D1242145

SCHLUMBERGER

The History of a Technique

SCHLUMBERGER
The History of a Technique

LOUIS A. ALLAUD
MAURICE H. MARTIN

A WILEY-INTERSCIENCE PUBLICATION
JOHN WILEY & SONS
New York • London • Sydney • Toronto

Library of Congress Cataloging in Publication Data

Allaud, Louis A.
Schlumberger.

"A Wiley-Interscience publication."
1. Electric prospecting—History. 2. Oil well logging, Electric—History. 3. Petroleum engineering—History. 4. Schlumberger Limited. I. Martin, Maurice H. joint author. II. Title.

TN269.A4913 622'.18'282 77-23566
ISBN 0-471-01667-5

Printed in the United States of America

10 9 8 7 6 5 4 3 2 1

To the memory of
Conrad and Marcel Schlumberger

Foreword

This book presents and explains the science-based techniques which, over the last half-century, have so vastly enhanced the prospecting for and the production of petroleum. The volume is a fitting tribute to the prominent part played by the ideas and work of Conrad Schlumberger in the development of these techniques, and in the creation of a large service company ranking at the very top of the specialized organizations devoted to the needs of the petroleum industry.

Between 1912 and 1926 Conrad Schlumberger, at first alone and then with a small, slowly growing number of associates, invented an original mineral-prospecting method based on the possibility, which he verified, of gathering from electrical measurements taken at the surface of the earth applicable information on the geometric and physical structure of subterranean geological formations. His success required not only a clear understanding of the physical phenomena and their potential utilization, but also a deep sense for experimentation in the laboratory and, above all, a genius for observation in the field. From the observation of diverse and complex facts he knew how to draw valid interpretations which basic research on necessarily simplified models, or even experiments based on this research, could only suggest. It is noteworthy that from the beginning his appreciation of the reasons for field test failures made such failures almost as valuable as successes.

Yet for Conrad Schlumberger there was a purpose broader than merely establishing principles for new methods of prospecting: what he wanted was to prove their practical value under the most diverse conditions and so to pave the way for their worldwide application. To this end, he promptly recognized the need for designing sturdy tools with which delicate electrical measurements could be made quickly in the field, even under difficult conditions. In 1919 Conrad's younger brother Marcel joined him, contributing his own remarkable inventive mind and his special skill for solving mechanical problems.

The establishment in 1926 of a small company, the Société de Prospection Électrique, marked the beginning of industrial development. A crucial factor in shaping the Company's future was the association with the Schlumberger brothers, in the same year, of Henri Doll, a young engineer. Within a year he was taking a large and gradually leading part in a new orientation of the research program, aimed at the employment in the oil wells themselves of the principles of electrical surface prospecting. In a few years this new field of electrical prospecting was to expand rapidly under the name, first, of "electrical coring" and, later, of "electrical logging." The great technical obstacles encountered required the development of highly sophisticated and dependable electromechanical tools; ensuing improvements in turn led to the delineation by electrical measurement of additional parameters such as borehole temperature or inclination. The recording of all these parameters as a function of depth constitutes logging in its broadest sense.

World War II seriously disrupted the Company's operations. Only in the United States, under the management of Henri Doll, could there continue to be any significant progress. One achievement of this period was the successful extension of the recruitment policy which had prevailed from the beginning of the work in America: the association, particularly in the research teams, of French and American engineers with widely different backgrounds.

After the war, research and development activities were concentrated in the United States under the continuing leadership of

Henri Doll; at the same time, close contact was maintained with the activities being resumed in France. During the postwar years, Henri Doll brought about a growing use of electronics in logging equipment and advances in the interpretation of measurements. One result was the electromagnetic induction method, which he had already thought of before the war. It came into standard use after 1949, and was to be the last original logging method introduced by Schlumberger.

I would like to take this opportunity to express my high regard for Henri Doll, whose outstanding human and professional qualities I was able to appreciate in 1943 when, as a visiting professor at Columbia University, I was invited by him to pay a few short visits to Houston to consult on certain aspects of the nonprofit organization established by Schlumberger in support of the Allied war effort.

Reading this book, scientists concerned with the penetration of the scientific spirit into the whole range of human endeavor will be happy to see that, even in the rather specialized field which is the book's subject, this spirit need not yield to the sense of business. Nor have Schlumberger's leaders lacked a keen business sense. Although the book is primarily addressed to engineers, who will find it a source of valuable information, I venture to state that readers who are not technically trained but who are eager for a better insight into the reasons for the prestige of certain industries will also find this modern epic unusually attractive even if they pass rapidly over the more technical paragraphs.

FRANCIS PERRIN
Member of the Institut de France and
Professor Emeritus of the Collège de France

Acknowledgment

We express our deep appreciation to Annette Gruner-Schlumberger. It was her idea that a book be written about the history of the Schlumberger technique. Without her guidance, encouragement, and support, this book could never have been published.

We also thank Marcel Schwob for his translation of the French text, and Frank Davidson, W. J. Gillingham, and R. R. Rieke for their very helpful suggestions, revisions, and corrections in the preparation of the English text.

LOUIS A. ALLAUD
MAURICE H. MARTIN

Paris, France

Contents

Figures

SCHLUMBERGER

The History of a Technique

Introduction

Schlumberger is a large concern belonging to what economists call "the service sector" of the business world. Although engaged for some 15 years in marketing products of its own manufacture, Schlumberger remains essentially what it was at the outset—a company offering specialized services to industry. The field to which these services are most applicable is petroleum exploration and production; the discipline is geophysics, a technology developed only during the last half century.

Comparatively recent developments in geophysical techniques include improved procedures for the measurement of certain physical properties of underground rocks. Great advances in mineral detection have resulted from applying these methods to rock density, electrical resistivity, sound transmission, radioactivity, and magnetism. By using special equipment, the measurements can be made either on the surface of the land or inside boreholes. In the first case, they help to outline broad features of the subsurface; in the second, they provide information on the characteristics of the formations penetrated by the drill.

Formerly called "electrical coring,"[1] and later known as "logging,"

[1] The term "electrical coring" (in French, *carottage électrique*) was adopted in 1927 because of its anology to "mechanical coring," which denotes the analysis of rock samples ("cores," in mining terminology) extracted from a formation by drilling. The French word *carottage* ("coring") was introduced as such into the Russian language in 1930 to designate the new technique. The English translation "electrical coring," under which the process became known in the United States, was replaced in 1933 by the term "electrical logging," now in use throughout the petroleum world. The word "log" means a strip of paper or film, on which measurements are recorded as curves (diagrams) in terms of depth. The word is borrowed from nautical science, where it refers to the plotting of the data that determine the motion of a ship in terms of time. After 1945 French-speaking oil people translated

the measurements in boreholes are the most important of the services provided by Schlumberger. It is essential, therefore, to grasp the meaning of logging as it applies to the petroleum industry.

The first objective of a drilling operation is to obtain as much information as possible on the various (and varied) formations penetrated, particularly those that may contain and yield oil; the second is to utilize the drillhole to produce oil from a reservoir found to be oil bearing. Logging is a convenient tool for obtaining much of this necessary subsurface information. It is performed by lowering an appropriate and highly sensitive device, by means of an electrical cable through which the physical properties mentioned above (except certain magnetic ones) can be measured and recorded in a detailed and continuous manner over the entire length of the open borehole. Through an analysis of the information obtained, the drilled formations may be characterized, their limits accurately determined, and the oil- or water-bearing levels identified. The basic physical property which the original Schlumberger log sought to delineate was the electrical resistivity of the rocks. Eventually, measurement of other properties increased the efficiency and scope of the method.

Schlumberger owes its existence to the electrical measurement tests applied to the surface by Conrad Schlumberger in France, a few years before World War I. This was the first time that such tests had ever had practical significance; they led to the development of innovative mineral-prospecting methods that came to play a major role in the application of geophysical concepts.

In 1919 Conrad and Marcel Schlumberger, with a handful of assistants and very little capital, launched a small venture for the purpose of applying these methods to the detection of metallic minerals, to geological surveys, and to petroleum exploration. Foregoing any idea of self-enrichment through personal speculation in promising mineral areas, the founders committed themselves to providing contract services of the highest scientific and technical standards. This commitment has remained a basic tenet of the Company.

After having contributed to the birth of geophysics, Conrad

"logging" into *diagraphie,* a new word derived from "diagram." In certain works (mostly textbooks), the expression "borehole geophysics" appears in all languages where logging is described. For the sake of historical accuracy, the term "electrical coring" will be used here for the technique prevailing until 1933, and "logging" thereafter.

Schlumberger in 1927 invented electrical coring. After 1931, when the firm's surface-prospecting methods were taken over by the Compagnie Générale de Géophysique, the Schlumberger group devoted itself principally to measurements and operations in boreholes. Then, in 1934, after 15 years of relatively slow growth, the company experienced a vigorous expansion extending its activities to most of the oil-producing regions of the world. In addition, in the late 1950's, while continuing to diversify its services to the petroleum industry, Schlumberger made broad inroads into the fields of electrical and electronic measurement and instrumentation.

At the present time this multinational enterprise ranks 148th among world firms (excluding those of the United States). Its 17 companies, with operations in more than 70 countries, employ a staff of 40,000. The group operates 18 research centers; the 3 major ones are devoted to logging.

Such growth is comparable to that experienced by certain pioneers operating in new countries. Indeed, the founders of Schlumberger were pioneers; the territory into which they ventured was an almost virgin area for mineral exploration. A continuous flow of creative ideas shaped their work. Sparked by the desire to meet the challenges of a rapidly evolving petroleum industry and to anticipate its requirements, the original ideas have bred more and more refined and complex techniques. The Company has been served by men who have efficiently organized and led industrial development under the most diverse conditions of geography, climate, infrastructure, law, and communications.

The purpose of this book is to provide an historical account of Schlumberger techniques from their inception to their current application, and to describe where and how they were employed. Lengthy technical descriptions will be avoided to spare nonspecialized readers.

The contributions made by Schlumberger in the field of surface geophysical prospecting were developed during the first part of this century, when the Company was functioning on a small scale. Today these accomplishments are almost forgotten in spite of their far-reaching effects: it was from them that logging, now in worldwide use, was developed; and they included enough scientific discoveries and inventions to justify a rather detailed historical review in the forthcoming pages.

For its first 15 years the company resembled a family concern with a modest staff; in many respects this was the most attractive period of its history. The founders had a hand in every detail; everyone knew everyone else personally. No two pieces of equipment manufactured for the same purpose were ever identical when they left the workshop. As Conrad liked to say, everyone was involved "for better or for worse." More space will be devoted to details of growth and development in this period than to subsequent periods when the staff grew to hundreds and then thousands, and when services multiplied and spread throughout the world. After 1934 the emphasis on personal recognition will necessarily diminish because the list of those who contributed to the common endeavor becomes too long—this is not to imply that these later associates deserve any lesser acknowledgment than those who formed the initial team between 1920 and 1933.

Although Schlumberger has held first place in a discipline of its own creation, other companies have offered similar if not identical services to the petroleum industry. Many engineers, geologists, and geophysicists in industry and in the universities have undertaken studies on specialized aspects of logging, particularly interpretation. The results have often provided Schlumberger with useful insights for the development and application of its processes. Since a listing would be too long for inclusion in this book, the reader may wish to refer to the specialized publications.[2]

To indicate the originality and importance of Conrad's first studies, this introduction will outline the field of mineral exploration as it existed at the beginning of the century. Some data will also be provided on the electrical resistivity of rocks in the crust of the earth. The measurement of this parameter is basic to electrical surface prospecting, and it constitutes one of the truly characteristic features of logging.

Mineral exploration in 1910

For many metallic ores (e.g., iron, copper, zinc, lead, manganese) the exploration devices available in 1910 were much the same as they had been centuries ago. These ores, whether in the form of lodes or intrusives in older rock formations, are mostly eruptive; prospecting for them basically consisted of searching for visible signs on the surface, either in

[2] A substantial, if not exhaustive, bibliography can be found in "Well Logging Techniques," by J. Riboud and N. A. Schuster, *Transactions of World Petroleum Congress,* VIII, 1971.

already productive areas or in other areas with similar geological features. Occasionally, thanks to geological experience, a deep, invisible deposit was found; but since the distribution of such lodes is generally erratic, their discovery was often due more to chance than to the prospector's educated intuition.

Other ore bodies (e.g., certain iron ores, coal, salt) have deposits formed similarly to sedimentary rocks. These are found in layers interspersed among ancient formations. For a long time their exploration depended on surface studies of the structure of the sedimentary series which presumably contained them. Geological exploration gradually became more reliable and significant. From observations and measurements taken where the subterranean layers outcropped, reasonable appraisals of their underground structural patterns, such as folds, breaks, overthrusts, and faults, could be made. The methods employed were those of tectonics, still a young science at that time.

On the other hand, actual physical inspection of the economic mineral—the ultimate target of exploration—no longer necessitated the digging of man-sized pits and tunnels until the time of exploitation. For several decades it had proved faster and less costly to drill small-diameter coreholes for cutting deep rock samples from the formations penetrated. Methods of examining the samples had been perfected, and advances in the new science of paleontology helped to determine the age of rocks, thus narrowing the scope of exploration.

Despite the persistence of the widely shared belief that petroleum gushed out of vast underground caverns (an idea popularized by many publications, including highly scientific ones), most professionals had long known that petroleum, liquid or gaseous, was to be found in the pores of permeable rocks such as sandstones and limestones. Since the first oil well was drilled by Colonel Edwin Drake in Pennsylvania in 1859, studies by English and American geologists had verified the close relationship between oil-bearing formations and deep geological structures. It was known that oil, being lighter than water, may accumulate in the upper parts of porous strata folded in the shape of arches or domes, and overlain by impermeable layers such as clays or shales (Figs. 1 and 2). This knowledge led to the "anticlinal theory," which was to become a key tool in petroleum exploration.[3] To be sure, not everything had been clarified,

[3] An anticline is a geological structure in the shape of a vault or an arch. A syncline has the

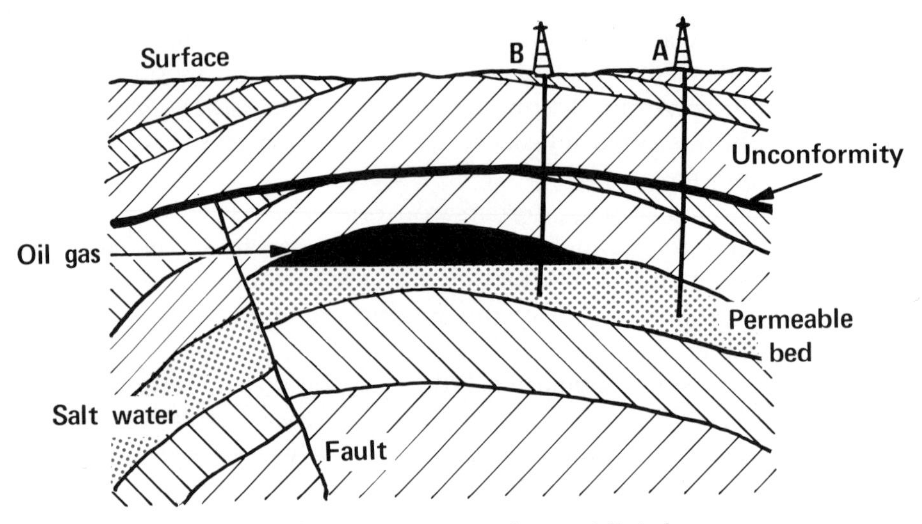

Figure 1. Schematic cross section of an anticlinical structure.

and many concepts held at that time were to be refuted or modified by experience. Anyway, in 1910 the geologist was not too well regarded by the majority of operators; the most popular way of prospecting for oil was to drill a borehole near surface signs such as gas seepages, eternal fires, asphalt lakes, and mud volcanoes.

Although the contributions of geophysics to the field of mineral exploration remained limited, several centuries of observing the magnetic and gravitational fields, and the more recent study of seismic tremors, had enhanced the knowledge of our terrestrial globe.[4] As early as the seven-

shape of a ditch. The word "dome" is self-explanatory. Many anticlines and domes have resulted from the thrust of rock salt, which is relatively plastic as compared to other rocks. Under the weight of the sedimentary overburden, wherever a passage is available, the salt formation shoots up extensions, often nearly vertical, which take the shape of gigantic ridges or mushrooms and either push their way through the surrounding layers (Fig. 2), bend them back, or uplift them. Salt ridges and domes are prevalent in certain regions such as Alsace and the Landes in France, the Gulf Coast (Gulf of Mexico) in the United States, the Emba region in the U.S.S.R., and particularly parts of Rumania and Germany.

[4] This area of learning was then called "geophysics" or "physics of the globe." Later, when measurement of the physical field would serve the purposes of mineral exploration, the term "applied geophysics" was introduced. Nowadays "geophysics" refers mainly to the scientific and technical studies involved in the exploration for minerals.

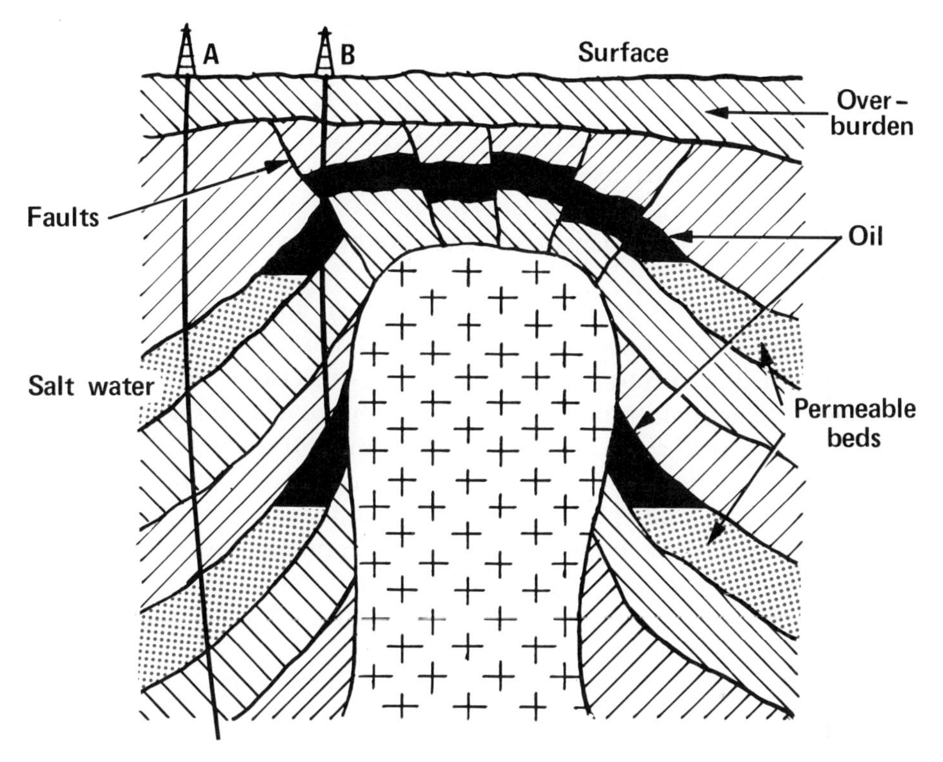

Figure 2. Diagram showing the uplife of the strata by a saline surrection.

teenth century, thought had been given to using the compass to detect, among other things, magnetite bodies; yet it was not until the middle of the nineteenth century that prospecting instruments of much greater sensitivity and accuracy (Thalen and Tiberg compasses) led to important discoveries in Sweden and the United States. In 1896 the Hungarian scientist Roland Eötvös designed and built his torsion balance, an instrument sufficiently sensitive to detect the variations in gravity above the cellars of his Budapest laboratory.[5] However, its only practical application at that time was to measure the depth of Lake Balaton. Not until the 1920's was the torsion balance used commercially in the search for salt domes and oil fields.

In 1830 a British scientist, R. W. Fox, used a rudimentary apparatus

[5] These measurements are the backbone of the gravimetric prospecting method.

to measure the electrical potentials generated by metallic sulfide deposits in Cornwall.[6] In 1880 an American, Carl Barus, published his observations in the same field while studying mines in Nevada. Elsewhere, some prospecting based on the electrical resistivity of rocks and utilizing high- or low-frequency currents had been going on for almost a century; the specialized literature fails to show, however, that these operations brought about the discovery of any new ore bodies or the extension of a known one.

In short, as late as 1910, geophysics had made scant impact on mineral exploration. There appeared to be little appreciation of the role that it might play in the study of geological structures; no attempt had been made to apply it to the search for iron ore or coal. As for petroleum exploration, which after 1925 would account for over 95 percent of geophysical activities, those involved knew very little about the applicability of geophysical principles. Professor de Launay at the École des Mines in Paris was one of the few who sensed these potentialities. After having briefly appraised the efficiencies of the magnetometer and other instruments of that time for measuring the physical field, he wrote:

> While one would think that the above-mentioned instruments would be of such great value, their use . . . has been restricted to a few rare instances and it does not appear that the spirit of scientific investigation, so active in every direction, is here applied with an intensity commensurate with the practical results that could be expected from a discovery.[7]

This, for geophysics, amounted to a statement of failure. There lay ahead a vast and relatively unexplored field, open to those keenly investigative and scientific minds that de Launay regretfully found wanting. The field of mineral exploration was ready for the studies of Conrad Schlumberger.

The electrical resistivity of rocks

Rocks encountered in sedimentary basins are composed of mineral grains more or less cemented. Most often, these grains do not fill the

[6] At that early date Fox predicted that methods of the electromagnetic type could help in prospecting for metalliferous lodes.
[7] De Launay, *La Conquête Minérale,* Flammarion, Paris, 1908.

entire space; they are separated by pores which vary in size with the type of rock.

In the finest sediments, such as clays and shales, the size of the pores can be as small as 1/100,000, even 1/1,000,000, of a millimeter. In sands, sandstones, and many limestones, sizes vary from 1/1000 to 1/10 of a millimeter, and in gravels and some vugs they reach 1 or more millimeters. Yet rocks are found with crystals so tightly joined that there are no pores at all. This occurs mainly in deep-hardened sandstones or limestones, or in matter produced by evaporation in deep water such as rock salt or anhydrite. Igneous or crystalline rocks are, in general, similarly compact.

In nearly all subsurface rocks, the pores are entirely filled with water varying in saline content from pure water, produced by rain or snow, to salt-saturated water (often encountered in oil fields). The highly aerated formations close to the surface are an exception. On the other hand, in hydrocarbon deposits the pores of the rock contain both water and oil, both water and gas, or the three fluids together. With a few exceptions, which will be considered later, mineral grains are electrical insulators. The rocks they form allow electric current to flow because they are permeated with water. Therefore the resistivity of a rock is of the electrolytic type; it is proportional to the resistivity of the water and decreases steadily when the temperature rises. The smaller the amount of water contained in a given volume of rock—in other words, the less porous the rock—the higher the resistivity. Resistivity is also contingent, although to a lesser extent, on the shape and arrangement of the pores, rather than their size. Thus there is only a loose relationship between the resistivity of a rock and its permeability.

Rock porosity varies widely, from a few percent in consolidated limestones, to 30 and 40 percent in sands and unconsolidated clays, and up to 80 or 90 percent in the silt of lakes and oceans. Since the salinity of waters also varies widely, the resistivities encountered underground vary considerably too, from a fraction of an ohm-meter[8] in unconsolidated formations filled with salt-saturated waters, to several thousand ohm-meters and on to infinity in compact rocks. No other physical parameter

[8] The resistivity unit adopted by Schlumberger and used almost universally in electrical prospecting is the ohm-m^2/m or, more simply, the ohm-meter; that is, the resistivity of a cube 1 meter on a side, the resistance of which, to a current parallel to an edge, equals 1 ohm.

(e.g., density, acoustical propagation velocity, magnetic susceptibility) presents such a wide range of values.[9]

When exploration is undertaken in a limited zone, where each sediment maintains a sufficiently uniform structure over a fairly large area, and the salinity of the water does not change appreciably, resistivity becomes a sensitive and accurate factor in identifying the formations. Furthermore, when the water in the rock is partially replaced with oil, gas, or both, the resistivity is much higher than if the pores are entirely filled with water of even salinity. Depending on hydrocarbon saturation,[10] the resistivity of an oil–gas formation may be 5, 10, or 100 times higher than that of a similar water-bearing formation. This property has far-reaching implications in logging since it offers a sensitive and dependable indicator for the differentiation of productive and barren zones.

Finally, certain ores, mainly metallic sulfides (e.g., pyrites), are made of crystals that conduct electricity as do metals. This metal-like conductivity ("electronic" is the term now used) is much lower than the conductivity of ordinary metals, yet much higher than that of other rocks, even those having the lowest electrolytic resistivity. A characteristic feature of this type of ore, it has proved useful in prospecting.

When Conrad Schlumberger began his research in 1911, only very fragmented data on the resistivity values of rocks were available to him. Although he might have intuitively foreseen the significance of formation resistivity for mineral exploration, there was no way he could have envisioned the impact it would have on geophysics and, more specifically, the petroleum industry.

[9] The density ratio of the heaviest to the lightest rocks does not generally exceed 3; the same ratio applies to the velocity of sound waves through the subsurface; and for magnetic susceptibility there is a marked difference only between crystalline and sedimentary rocks (with the exception of certain iron ores).

[10] In the language of the petroleum industry, saturation designates the fraction of pore volume occupied by a fluid: water, oil, or gas.

PART
ONE

Surface Electrical Prospecting

Phase One: Before 1914

First laboratory studies

September 1912. At the family estate, the Val-Richer, in Normandy, Conrad Schlumberger was putting the finishing touches on lines drawn on a large sheet of paper, with a few reference marks indicating lawns, ponds, driveway, the outline of a mansion, and farms. The lines appeared to mark topographic contour levels; however, instead of connecting points of equal elevation, they connected points of equal electrical voltage and were called equipotential curves. Conrad had before him the plotting of the first on-site measurements he had made to test, under actual conditions, his ideas about the exploration of the subsurface through the transmission of an electric current.

These field tests were the result of many months of meditation, theoretical work, and small-scale experimentation. During that same summer he had developed, almost entirely by himself, the apparatus and methods that would enable him to put into practice his theories. However, since there was no mineral wealth buried in the grounds of the Val-Richer, his tests served merely as a feasibility study. The results were strikingly clear; the measurements ranged neatly along regular equipotential curves, whose shape and spacing were easily explained by the ground relief. Although others before him had sent current into the ground, they had become bogged down in insurmountable difficulties; no one else had succeeded in making such precise, cohesive, and significant observations: a method had truly been "discovered."

What were the underlying principles of this discovery, and how did they give birth to a method? Looking back almost 10 years later, Conrad would answer that the same circumstances that had earned him a professorship at the École des Mines in Paris* had later compelled him to devote his holidays to researching the application of physics to the field of mining, with particular emphasis on prospecting. It did not behoove him to add that these "circumstances" were strongly helped by a rare intelligence, exceptionally gifted for the abstract and practical, a fertile imagination, a varied cultural background, and a lively curiosity controlled by a rigorous scientific approach. To a man so endowed, who was little more than 30 years old, an academic career, however prestigious, could hardly bring fulfillment. Teaching a course in physics at the École des Mines was not, however, without its advantages; it provided a quiet occupation, scientific surroundings, a vast library, and the stimulation of contact with a young audience. This was the "environment" that became the fertile ground of creativity for one whose mind was shaped by mathematics and physics. Thoroughly aware of the problems of the mining industry, and having a keen interest in the sciences of the earth, Conrad was inevitably attracted by the challenges of the field later known as geophysics. Furthermore, his fondness for research and his desire for orderly and efficient work, as well as his concern for economic and social factors, led him toward efforts to improve the methods of mineral production.

The search for metalliferous deposits was expanding during this period, and Conrad's first thoughts were directed toward methods for improving reconnaissance. The problem was to decide on a physical parameter with which one could accurately differentiate the deposits themselves from the rocks in which they were embedded.

According to information available at the time, electrical conductivity appeared to be a more significant parameter for the exploration of metalliferous deposits than the rather uncharacteristic density, or magnetic properties, limited to certain types of iron ores. However, while a magnetic mineral body may be revealed by the disturbance it creates in the magnetic field of the earth, there is no similar natural electric field that can be scrutinized in the search for conductive ores.

Conrad thought, therefore, of creating an artificial electric field. The simplest way was to transmit a current between two electrodes (two series of pegs) driven into the ground, and to plot the equipotential lines reflecting the distribution of resistivities of the subsurface.

* Graduate institute of mining engineering. (Translator's note.)

In the case of a homogeneous underground, the path of the current lines and the shape and the arrangement of the surfaces of equal potential were well known. A simple calculation shows that the current does not flow inside the narrow channel connecting the two electrodes, the most direct route (as certain experimentalists believed), but spreads in breadth and in depth across the subsurface. The surfaces of equal potential intersect with the plane surface of the ground, forming equipotential lines (Fig. 3). Should an ore body with a higher conductivity than the surrounding formation (always assumed to be homogeneous) occur, these lines would be distorted, as revealed by the plotting.

It appears that this reasoning was somewhat oversimplified because the strata around a mineral bed are never homogeneous. Certain difficulties, which Conrad hoped to overcome in future experiments, had to be expected. Still, no serious theoretical objection could be raised against the apparatus he had in mind, whereas the system unsuccessfully operated a few years earlier by other researchers (Brown and MacClatchey)[1] could not withstand close criticism.

Early in 1911 Conrad conducted a study of the various elements involved in his initial thinking. The laboratory of the École des Mines afforded him the opportunity to examine ore samples and to measure their electrical resistivities in order to supplement, correct, or confirm the published data. He also extended his observations to geological formations of sedimentary origin, especially argillaceous sands in recent alluvial deposits. Small-scale devices were set up in the basement of the École des Mines. At first, wooden crates were filled with a mixture of sand and clay impregnated with water that varied in saline content. In these crates a current flowed between two current electrodes (hereafter referred to as A and B), and its effect was explored by means of two measure electrodes (hereafter called M and N), with which the carefully leveled surface was scanned. The alternating current used was of audible frequency. Silence, or reduced noise, in a pair of earphones connected by loose wires to electrodes M and N indicated that they were at the same potential.

Step by step, Conrad plotted grids of equipotential curves, checking that they confirmed the theory. Other experiments using various electrode

[1] Brown and MacClatchey had been concerned solely with measuring resistance through the ground between two ground electrodes conducting the current. They were unaware of the fact that 95 percent or more of the resistance is concentrated very close to the current electrodes and remains the same whether or not an ore body is buried between them.

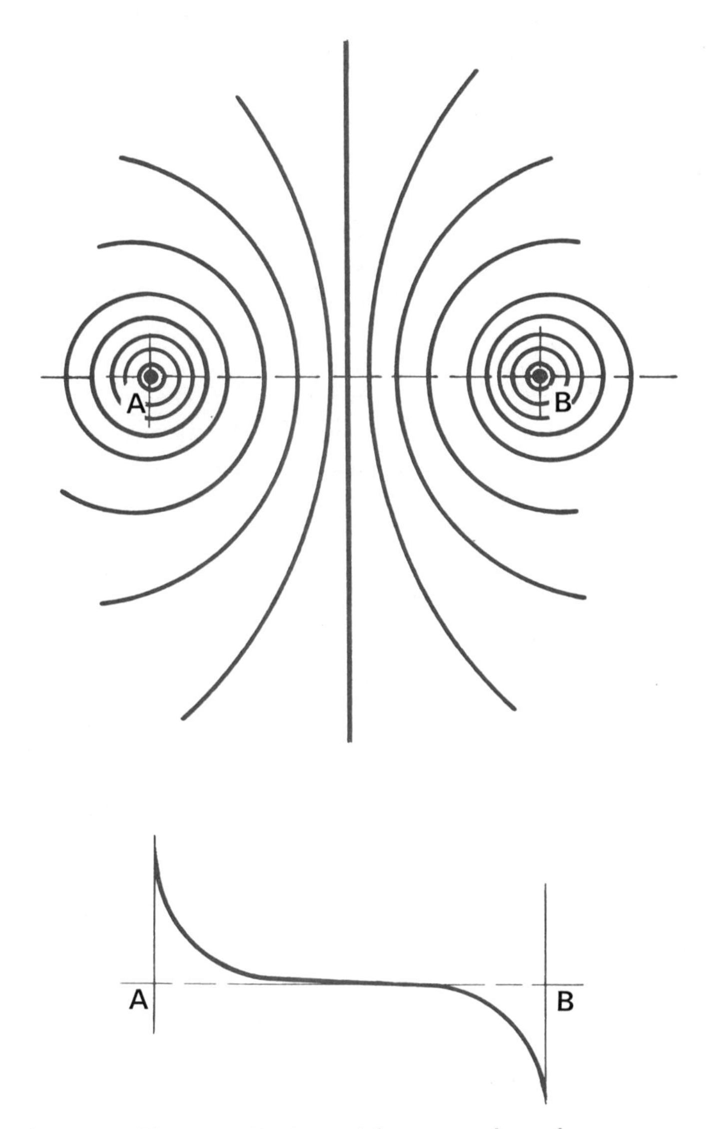

Figure 3. Equipotential curves at the surface.

arrangements followed, but the wooden crates were no longer adequate. A watertight, metallic tank was needed to hold water of various salinities and to set a zero potential. The small copper bathtub that had been used by his children approximated these conditions and became the indispensable item of equipment in Conrad's experiments. To this day the term "bathtub" is used in the idiom of "La Pros" (a nickname of the Company originated by French engineers; see p. 37) to designate a metal tank through which the passage of a current reveals a variety of complex and informative data.

With these items of equipment it was possible to answer a question that was basic to the practical application of the method: would a deeply buried body of metalliferous ore reveal its presence by sufficiently distorting the equipotentials? Conrad devised experiments wherein a parallelepipedic slab of potter's clay was placed horizontally inside the clayey sand of a crate and the effect on the curves observed. Repeated on several crates, each one with its slab of clay at a different depth, these seemingly simple experiments required meticulous attention to accuracy and detail in order to achieve two media of natural materials that were sufficiently homogeneous and had a resistivity contrast such that the measurements were representative.

These experiments showed that the suggested method would work: it was possible to detect a buried object provided that it was large enough and its resistivity contrasted sufficiently with that of the surrounding formations. The idea of adjusting the distance between A and B (hereafter termed current electrode spacing) gave rise to a special advantage of the current transmission method: the possibility of altering the depth of investigation to meet the desired objective.

First field tests

After a few months of preliminary testing, done on a relatively low budget during free time between teaching assignments, Conrad had developed his initial ideas to the extent that he felt ready to undertake full-scale tests.

In 1912, a summer vacation provided the necessary leisure, and the park of the Val-Richer a field for experimentation (Fig. 4). The undertaking was not an easy one. It was no simple task to shift from laboratory-

Figure 4. Conrad Schlumberger performing a field measurement (Normandy, 1912).

scale models to actual field conditions. The earphone method had proved disappointing. Mutual induction between the current-emitting (AB) and measure (MN) circuits was negligible in the laboratory, but in the field, with lines of several hundred meters, it became strong enough to jam observations.[2] Conrad then switched to direct current; experience confirmed its superiority.

Within a few weeks he procured, tested, and adapted his instruments and perfected the operating methods. Several months later, in the early part of 1913, he summarized this first experiment on the same sheet of heavy drawing paper on which he had originally plotted his measurements. He described the crude cables and reels borrowed from the army, the small generator driven by the engine of the truck, the ground stakes and the way they were set into the soil to get good grounding, the practical maneuvering of the electrode system, the trumpet signals used for field communication between himself and his assistant, the deviations of the needle, and finally the galvanometer itself. After testing several galvanometer models, he finally found, in Frankfurt, one of adequate sensitivity, accuracy, and convenience, made by Hartmann and Braun. He would add to this galvanometer the components (batteries, resistances) that would enable him to measure potential differences by the balance (or null) method (see p. 47).

Another difficulty, arising from the use of direct current, had to be solved. If M and N are plain pegs of copper, for example, electrochemical phenomena take place at the point of contact of the metal and the water present in the soil: the ground electrodes are said to "polarize." As all ground electrodes do not polarize in an identical manner, this imbalance generates an electromotive force of up to several tens of millivolts, more than enough to drive the needle of the galvanometer out of scale, making any measurement impossible. Conrad conceived and built a nonpolarizing type of ground electrode that was well suited for fieldwork (Fig. 5).

The subsurface of the Val-Richer testing ground was simple: a thick layer of moist, uniform, clayey soil covered the ancient rocks, which were embedded too deeply to affect the measurements. Electrode A was staked at the center of a well-leveled lawn and connected to remotely located

[2] Conrad has explained by these disturbance effects the failure of the earlier attempts of L. Daft and A. Williams, who were sending alternating current into the soil and measuring the potentials with earphones. Various other tests employing the same technique, especially in Sweden, had not been followed up.

Figure 5. Paraphernalia of Conrad Schlumberger. 1: Tripod with compass, a military instrument used for topographic survey of measuring stations. 2: Nonpolarizing electrode, the handle of which features a coil used in the measuring circuit and the wooden box crating the potentiometer. 3 to 7: Electrodes of various lengths. 8: Horn to call the helpers. 9: Earphones. 10: Stake and large coil for power circuit. 11 and 12: Current generators. 13 and 14: Storage batteries. 15: Geologist hammer. 16 to 19: Commercial ammeters and voltmeters.

electrode *B*. Thus *B* had practically no effect on the shape of the curves plotted around *A*; it was as if *B* had been removed to infinity. On the lawn the equipotentials formed almost concentric circles, as stated in the theory; to the northwest, where they spread out and tightened again, the pattern indicated the presence of a slope (Fig. 6).

The test was conclusive, and the technique dependable.

First surveys of geological structures

The second experiment took place slightly to the south of Sassy, Calvados. The main features of deep geology in this region are known, thanks to numerous iron mines. The subsurface is made up of steeply dipping, occasionally almost vertical (Silurian) formations. The thick basement is composed of three types of rocks: Armorican sandstones, Calymenae shales (with a sedimentary seam of iron ore between them), and May sandstones (actually a stack of sandstone and shale beds). These ancient formations outcrop in places, depending on their hardness and on the relief of the ground, but they usually remain hidden beneath younger, softer, and nearly horizontal Jurassic marls and limestones. Conrad's purpose in undertaking the study was not to detect the iron ore, made of hematite and carbonate and having no particular electric conductivity; what he wanted to ascertain was whether the resistivities of various types of rock were sufficiently distinct to reflect their shapes through the equipotential lines, and, if so, to what deductions this might lead (Fig. 7).

As previously noted, Conrad had, at that time, little information on the resistivity of the rocks constituting the ancient basement of the underground terrain. Laboratory measurements on samples had revealed much higher values than those for recent alluvial sands and clays. Furthermore, it was known that sandstones are harder than shale and marly Jurassic limestones. This fact was confirmed by the outcropping of sandstones south of Sassy (Fig. 8). Shales covered with a thick limestone, on the other hand, showed that during a certain era they had been more intensively eroded. The Armorican sandstones, in particular, formed a slight, east–west oriented hillock, at the foot of which ran the trace of their contact with the limestones, and, below, their contact with the shales. Therefore it could be assumed *a priori* that such differences in hardness would be reflected electrically, and that the resistivity of the sandstones ought to be higher than that of the other formations. Unlike the Val-Richer experiment, the one at Sassy involved a subsurface composed of several formations with different resistivities.

The measurements were carried out over a few days in October, and the dependability of the technique was confirmed. In Conrad's experiments, the position of A varied, whereas B was fixed at a remote location. The tests showed that the equipotential curves no longer appeared as concentric circles: they clearly reflected the subsurface formations and the locations of their contacts. To analyze their arrangement even more accurately, Conrad resorted to a method that geophysicists were to employ in the future: a model composed of several formations was devised. It was simple enough to lend itself to mathematical treatment, and similar

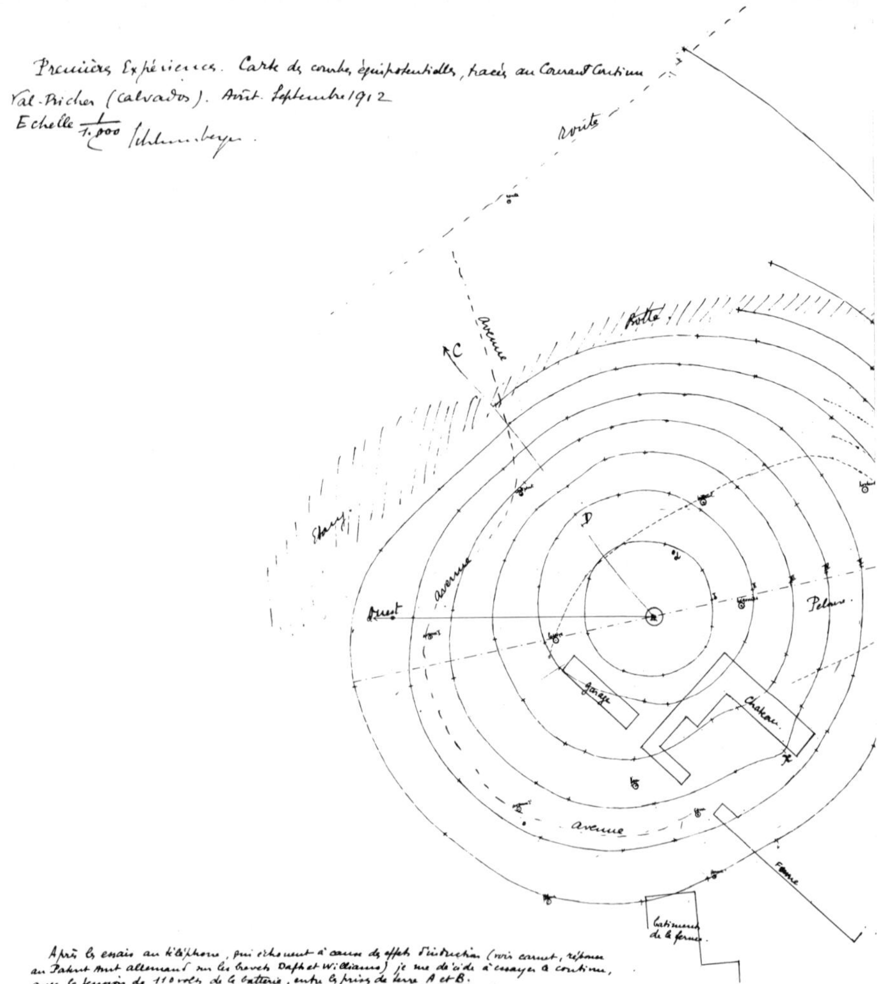

Après les essais au téléphone, qui échouent à cause des effets d'induction (voir carnet, réponse au Patent Amt allemand sur les brevets Daft et Williams) je me décide à essayer le continu, avec la tension de 110 volts de la batterie, entre les prises de terre A et B.

Résistance de ligne. La résistance totale est de 23,5 environ ; résistance du cable 7,5 (chiffre exagéré me semble-t-il vu la faible distance des prises de terre (320m) je me décide à commander du cable de résistance moindre, de 8 au klm par ex., ce qui amène d'autre part une meilleure résistance mécanique ; le cable employé alors avait 16 au klm environ) Les prises de terre sont composées chacune d'une circonférence de 6 mètres de diamètre, avec 9 piquets galvanisés type militaire. La résistance de chaque prise de terre est de 23,5 − 7,5 = 8. La résistance de chaque piquet est de 9 × 8 = 72. Quand je prends seulement 5 piquets par prise de terre, la résistance moyenne par piquet tombe à 55. Elle tombe à 50 avec 4 piquets par prise, à 46 avec 3 piquets, à 47 avec 2 piquets, à 45 avec 1 piquet par prise. J'en conclus qu'avec 9 piquets, (soit les piquets à 2m les uns des autres sur la circonférence) les conductivités des piquets ne s'ajoutent que très imparfaitement, mais qu'avec 6 piquets ces conductivités s'ajoutent déjà suffisamment (puisque la résistance par piquet ne doit guère dépasser 55 et que les piquets isolés donnent 45). Mettre 6 piquets, c'est les placer à 3m les uns des autres, sur une circonférence de 6 mètres de diamètre. J'énonce la règle pratique suivante : Dans les sols à couches terre profonde (ce qui est le cas au VR), prendre autant de piquets par prise de terre qu'il y a de mètres dans le diamètre de la circonférence de la prise de terre. Les piquets placés plus près se gênent mutuellement trop et leurs conductibilités s'ajoutent très mal ; d'autre part on ne gagne pas beaucoup (20%) à les écarter d'avantage. Dans les terres peu profondes, où la répartition du courant à partir des prises de terre est bien plus superficielle, ces résultats ne sont plus valables. Il faut probablement écarter plus les piquets.

J'ajoute que le terrain du V.R est particulièrement favorable (argileux et humide). Dans les expériences ordinaires de Normandie (Sommont etc) les résistances totales ont varié de 70 à 100, avec il est vrai 20 de cable. Quand les terrains sont peu profonds, on atteint facilement plus en cas de sécheresse.

La résistance varie très notablement (10%) quand on inverse le courant. Il ne s'agit pas de la force électromotrice de polarisation due aux gaz (O et H) dégagés au piquet, celle-ci ne dépassant pas 1% des 110 volts employés, mais bien d'une augmentation de résistance due aux gaz. L'O et l'H ne semblent pas agir de même. Il faudrait étudier la chose, pour savoir, dans le cas de prises de terre très inégales, celle qu'il y a intérêt à choisir comme anode. Si c'est la couche d'Oxygène qui augmente le plus la résistance, c'est la meilleure des prises de terre (piquement métallique par ex) qu'il faut choisir comme anode, pour y former le dépôt d'Oxygène.

Ces variations dans la ~~mesure~~ résistance ~~pinent pour la~~ empêchant de prendre pour le numérotage des courbes, les prises de terre comme points 0 et 100 (voir brevet).

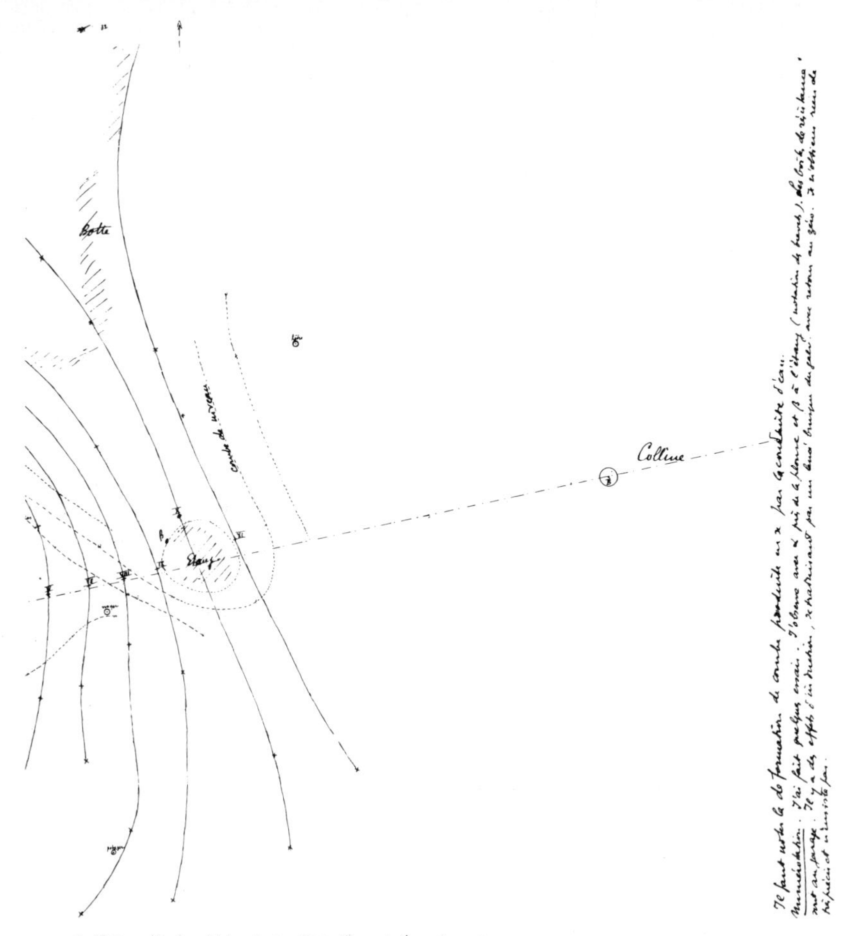

<u>Matériel employé</u>. <u>Ligne volante</u>. Fil militaire extra léger en bronze. Solide, mais vieille et est mal isolé. <u>Déroulement</u>. Petits rouleaux [illegible]; peu satisfaisants. <u>Galvanomètres</u>. J'essaie: 1°. de voltmètre Carpentier, sensibilité [illegible] (après divers galv. à miroir [illegible] à microscope de Carpentier [illegible]). 2°. Un galv. Siemens à aiguille à vis calante. Sensibilité 1 div pour 3.10⁻⁷ amp. C'est une révélation, mais il faut mettre le galv. en station. J'emploie le pied du tachéomètre. Ou bien je fais observer le galv. par l'aide qui commande le courant et je me fais annoncer à la trompette le sens de l'écart. Dans ce cas la ligne volante est très longue, passe près des prises de terre, il y a des phénomènes d'induction appréciables. 3°. Galv. à aiguille Hartmann Braun, sur pivot. Portatif. Excellent. Progrès immense. <u>Électrodes</u>. Essais: bougie de filtre (trop poreux, perd le liquide); terre peu cuite (pourrit vite sous l'action de SO_4Cu); terre très cuite (bonne). Formes variées de vases. Fixation variée du tube de cuivre par caoutchouc, je crains l'action du soufre, par mastic Golaz (bien, mais se fendille assez vite). Je laisse de côté les premiers balbutiements (plaque de cuivre, entourée de chiffons mouillés avec SO_4Cu, vase poreux de pile entouré de chiffons).

<u>Technique</u>. Inversion par commutateur à la main. Le commutateur mécanique est introduit pour Saint-Bel. A partir de Oct. 1912, je me borne à une inversion de peu de temps, au lieu de l'établissement, l'inversion et la suppression. Il est vrai que l'on polarise les piquets de prise de terre, mais cela n'a guère d'inconvénient (pas d'attaque notable), et l'avantage est de faire travailler régulièrement le moteur, au lieu de le faire travailler par à coups. La consommation d'essence est probablement améliorée et la commutation mécanique est très simplifiée, puisqu'il n'y a plus lieu d'agir sur l'accélérateur.

<u>Résultats</u>. Résumés dans la carte ci-dessus, relevée au tachéomètre. Les courbes sont assez rondes; elles le sont parfaitement dans les endroits horizontaux et réguliers (pelouse). Les courbes sont affectées par le relief du sol. Elles tendent à longer la Botte. — Considérons le profil AC. Les courbes, qui ont été prises équidistantes sur la pelouse (ligne AB), sont écartées entre A et D, puis serrées entre D et C. Voici mon explication (voir coupe du terrain). Vers la crête D, la densité de courant, donc la chute ohmique par mètre, est faible, car cette crête constitue une sorte d'angle mort pour le courant. Les courbes équipot. doivent donc être relativement très écartées les unes des autres. Vers le rentrant formé par la Botte, les filets de courants au contraire se resserrent; la densité de courant augmente, la chute ohmique par mètre augmente, les lignes équip. se resserrent. A côté de cet effet, peut-être y a-t-il des perturbations provoquées par des hétérogénéités de conductibilité (plus grande humidité dans le voisinage de la Botte et dans la vallée).

<u>Polarisation</u>. Je l'observe près des tuyaux métalliques des conduites d'eau, notamment en x (Sud du Château). La courbe n'est pas au même point suivant que le courant est dans le sens A vers B (A anode) ou dans le sens inverse. Il y a en x, sur l'aplomb d'un tuyau connu, un écart de 1 à 2 m entre les 2 courbes. Quand l'électrode mobile est en y, entre les 2 courbes, on observe: 1°. une déviation à gauche, indiquant qu'on est par ex. à l'extérieur de la courbe. 2°. après l'inversion une seconde déviation à gauche (je suppose celle-ci plus forte que la première pour qu'elle soit perceptible), indiquant qu'on est à l'intérieur après inversion. En z la 1ère déviation est nulle et l'on ne voit que la seconde. En w la 1ère dév. passe à droite, la seconde restant toujours à gauche et étant plus grande que la 1ère. La non coïncidence des courbes équipot. se traduit donc par l'inégalité des déviations, les déviations étant de même sens, lorsque l'on est entre les deux courbes. L'explication de la non coïncidence, c'est [illegible] les f.é.m. de polarisation, qui se superposent aux chutes ohmiques, sont dissymétriques et inégales suivant le sens du courant.

Figure 6. Experiment at the Val-Richer.

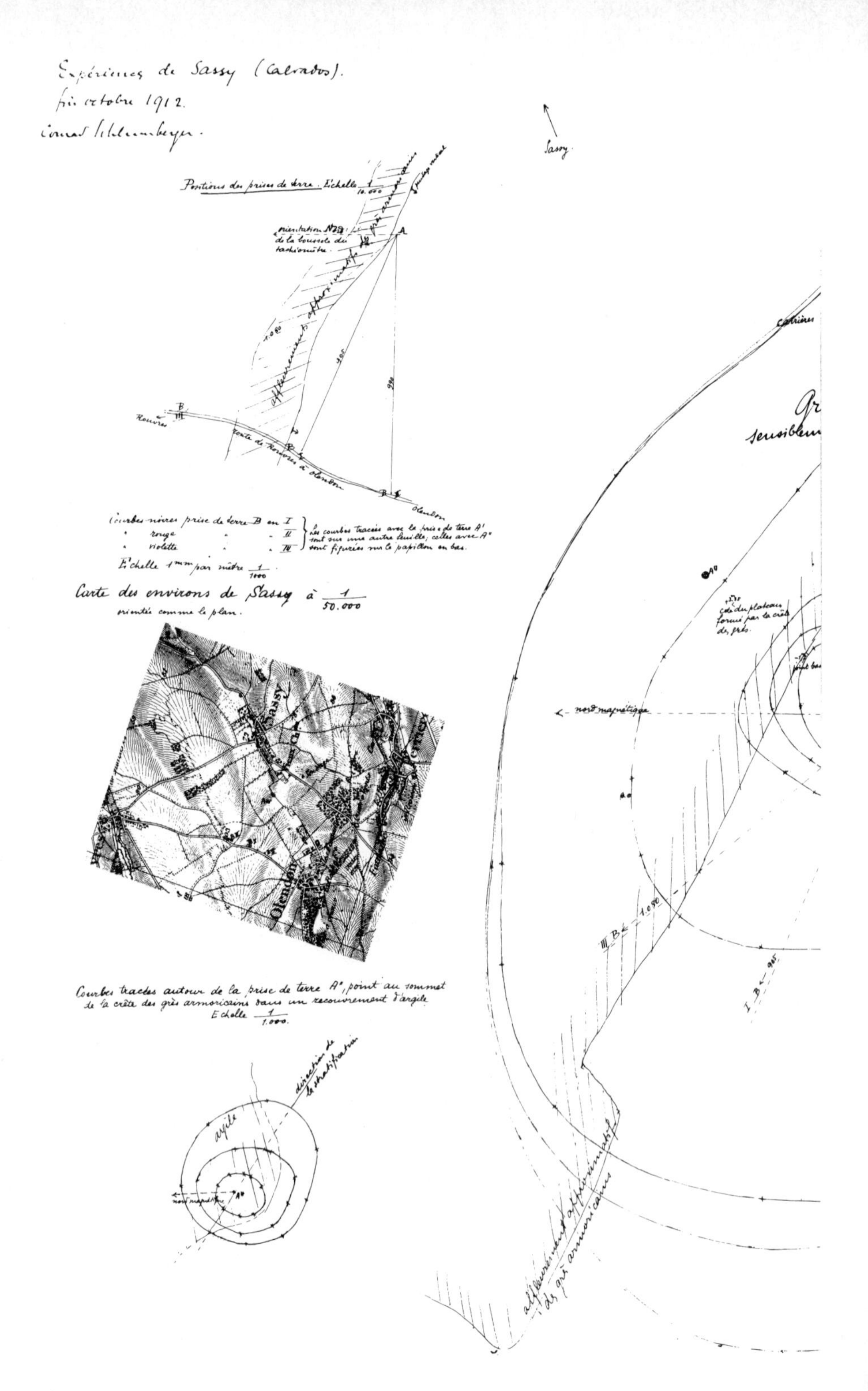

Expériences de Sassy (Calvados).
fin octobre 1912.
Conrad Schlumberger.
Sassy
Positions des prises de terre. Echelle 1/10.000
Carte des environs de Sassy à 1/50.000
orientée comme le plan.
Courbes tracées autour de la prise de terre A°, point au sommet de la crête des grès armoricains sous un recouvrement d'argile
Echelle 1/1.000
argile
direction de la stratification
nord magnétique
affleurement des grès armoricains

Figure 7. Experiment at Sassy.

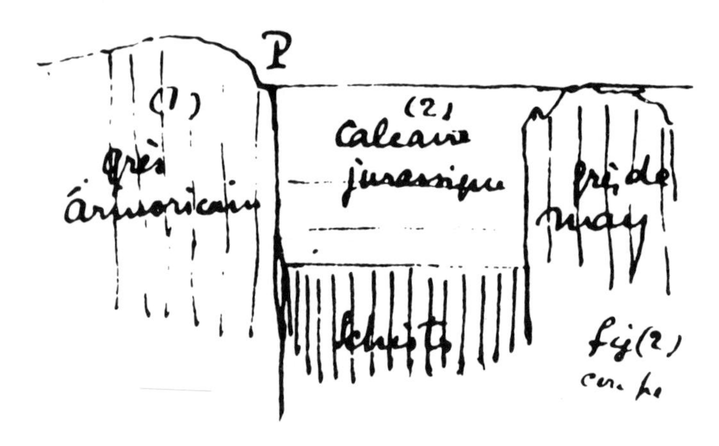

Figure 8. Experiment at Sassy: enlarged detail.

enough to the actual structure of the subsurface so that a plot of the field measurements could be unequivocally explained by means of the calculated model. In Figure 9, the drawing on the left shows the calculated equipotentials, separated by a vertical plane at the contact of two homogeneous formations, one resistant, the other conductive. The sketch on the right refers to a conductive medium between two resistant ones. The curves plotted on the map appeared quite similar to those of the drawings. The large equipotentials were elongated in an east–west direction, that is, alongside the trace of limestone and shale layers, which are softer and electrically less resistant than the flanking sandstones. Finally, the breaks they showed at the contact between sandstones and limestones provided experimental verification of the refraction phenomenon as predicted by theory.

Thus the potentials map can be informative even in the absence of metallically conductive ore bodies, inasmuch as it can reveal certain features of underground geological structure, such as the strike of the strata and the contacts between formations. It is true that Conrad was experimenting at Sassy on a known structure, visible over almost the entire surface—a condition necessary to check the accuracy of the method; yet it was becoming apparent that similar studies could be attempted in regions where the deep structures were completely concealed. Such was the case in Normandy, where iron ore exploration was based on the identification of contacts between generally hidden sandstones and

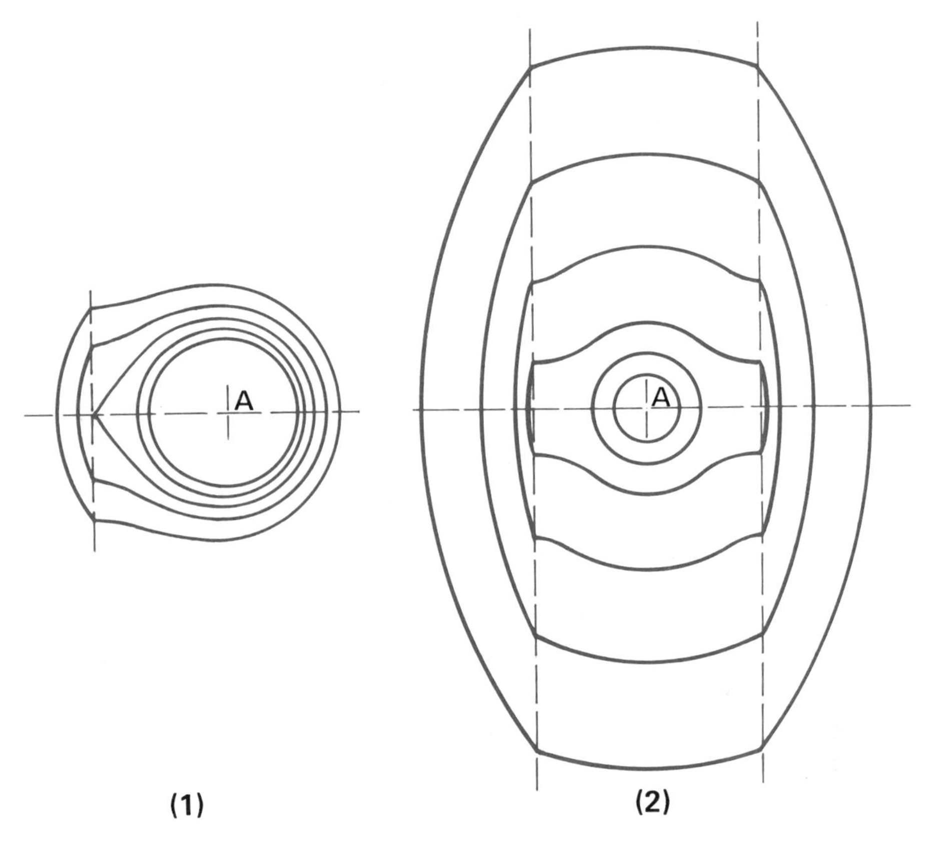

Figure 9. Left: *Calculated plot of the equipotential curves around an electrode* A *(with* B *very remote, practically infinite) and close to the vertical plan separating a more resistant medium to the left from a more conductive medium to the right. The curves are being refracted when crossing the contact.* Right: *Approximate image of the behavior of the equipotentials when* A *is located in the middle of a layer, the conductivity of which is higher than that of the flanking formations.*

shales. At that time a survey could be made only by drilling coreholes, and it appeared logical that the potentials map could complement the data obtained by drilling, if not become a partial or total substitute for it. After a few weeks of tests, prospects for industrial application were in sight.

The third experiment took place in January 1913 at Fierville-la-Campagne, in the Calvados iron ore basin (Fig. 10). The composition of the subsurface there is similar to that in Sassy, that is, nonoutcropping, very steep sandstones and shales totally overlain by 60 to 90 meters of

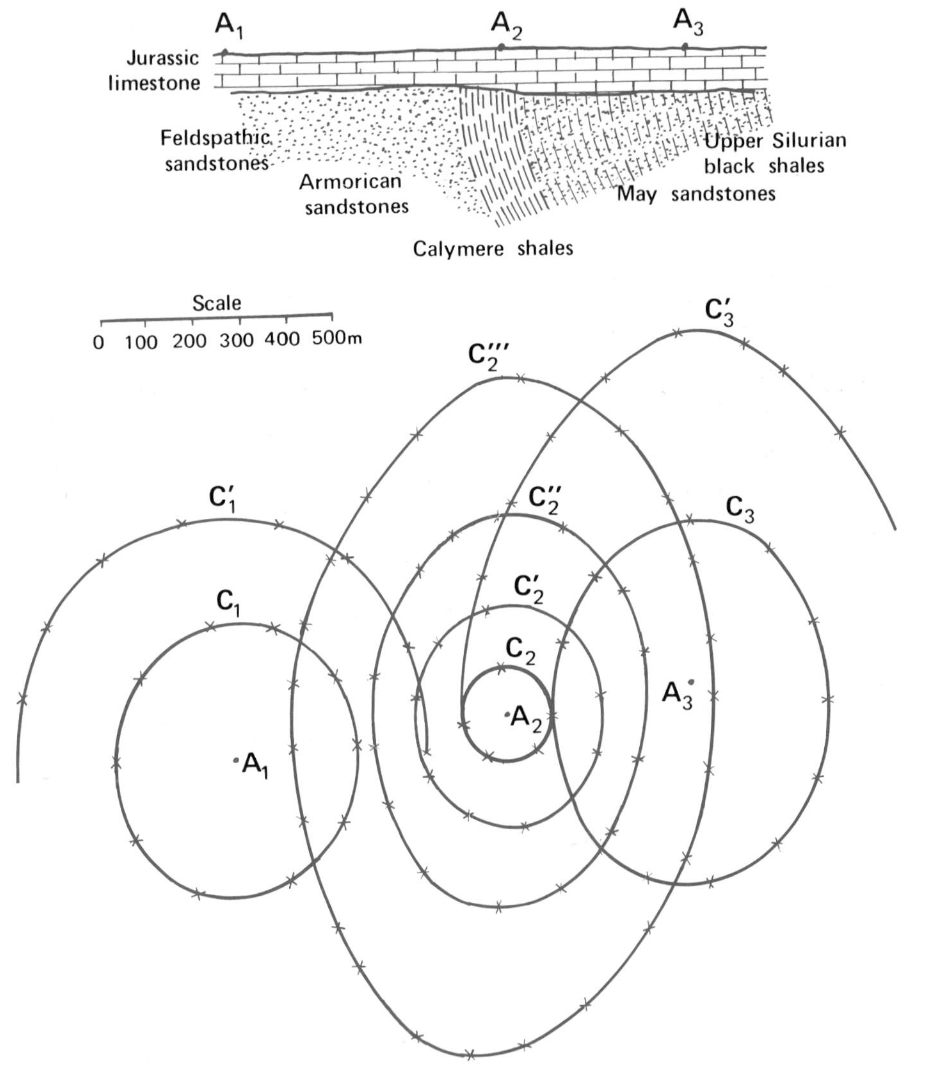

Figure 10. Test at Fierville-la-Campagne, Calvados, 1913. (Reproduction from C. Schlumberger, Etude sur la prospection electrique du sous-sol, Gauthier-Villars, 1920.)

marly Jurassic. Seven drillholes were required to outline the ore body and thoroughly explore the formations. As in Sassy, the strike of the ancient strata was clearly shown by the elongated shape of the equipotentials surrounding electrode *A,* which was planted straight above the shales, while *B* was placed a sufficient distance away so that it would not influence the shapes of the curves. However, the contact between sandstones and shales appeared less clearly, because of the blurring effect of the thick, overlying formations.

The flexibility of the electrical method was now to compensate for this lack of precision. As shown in Figure 3, homogeneous ground produces very uniform potentials in the center of the drawing: the electric field, that is, the potential drop by unit of length, remains constant. But if any discontinuity should appear in the soil between *A* and *B* and separate two beds of different resistivities (e.g., the contact between shales and limestone strata), the electric field would vary at the level of this separation; it would be weaker above the more conductive formations, and stronger above the more resistant ones. This led Conrad to test what he referred to as "field profiles." Electrodes *A* and *B* were placed on a line perpendicular to the strike of the strata, each about 1 kilometer away from the presumed trace of the contact, and the electric field between two very close electrodes, alongside three profiles parallel to *AB,* was measured. The result was clear: each profile showed an abrupt change in the value of the field, making it possible to locate the contact within an accuracy of 50 meters.

Another new feature was apparent on the right-hand side of the map: the equipotentials surrounding an electrode planted directly above the May sandstones were similarly elongated, and ran in the same direction as the curves plotted above the shales. Therefore the current from *A* flowed more easily alongside the strata than across them. This occurred because May sandstones are composed of thin, alternating layers of varied resistivities, affecting the current with what is known as an anisotropic medium. The curves plotted on the map confirmed the theory that in such a medium the equipotentials are ellipses elongated in the direction of the strata. This was the first empirical demonstration in the field of the electric anisotrophy of sedimentary layers.

Somewhat later, during 1913, the Soumont survey was made in the same basin with the same Silurian structure hidden by Jurassic. During that study a practical mining problem arose: the ore body under production was obstructed by a fault that had thrust its extension outside the working face of the heading. It was decided to use the electrical method for relocating the ore.

After having plotted a few large-diameter equipotentials and several field profiles, all of which confirmed the conclusions reached at Fierville, Conrad tested two new variants of the electrical method that enabled him to determine the slippage of the fault. These procedures lasted for several days and kept the small team of five very busy. Eugène Léonardon, a young graduate of the École Polytechnique and one of the first prospectors to join the firm, was a member of the group. He and a young worker named Jules Carré were to become popular figures of "La Pros."

The Soumont success, which Conrad still called a test, in face represented authentic mineral prospecting. It was later followed by another study in the same region near Sées, Orne, from which additional valuable information was extracted and in the course of which the techniques and analytical processes were refined.

By the middle of 1914, before World War I interrupted his work, Conrad had succeeded in demonstrating that physical measurements made on the surface could provide useful data for mineral prospecting, even when the ore could not be detected directly from the surface. By a thorough study of the structure of the ancient sedimentary rocks of the subsurface, it became possible to track down deposits of geologically related minerals. Although at the time only iron ore was involved, a few years later geophysicists all over the world would apply the same principle in their search for a mineral that had eluded any instrument designed solely for surface exploration: petroleum.

The exploration of conductive ores

While pursuing his experiments in direct prospecting of high-conductivity metalliferous ore bodies, Conrad had noticed that, after the current sent into the soil was switched off, small but measurable potential differences continued for some time. It seemed to him that the ground as a whole was polarized by the direct current and then discharged. A local incident led to a comment that he jotted down on the potentials map of the Val-Richer: straight above an iron water-supply pipe buried close to the surface, the equipotential curve had moved approximately 2 meters when the current was reversed. He reasoned that, because the pipe had become polarized, it affected the current differently, depending on the direction in which the current circulated. From this observation he assumed that a metallic mass, or a metal-like conductive ore, could be distinguished from the surrounding rocks by a particularly intensive polarization. A new prospecting method became feasible: send direct current into the ground, interrupt it, observe the effects of the polarization, and then look for the spots where they were abnormally high.

The idea led to little experimentation at the time, and when it was taken up again at a much later date it was not intensively pursued because efforts were directed toward new projects that seemed more attractive. By 1913 the induced polarization tests[3] were sidetracked after the discovery of another phenomenon that was much sharper and easier to exploit. This discovery would lead to a prospecting method widely utilized in the geophysical industry: spontaneous polarization.

In April 1913, near Sain-Bel, Rhone, Conrad was developing a potentials map of an area above a pyrite lode in the shape of an elongated lens and about 100 meters deep. Before any current had been sent between A and *B,* he noticed, in the vicinity of the ore, potential differences between nonpolarizing electrodes spaced less than 100 meters apart. These differences reached hundreds of millivolts above the axis of the lens (an amplitude substantially higher than that usually observed in such tests), and faded away as the electrodes were moved farther from the ore. Multiple observations confirmed the phenomenon: the measurements would accurately reproduce themselves. A new type of map was drawn (Fig. 11); with potentials becoming negative directly over the lens, the map not only indicated the outline of the pyrite already mined in the area but also showed the location of an unknown lens whose existence was confirmed a few years later. This was an instance of a spontaneous phenomenon occurring in the ground without any artificial stimulation. The pyrite lens acted like a huge metallic mass, whose upper parts had become oxidized by the circulation of aerated surface waters,[4] while its lower parts had escaped alteration. The pyrite lode, once polarized, generated electric currents that flowed around it from bottom to top, creating a center of

[3] Induced polarization has now become an efficient process, widely used since 1950.

[4] Such circulation and the resulting oxidation are enhanced by the seepages into mine works; they can be sufficient to generate a measurable phenomenon for a virgin lens, provided that its top is close enough to the surface.

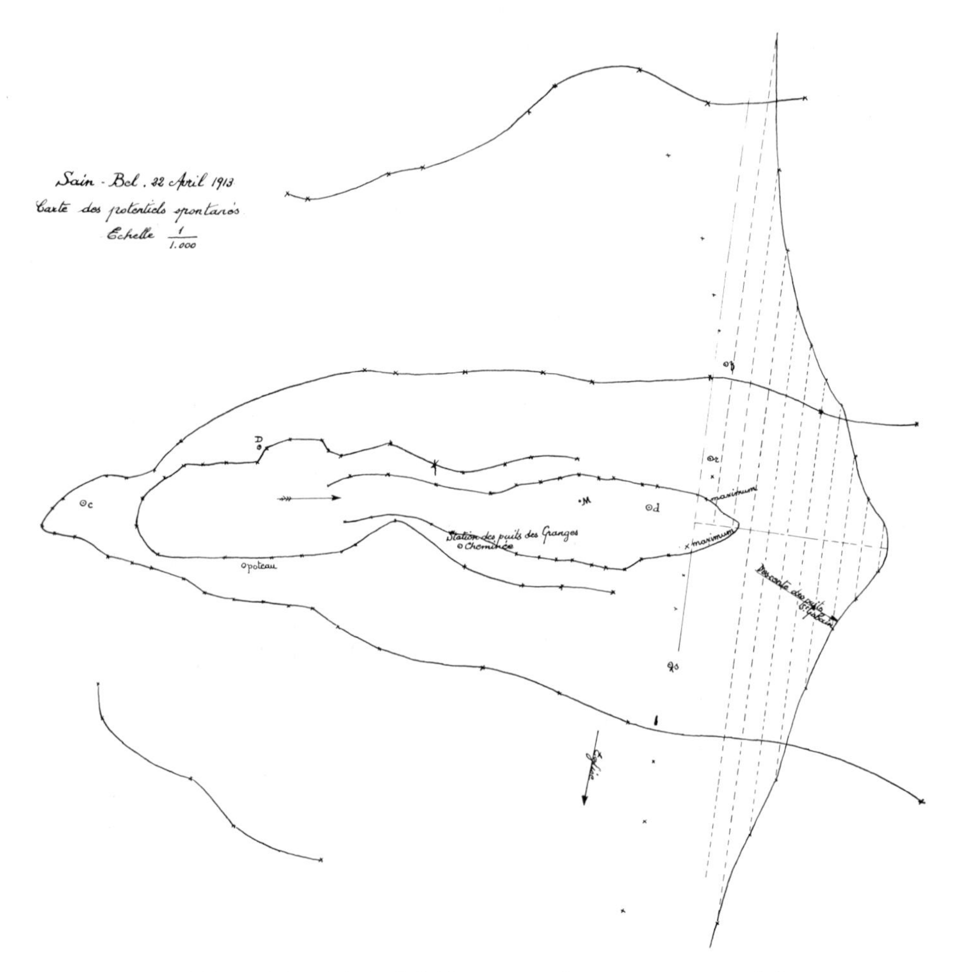

Figure 11. Sain-Bel (Rhône): map of spontaneous potentials.

negative potential at the surface above it. The validity of the assumption was demonstrated through imaginative laboratory devices and measurements. New studies at Vaux, Rhône; Saint-Félix-de-Pallières, Gard; Herrerias and Campanario, Spain; and Bor, Servia, all confirmed the Sain-Bel observations and demonstrated that spontaneous polarization (the term was eventually used to designate the mineral-prospecting process itself) characterizes all formations containing pyrites. As a result of the survey made there in the early summer of 1914, the Bor copper mine was expanded considerably. This was the first time in history that geophysics had been instrumental in the discovery of a nonmagnetic deposit.

During an earlier experiment at Bor, in September 1913, Conrad had succeeded in discovering a variant of the artificial current process: if a transmission electrode is planted in the mass of pyrite inside a mine heading, the whole ore body is raised to the same potential and becomes a huge electrode, accurately outlined by the equipotentials plotted on the surface. This so-called "grounding" method was subsequently used on several occasions.

By 1914, in the limited time available between his obligations as a professor at the École des Mines, Conrad had been able to advance the research work far enough to have in his possession an operational tool that had proved of value in several industrial surveys. However, he considered such studies to be of an experimental nature only and had sought no compensation from the mining operators who had benefited from his work. He continued to regard his experience, although encouraging, as too narrow for expansion into the commercial field. Only after World War I, in 1920, did he publish the first work describing his methods of electrical prospecting of the subsurface.[5] But he had been prudent enough to protect his inventions, and his first patent, *Procédé pour la Détermination de la Nature du Sous-sol au Moyen de l'Électricité* ("Process for Determining the Nature of the Subsurface by Means of Electricity"), had been filed in France on September 27, 1912. Several additional patents were to follow.

Out of a belief that nothing should be left unexplored, Conrad meticulously investigated the findings of other scientists and scrutinized whatever published material was available. He gave special attention to the experiments of the Swedish researchers Gunnar Bergstrom and Carl Bergholm, who had worked on the detection of metalliferous deposits through the transmission of current and the plotting of equipotentials by the earphone method. They had experimented for several years without producing any conclusive results, and only after 1914 did they contribute effectively to the field of mineral exploration. Moreover, it was a long time before they could achieve the accuracy

[5] C. Schlumberger, *Étude sur la Prospection Électrique du Sous-sol,* Gauthier-Villars et Cie, Paris, 1920.

assured by the use of direct current. Additionally, the phenomenon of spontaneous polarization was unknown to them. In this respect Conrad called attention to the work of R. W. Fox, described in an article in *Annales des Mines* (1837), and to the writings of Carl Barus (1880) and R. C. Wells (1914), whose field and laboratory observations he found most instructive. Yet nowhere could he find any account of their application to mineral exploration; in fact, in 1914, the director of the U.S. Geological Survey, George O. Smith, wrote in a preface to a work by Wells:

> It should be emphasized that the results thus far obtained afford no adequate basis for any method of electrical prospecting nor any promise of the development of such a method which would make possible the detection of an ore body by electrical measurements.

This was a few months after the Bor discovery. In all fairness to the director, however, it should be added that information about this discovery had not yet been published.

Phase Two: After 1919

Technical Developments

An overall view

During World War I some progress was made in the application of physical measurements to the search for minerals. Advanced magnetometers (inclinometer and Schmidt balance) became tools for mineral exploration. In addition to the contributions of the Swedes in the field of electrical exploration, Frank Wenner, a physicist of the U.S. Bureau of Standards, announced in 1915 the development of the four-electrode device bearing his name, for the measurement of formation resistivity. The possibility of using the Eötvös balance to locate potential oil-bearing structures had been recognized; however, thorough experiments conducted on salt domes in Czechoslovakia (then a part of the Hapsburg Empire), Germany, and Rumania had not resulted in any discovery. As to seismic methods, 5 years in the German artillery gave Ludger Mintrop the opportunity to meditate on the propagation of shock waves in the ground, but only after the war was he able to take out the patent that became the basis of the refraction method. Similarly, Conrad Schlumberger, then a French artillery officer, meditated on his side—whenever a lull gave him the opportunity—about his prewar studies, to which he could relate certain features in the military technique of ground telegraphy.

In 1919, with World War I at an end, Conrad decided to resume his work, placing a heavier emphasis on its application to industry. He teamed up with his younger brother Marcel, who brought with him practical experience in the mining business, as well as a creative mind and a talent for mechanics. The two brothers, with the moral and financial support of their father, established the Company that bears their name. In 1923 Conrad resigned from his teaching post at the École des Mines to devote himself fully to the new endeavor.

The Company premises at No. 30 rue Fabert in Paris consisted of five rooms. Three were converted into offices, while the other two became the workshop and the laboratory. Surrounded by maps, diagrams, and charts, the two brothers shared a desk, with a drawing board nearby. They supervised every aspect of the operation: conception and design of methods and instruments, manufacture of equipment, seeking out customers and drafting contracts, purchase orders, analysis of supplies, and invoicing. They directed and monitored the work of what they called the missions: the crews (usually consisting of two prospectors and a few assistants) in charge of carrying out the field surveys. For years they occasionally took part in the manual work required for the measurements: handling cables, generators, and electrodes, reading the potentiometer, and so on. They interpreted the results, discussed them with their clients, drafted the reports, drew up the budgets, and controlled the expenditures. Frugality was the rule.

In the beginning they had the help of a single engineer, Léonardon, who had rejoined them after the war. In addition, fieldwork had the unflagging assistance of two young men trained in vocational schools, Pierre Baron and Jacques Gallois, who were to have long and brilliant careers with the Company. In Paris J. Carré was the storekeeper, and R. Jacquin the mechanic and driver.

Thanks to the eagerness and dedication of its initial team, and also to the growing interest of the French mining industry, the small Company was able to expand, despite the unfavorable conditions of a period of economic reconversion following World War I. New collaborators gradually joined Conrad and Marcel. In 1922 Edward Poldini, a young Swiss geologist, became their part-time consultant. Priority was given to the recruiting of graduate engineers with university-level training. Mission personnel was provided mainly by the various écoles des mines[1] and the École Centrale,[2] with top consideration for grades received in geological

[1] Graduate institutes of mining engineering. (Translator's note.)

[2] École Centrale des Arts et Manufactures, a multidisciplinary graduate institute of technology. (Translator's note.)

and electromechanical training. The Écoles des Arts et Métiers[3] and the École Supérieure d'Électricité[4] provided the candidates for the engineering office, but there was never any sharp discrimination in this respect. In fact, prospectors available between missions could be assigned to laboratory work or instrument design. In any case recruiting in the first years drew on widely diversified backgrounds, from the Institut Agronomique[5] to the Écoles des Maîtres-Mineurs.[6] In the early years of the Company several employees were doubtful about its future and left, but in 1924, when an expanding world economy brought in long-term contracts, especially in Rumania and the United States, and stabilized the development rate of the Company, some of them rejoined. Also, new staff now joined with confidence. In the summer of 1926, after a few months of part-time collaboration, Henri Georges Doll, Conrad's son-in-law and a graduate of the École Polytechnique and the École des Mines, joined the Company. With his genius for solving electrical problems, he soon clarified and improved measurement procedures. During his long career he was responsible for a continuous series of new developments in both instrumentation and methods. Félicien Mailly joined the Company in 1925 and for 20 years was the mainspring of its electrical research laboratory.

On July 1, 1926, the Société de Prospection Électrique (S.P.E.) took over all of the Schlumberger brothers' operations, including personnel, equipment, and premises. In Paris and in the field as well, everybody was already calling the enterprise "La Pros." One year later the staff had grown to 16 engineers or equivalent technicians.

The concept of apparent resistivity

Two essential elements contributed by Conrad and Marcel enabled "La Pros" not only to survive but also to expand its operations: the concept of apparent resistivity, and the potentiometer. At this point it is appropriate to describe briefly the methods and techniques that evolved and were developed during the 1920's from these two elements.

[3] Technical schools emphasizing the practical training of mechanical engineers. (Translator's note.)

[4] Graduate Institute of Electrical Sciences. (Translator's note.)

[5] The highest agricultural institution in France. (Translator's note.)

[6] These schools produced originally mining foremen, but were upgraded to schools of mining engineering with emphasis on practical training. (Translator's note.)

To this day the notion of apparent resistivity has remained a basic concept in electrical processes used on the surface and inside boreholes. Although the potentiometer has been largely replaced by the photographic recorder since the 1940's, it is still used in certain kinds of surface work.

Before 1914 all the tests in which current was emitted into the ground had consisted of plotting curves of equal potential. The electric field profiles were nothing but a variant of this method. The surveys, resumed in 1920 for the reconnaissance of ancient, sharply dipping formations, showed that the potentials map technique was limited in its application. First, the electrical field becomes very strong when measurements are made close to the electrodes (Fig. 3); this phenomenon can distort the interpretation of the map when the medium is not homogeneous, as is almost always the case. Second, when the survey extends over a large area, a single map of potentials, referred to (stationary) electrodes *A* and *B*, becomes inadequate. The measurements must be repeated while moving line *AB*, and it is no easy task to link such a succession of readings. Finally, since the differences in the voltage between successive equipotentials are proportional to the intensity of the current transmitted into the ground, they reflect, with only approximate accuracy, the resistivity variations. They do not give any absolute values, even approximate ones, and therefore are not comparable from one area to another.

At this point it was decided that the reference parameter would be called apparent resistivity. In a configuration of four ground electrodes *ABMN,* where *MN* is located between *A* and *B,* a current is emitted into *AB* and the corresponding potential difference is measured between *M* and *N.* The apparent resistivity equals the ratio between the potential difference and the intensity of the current emitted, multiplied by a factor that is a function of the respective distances between the four ground electrodes. This coefficient is easy to calculate and is such that, if the system is placed in a homogeneous medium, the apparent resistivity must be equal to the resistivity of this medium.

The apparent resistivity reflects an average value, depending on the resistivities and spatial shapes characterizing the various formations included in the total volume of the subsurface being measured.[7] The vol-

[7] The word "apparent" is often omitted in the professional language without any resulting ambiguity. "Resistivity" designates a precise parameter, which characterizes certain mat-

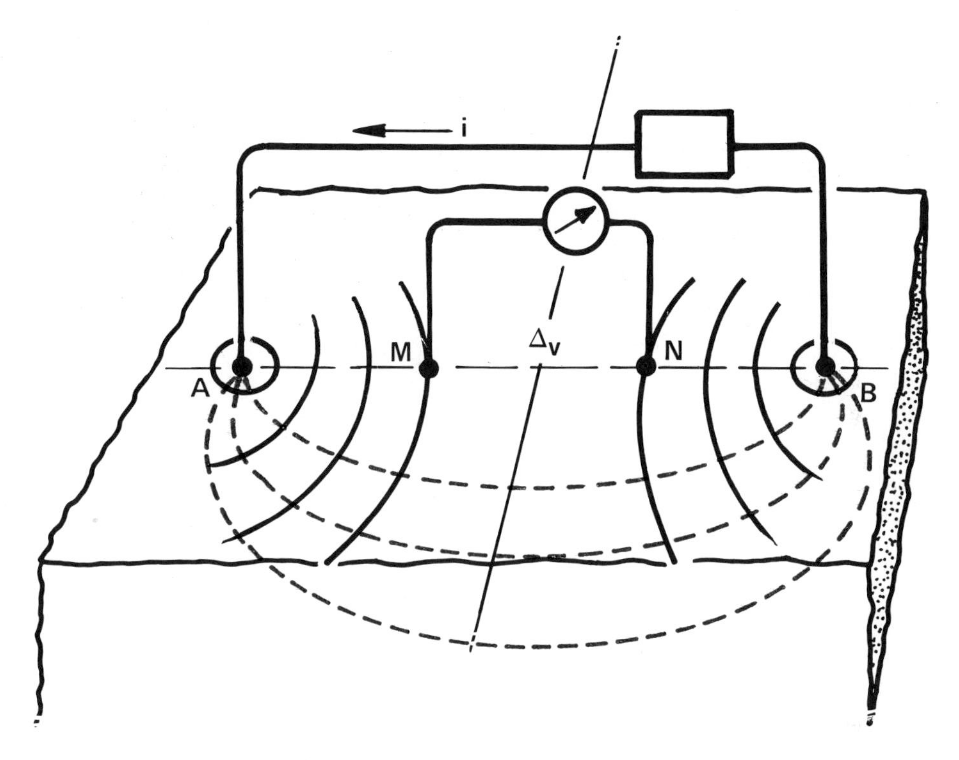

Figure 12. Measurement of apparent resistivity (schematic).

ume itself is not sharply delineated, because the density of the current is not suddenly reduced to zero at a certain depth, but decreases and disappears gradually in all directions as the distance from the electrode system increases.

Figure 12 shows a current being emitted along the line *AB* and, in a vertical plane, lines of the current expanding as a function of depth. These lines radiate around electrodes *A* and *B* when in the zones close to them. If electrodes *M* and *N* are close to *A* (Fig. 13), the voltage difference of two rather shallow equipotentials is measured. As the electrodes move further away from *A*, the equipotentials become deeper and deeper. In

ter (e.g., a rock) when measured in the laboratory on a representative sample. The same word also designates apparent resistivity, that is, an average, when the measurement takes place with an *AMNB* configuration stretched over the ground.

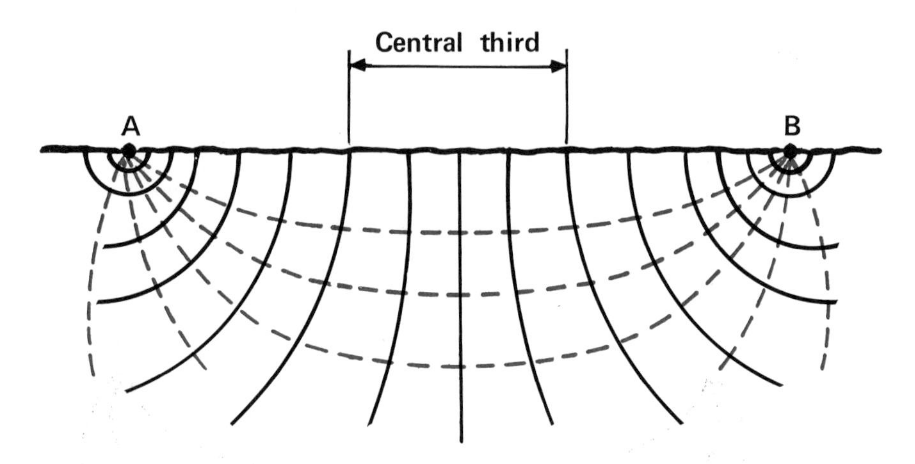

Figure 13. Variation of the depth of investigation according to the position of MN *between* A *and* B.

other words the depth of penetration of the system is in direct proportion to the remoteness of *MN* from *A*. This depth reaches its maximum, then remains practically constant when *MN* is within the central third of *AB*, and decreases again when *MN* moves closer to *B*. In the central third the equipotentials are nearly vertical surfaces (the current lines being approximately horizontal). Originally it was assumed that, when *MN* is within the central third of *AB*, the slice of ground below *MN* to a depth equal to *AB*/4 had a prevailing action on the value measured.

Thus the depth of investigation became quantified. Granted, this was a convenient and provisional approximation, reasonably correct only in a homogeneous medium. It was recognized that in real formations consisting of layers of different resistivities the distribution of the current lines would be more complex than indicated in this ideally simple case. It was a combination of empirical evidence acquired in the field and scientific data computed by mathematics and laboratory research that gradually defined the relationship between apparent resistivity and its various components, and increased the accuracy of depth of investigation.

The initial system consisted of a line several hundred meters long. Electrodes *M* and *N* were located 20 to 50 meters apart on line *AB*, and a series of measurements was made by displacing *MN* within the central third of *AB* (this was called the *AB* profile technique). The cables connecting *A* and *B* to the source of electric current were laid on the ground. Once a profile was completed, the *AB* configuration was moved either sideways or in the same direction, depending on the type of problem being solved. In the latter case, care was taken to achieve continuous measurements. A slightly more complex method first used in Rumania (hence the name "Rumanian method") consisted of making four series of measurements around *A* by moving *MN* into four different azimuths for each position of *AB*. The same was done around *B*. It would be tedious to describe other configurations used, though they clearly illustrate the difficulty of selecting the best technique.

Later, a more expeditious method was adopted. It employed a regular network of measuring stations with the same *AB* spacing, while electrodes *M* and *N*, always on the same alignment and within the central third, were symmetrical about the center point of *AB*.[8] Thus all measurements were made with the same average depth of investigation. The stations were equidistant from one another and positioned alongside alignments as straight as possible. The *AMNB* configuration stretched along the same alignments and was moved from station to station by a distance equal to *MN*, thereby resulting in adjoining measurements. Since the operation required that the cables be dragged on the ground, this resistivity profiling was called the "dragging" technique at "La Pros." Until about 1927 the *AB* spacing never exceeded a few hundred meters, and the rather lightweight equipment could easily be carried on a pickup truck.

Resistivity profiles were used for the first time in the fall of 1920, during a survey of an iron ore-bearing syncline at May St. André in Normandy. The problem was similar to those that had prompted the above-described tests in their early stage: reconnaissance of tilted Silurian formations, predominantly shales and sandstones, concealed by a nearly horizontal Jurassic caprock of marls and limestones 20 to 40 meters thick. However, the area to be surveyed was substantially larger. The survey involved a series of nearly parallel profiles, about 2 kilometers long, each

[8] This configuration was very similar to that mentioned in 1915 by Frank Wenner, who had introduced the term "effective resistivity" to designate the same concept as "apparent resistivity." The latter has come into common use. In the Wenner configuration, however, *MN* is always equal to one third of *AB* regardless of the length of *AB*. Wenner's measurements were limited to quite shallow depths, and he operated with very short lines (a few tens of meters), using alternating current. However interesting, his studies remain mainly theoretical.

with a south–north orientation, that is, perpendicular to the general strike of the Silurian. The depth of investigation was such that the shape of the Silurian formations was clearly revealed.

Figure 14 shows one of these profiles with the corresponding geological cross section: peaks appear above the sandstones, troughs above the shales. By correlating the peaks of these profiles, it was possible to plot the trace on the ground of the various formations hidden under the Jurassic caprock.

Shortly thereafter, a cartographic representation, clearly indicating the values plotted, became the simple complement of the profiles. The resistivity values plotted at each station were transferred to a topographic map, and curves of equal value were drawn by interpolation. The result was a "resistivity map" where the equal-value curves showed the limits between various zones of apparent resistivity. The zones were marked by conventional colors: red for higher conductivities, blue for higher resistivities, and so on.

When the purpose of prospecting is the reconnaissance of steeply dipping ancient formations overlaid by a caprock, the *AB* spacing is

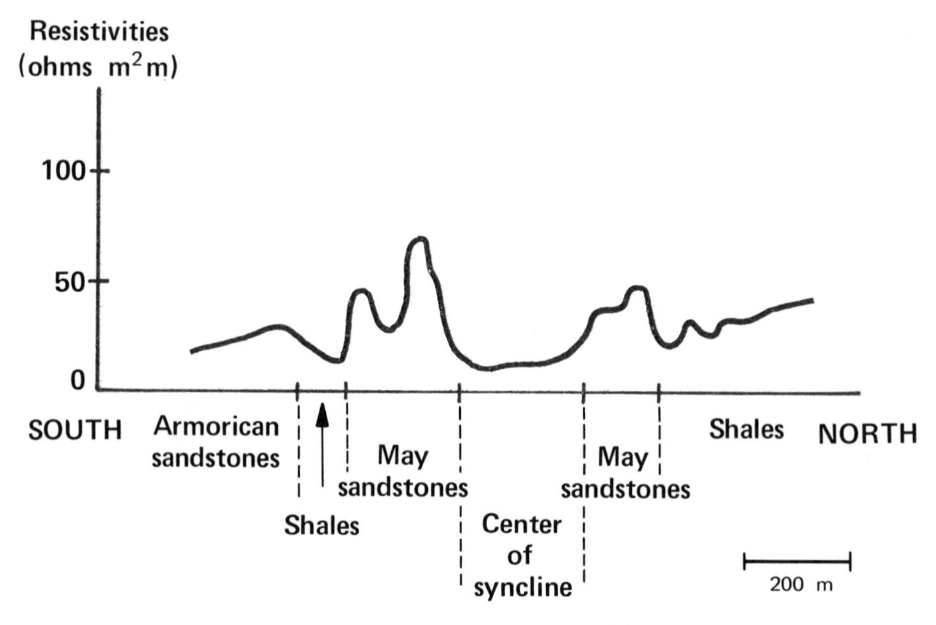

Figure 14. First resistivity profile (May St. André, Normandy, 1920).

adjusted commensurately with the thickness of the caprock, so that the resistivity measurements clearly reflect the formations being surveyed. When the caprock is rather uniform, the resistivity map looks like a geological map plotted as if the caprock had been pulled off; this is called "geological skinning." It is important to remember that in such cases the resistivity map reflects only rather gross features and gives a more global or less detailed description of the formations than does the geological map. On the other hand, the method has the advantage of spotting elements invisible on the surface: faults, stratigraphic contacts, and the like. For the Normandy synclines the result was a sketchy outline of the upturned Silurian strata at the beginning of the Jurassic period, when the surface of the earth had been eroded into a peneplain, and an invasion of the sea had begun to bury these strata under new sediments.

The application of the resistivity method to the study of low-dip, nearly horizontal formations came a little later, in 1923, under the stimulus of oil exploration. At one time it may have been thought that oil could be located directly through surface resistivity surveys; but this proved impossible since, in spite of a volume reaching millions of cubic meters, oil deposits represent only a minute fraction of the formations containing them. On the other hand, it seemed reasonable to expect that the resistivity map would furnish clues regarding the position of the subterranean structures suitable for the accumulation of oil, and thereby provide indirect guidance for its discovery. This was also the objective of the gravimetric and seismic methods, which at that time were undergoing their first tests.

Figure 15 represents schematically a hypothetical vertical cross section of formations across the axis of an anticline. The porous and permeable layer where oil may have accumulated—a potential reservoir—belongs to a very thick sequence of mainly argillaceous sediments with a rather low resistivity (about 5 ohm-meters), capped by formations of higher average resistivity (30 ohm-meters). Since the surface is approximately horizontal, the thickness of these formations above the top of the anticline decreases. An *AMNB* configuration of appropriate length, when moved on the surface in a direction perpendicular to the axis of the structure, will record a minimum of apparent resistivity above the top of the anticline, where the thickness of the resistant formations is lowest. If now a series of profiles is plotted at suitable intervals, the result should be a resistivity map on which contour lines of equal resistivity will reflect the

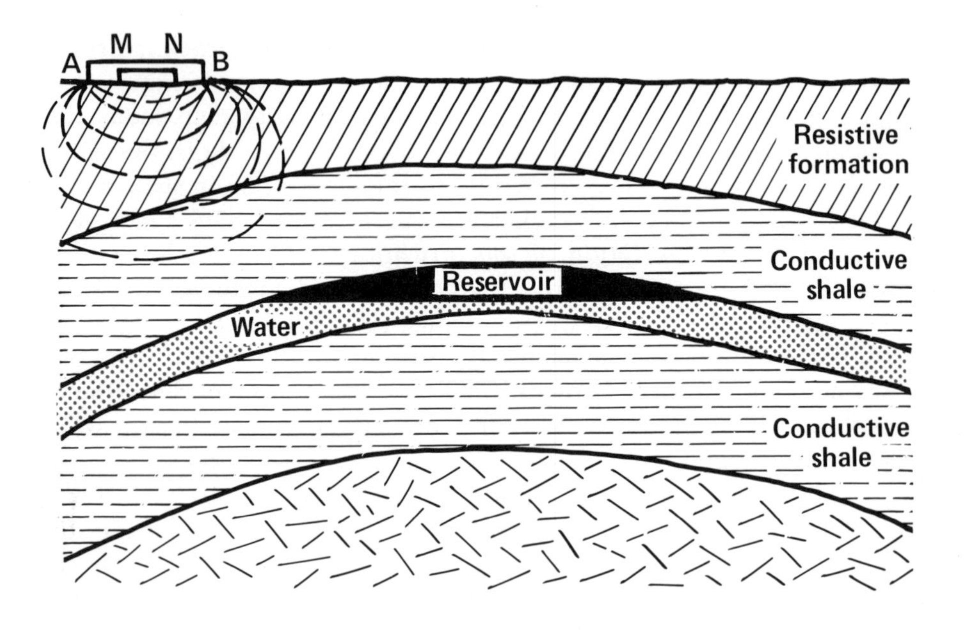

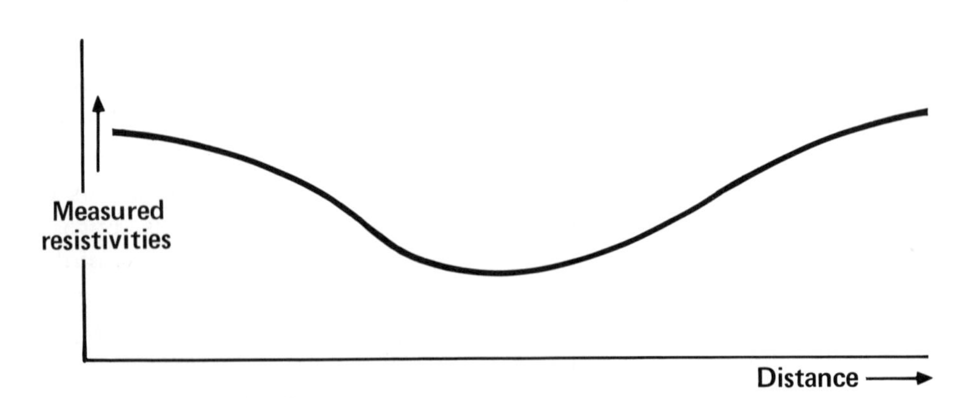

Figure 15. Hypothetical cross section of an anticline and corresponding resistivity profile.

isobaths (lines of equal depth) of the top of the shales, as well as the general shape of the structure. The reasoning is identical if the deep underground is more resistant than the shallow formations.

The analysis of sedimentary series is usually less simple. If the composition and hence the average resistivity of the upper formations vary laterally, this variation will affect the resistivity map and make its interpretation more complex. In fact, to distinguish between the respective influences of the shallow and deeper formations, it is necessary to use two different spacings for *AB*, corresponding to two different depths of investigation. For example, by comparing the readings it is possible to determine whether a portion of the resistivity map indicates a deep tectonic feature such as an anticline, or merely a change in the lithological structure or thickness of the shallow formations. As a general rule, the resistivity map will be plotted only with a long *AB* spacing; the short spacing serves mainly as an aid in its interpretation.

"Dragging" with two spacings, whether for low-dipping or for upturned formations, has remained to this day one of the basic techniques of electrical prospecting. Although it seems simple enough, in the 1920's it was such a valuable innovation that Conrad and Marcel, and particularly Eugène Léonardon, endeavored for several years to keep it secret. In fact, until protected by a patent, the very terms "resistivity map" and "apparent resistivity" were carefully omitted from all communications, such as reports and publications outside the Company. Instead, vaguer terms such as "electrical profiles" and "parameters" were used.

It frequently happens that geological data fail to indicate whether or not a dip is steep. Usually greater accuracy in this regard may be obtained from the resistivity map. A rapid variation of values, that is, a tightening of the equal-resistivity curves, reveals a feature characteristic of two different formations making contact alongside a steeply dipping plane, be it a stratigraphic boundary, a fault, or a sheer fold. Widely spaced curves are indicative of slow variations and are generally evidence of formations that have a smooth tectonic trend with very low dips. To complete and refine the data of the resistivity map in such zones, a new method was introduced: "vertical electrical sounding," a procedure that has remained basic to surface electrical prospecting. Although the method is suitable mainly for the reconnaissance of strata composed of horizontal layers, it is also applicable wherever the dips do not exceed a few degrees. It consists of a series of measurements carried out with increasing *AB* spacings, while the center of *AB* remains in a fixed position. Thus each station records apparent resistivities pertaining to slices of ground of increasing depth. Eventually, the rather short spacings used initially reached several hundred meters.[9] The results of the measurements were then plotted on a graph with apparent resistivities as ordinates and the successive values of *AB* as abscissae. The coordinates were linear. The curve reflected the depths of the formation boundaries, provided that the formations remained reasonably homogeneous horizontally, at least within the range of the investigation.

In most cases vertical sounding curves are representative of an area where the subsurface is divided into several thick layers, each distinguished by its average resistivity. In fact, each of these layers is, geologically speaking, a stack made up of many thinner layers. Whatever the contention of certain researchers of that time, electrical soundings are insufficient for identifying every single layer of a stratigraphic series.

The potentiometer and its accessories

As has been described, before 1914 Conrad employed a rather rudimentary measuring apparatus for his testing and research. Now the time had come when it was necessary to tool up for industrial operations for which dependability, efficiency, and cost were as important as the accuracy of the measurements. What was needed was a device simple to operate, of limited weight yet sturdy and highly sensitive; it also should have a rather short time constant and adequate electrical insulation. Whatever the terrain or the climate, the measuring tool had to be easily transportable by man or vehicle, without risk of damage that would inhibit a rapid resumption of the readings.

Once the exact specifications had been drawn, the Hartmann and Braun firm was selected to manufacture the device. It took only a few months to produce the famous "black box," which for the next 20 years was to remain the basic tool of prospecting and is still used for certain types of surface work. Compact, lightweight, and watertight, the box screwed onto a folding tripod and could be carried on the shoulder like any topographic surveying instrument. The galvanometer it contained was

[9] After 1927 the spacings were extended to several kilometers.

as sensitive as a laboratory instrument,[10] but once locked (at the touch of a button), it could be moved about safely.

The measurements determined a potential difference between two ground points through electrodes, the resistance of which usually changed from one station to the next. The reading of such differences on the dial of a galvanometer, serving as a millivoltmeter, would have required the insertion of a high additional resistance into the circuit to render negligible the resistance of the ground electrodes. This, however, would also have reduced the sensitivity of the circuit to a point incompatible with a precise reading. For this reason, the instrument included a potentiometer,[11] which made it possible to use the null method with the assurance that the measurement was independent of the total resistance of the circuit. In this method the potentiometer circuit consisted of an ordinary 1.5 volt dry battery connected through appropriate resistors in opposition with the potential difference being measured. The resistors were adjusted through knobs graduated in millivolts (one knob for tens and one for units) until the two potential differences reached equilibrium and the reading in millivolts thus indicated the value of the measurement. The potentiometer had four sensitivity ranges.

Nonpolarizing electrodes of the kind built by Conrad for his prewar tests were required for spontaneous polarization measurements. Since the potential difference between these electrodes was very low, the needle remained near zero when the station was far from the ore, provided that there was no noise from stray or telluric currents (see p. 85). When approaching a deposit, the spontaneous currents deviated the needle; by bringing it back to zero, the millivolts could be read as indicated above.

Nonpolarizing electrodes were first used for resistivity measurements; later, they were replaced by copper pegs whose polarization was compensated for by a small auxiliary instrument. Under such conditions, with no current flowing between *A* and *B,* and on the assumption that the noise level was weak enough not to interfere, the needle of the galvanometer would remain stable and near zero. When the emission of a current generated a potential difference between *M* and *N,* it deviated the needle, sometimes out of scale. The measurements were made by succes-

[10] From 25 to 150 x 10^{-8} ampere per scale division.

[11] As a matter of fact, the name "potentiometer" has always been extended to include the whole instrument ("the pot" in "La Pros" language).

sive approximations. The circuits of current transmission and of the potentiometer's dry battery were closed and opened simultaneously and repeatedly, the duration of each closing varying from a fraction of a second to several seconds, until two successive current emissions, corresponding to two adjoining graduations of the millivolt knob, slightly deviated the needle in opposite directions; the measurement was then bracketed between the two graduations, the needle of the galvanometer serving merely as a mark.

To obtain accurate resistivity values with an unreliable standard like a portable flashlight battery, the technique consisted of using the same potentiometer, and hence the same battery, to measure the potential difference between *M* and *N* as well as the intensity of the current transmitted, the apparent resistivity being proportional to the ratio between these two quantities. A precisely calibrated shunt of 0.1 ohm was placed in series in the *AB* circuit. To measure the current, the shunt was connected with the input terminals of the potentiometer; from that point on, the operation was the same as in the preceding case, the intensity of the current being read on the knobs.

With the *AB* profile method the prospector moving the potentiometer from one *MN* location to the next would also move away from the generator (initially a motor–generator unit, but soon thereafter dry batteries) and from the switch. The latter was operated by an assistant who closed and opened the circuit in response to horn signals from the operator, who, in turn, synchronously closed and opened the potentiometer circuit. The switch was contained in a box, together with the shunt and an ammeter, and placed on a tripod. At the beginning and the end of the profile, the prospector measured the value of the current with his potentiometer by connecting the input plug to the terminals of the shunt; meanwhile, the assistant checked to see that the current had not varied on the ammeter, as occasionally happened with worn-out batteries. Later, the operation became more practical when the box containing the shunt and the ammeter was fastened on the right-hand side of the potentiometer; the prospector could then control the current transmission from the very point where he was making his measurements, and the horn signals could be eliminated. The switch was operated by a button located next to the button that opened or closed the potentiometer circuit. Through a clip resting on both buttons and a small system of springs located under the instrument (the arrangement looked like the hand control of a bicycle brake—hence the name "Bowden"), the operator was able to open and close the two circuits simultaneously by a pressure of the hand.

Whereas nonpolarizing electrodes were essential for the measurement of spontaneous polarization, several years passed before it was realized that they were not required for resistivity measurements. The discovery was made in April 1925 in the Haguenau forest, where resistivity profiles ("draggings") were made with short spacings (250 meters for *AB*, 50 meters for *MN*). Conrad was visiting the team, and while taking part in the measurements, he noticed that the needle deviated every time Roger Jost, the young prospector, moved his foot in the vicinity of an electrode.[12] It might be, Conrad thought, that if a dry battery were connected to the terminals of the potentiometer, the needle would return within scale. A battery purchased from a local store was found to be suitable. It was round and yellow and looked somewhat like a hunting cartridge; hence the assembly (i.e., a few solders and resistors) was christened "the millivolt cartridge." This also became the name of the new black box, now built with a circular rheostat on the left-hand side of the potentiometer. The nonpolarizing electrodes were replaced by copper pegs, and the electromotive force generated by their polarization was compensated for by the cartridge. In addition to the advantages of simple assembly and operation, this device substantially decreased the resistance of the ground electrodes, thus increasing the sensitivity of the system.

Finally, the potentiometer (Fig. 16) was completed by the addition, on a third side, of yet another black box containing an induction corrector. This was a small, adjustable transformer, with its primary winding in series in the current circuit and its secondary in the measuring circuit. Its purpose was to correct the mutual induction between *AB* and *MN* circuits, which, when long spacings were used, caused the needle to jump every time the current was switched on or off. The fourth side belonged to the operator; it had to remain open since it carried the input terminals: the connectors plugged into the cartridge on the left and into the terminals of the shunt or into the double jack leading to *M* and *N* on the right.

There is little to be said concerning the other accessories. Since no

[12] Probably a phenomenon of electrofiltration: when more or less mineralized water flows through a porous body, the motion generates an electromotive force proportional to the pressure producing the flow. In the case above, when Jost moved his foot, the changing pressure on the loose soil (humus) resulted in changing the electromotive force and hence the stray current flowing through the electrode, the line, and the potentiometer.

Figure 16. Potentiometer with switch and shunt on the right, polarization compensator (millivolt cartridge) on the left, induction corrector facing. The Bowden cable can be seen between the legs of the tripod.

polarization problem arose for the ground electrodes A and B, they were made of wooden, iron-sheathed pegs, while copper was used for M and N. They were driven into the ground with steel-hooped mallets. The cable was made of stranded copper wire, later reinforced with steel wire, and came in 500 meter spools. It was dispensed by hand-operated reels mounted on tripods. Convenient as the equipment was, its operation sometimes caused difficult problems. Roads were not always available for the pickup trucks. More than once the cables had to be unwound by hand across woods, dunes, or marshes, and batteries and pegs had to be carried by the team in all kinds of weather. In humid or rainy weather, insulation defects were irritating because leaks in the AB circuit had an ill effect on the measurements.

For years the effect of leaks could be only roughly estimated, and since nobody knew exactly how they occurred, it was impossible to obviate them. One of Doll's first contributions was to analyze these leaks and to show how to identify and to counteract them. He particularly stressed the importance of connecting the metal box of the potentiometer to the potential of the shunt when the current was being measured; this prevented leaks and electrostatic effects on the winding of the galvanometer. He also shed light on the effects of the resistance of the ground electrodes on the errors arising from the leaks. A small special device, called relay resistance, was connected in series next to each ground electrode *A* and *B*; by reversing the current, it was possible to multiply the resistance of the ground electrodes by an average of three to five times. The necessary control was provided by transmitting the current in both directions and comparing the measurements.

Taking the readings themselves often proved difficult. With very low resistivities and very long electrode spacings, the potential differences could become as low as a fraction of a millivolt and require the highest sensitivity range of the potentiometer. The total resistance of the measuring circuit could be so high that at the closing of the current the deviation of the needle was hardly visible. A tiny microscope fixed above the dial helped in such cases. Additionally, should the needle become erratic because of stray currents, it was very time consuming and difficult to achieve an accurate reading (Fig. 17).

In the case of long spacings and low resistivity formations, another complication arose from the phenomenon known as skin effect. Since at the closing of the circuit the current concentrates near the surface before spreading in depth, the potential difference between *M* and *N* was temporarily higher than after stabilization. As a result the needle deviated beyond any possibility of correction, and a delay of several seconds was necessary before equilibrium could be reestablished and the reading taken.

When a survey takes place close to industrial installations such as electric railroads or power stations, stray currents can reach such an intensity that they cause erratic needle fluctuations which preclude any measurement with direct current. On the other hand, alternating current of the sinusoidal type is also unsuitable because of the mutual induction between *AB* and *MN* and, in the case of long spacings, because the skin effect reduces the penetration of the current into the ground and, hence,

Figure 17. A painstaking measuring with the potentiometer (1928).

the depth of investigation.[13] As a remedy Conrad devised a variant of the direct current technique, consisting of rapid periodic reversals,[14] and the first "pulsator" was born.

Basically this instrument was made of two sections operating synchronously, one in the current emission circuit, the other in the measur-

[13] The reduction becomes larger as the frequency of the current increases, the length of the transmission line increases, or the resistivity of the ground decreases.
[14] The idea was proposed at the same time (1926) by O. H. Gish and W. J. Rooney.

ing circuit. Switched off and reversed 15 to 30 times per second, the direct current was sent into the ground electrodes *A* and *B*, thus creating interrupted and reversed differences of potential, superimposed on those generated by the stray currents circulating between *M* and *N* (Fig. 18). The pulsated reversed component being rectified, the potentiometer received 15 to 30 continuous impulses per second, separated by pauses too short to be perceptible; and since the parasitic stray component reached the potentiometer with the same pulsated-reversed frequency, the result was only a very slight wobbling of the needle about a fixed position: the inconvenience of an erratic needle was eliminated.

In practice the pulsator was made of two cylindrical collectors mounted side by side on the same shaft, and connected by brushes to the emitting and measuring circuits. The first model designed by Conrad was faultless in terms of physics, but not with regard to its mechanical functioning. Two bronze bearings, in particular, required a most delicate and rather precarious adjustment. This made the model impractical for handling in the field, and as a result it had a short life span.

Marcel then tackled the project, on the basis of one of those simple and practical ideas that were his forte. What was needed was a device with a double collector that rotated much faster than its driving crank. He thought of a tool available in any hardware store, namely, a hand grinder. All he had to do was substitute the collector for the grinding wheel, with the necessary brushes and wiring. The result was an instrument that served for many years in surface prospecting; over several decades it was gradually improved and became part of logging equipment until it disappeared around 1955 to 1960 with the advent of electronic instruments.

A rather long idle time between phases had been provided in the measuring part of the pulsator to eliminate the effect of the transitory phenomena occurring at the closing and opening of the current pulses. This was satisfactory for surveys with relatively short spacings on rather resistant ground, but it became inadequate with long spacings and on conductive ground where the transient from the skin effect can last for 1 second or more. At a frequency of 15 to 30 cycles per second, the current no longer had time to penetrate to the depth it would have reached under a steady state. In other words, the depth of investigation was substantially reduced to about the same extent as if a sinusoidal current of the same frequency had been used. For this reason, surveys with long spacings were restricted for many years to the direct current technique with all its

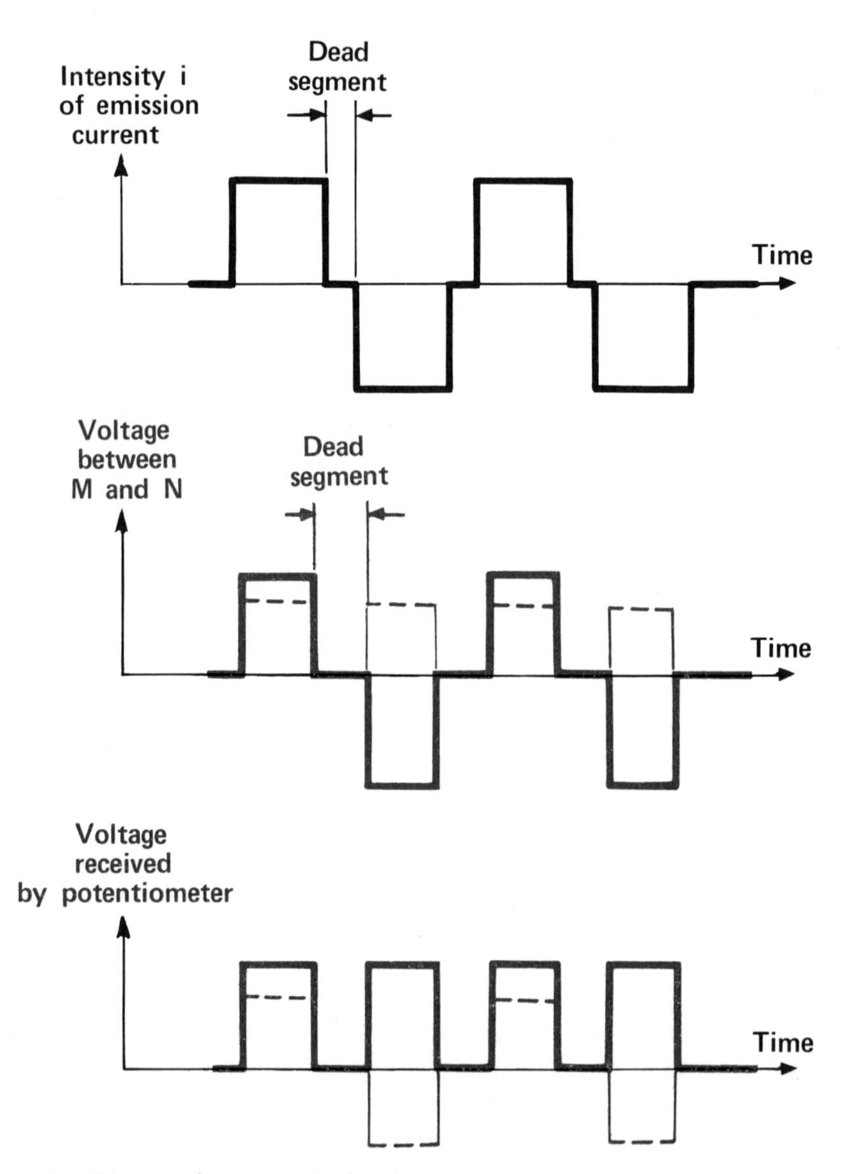

Figure 18. Diagram of operation of pulsated-rectified current. Center: *Solid lines show voltage created between* M *and* N *by the current emitted. Dotted lines show voltage from stray currents, with variations very long as compared with emission frequency.* Bottom: *Solid lines show rectified voltage sent to potentiometer. Dotted lines are pulsated-inverted stray voltage, producing a mere vibration of the needle.*

shortcomings. For the prospector this meant spending a long time at each station, bent over his microscope and pressing the Bowden as many as 20 or 30 times, with 2 or 3 seconds for each closing, while endeavoring to notice some acceleration or slowing down of an erratic needle—all of this under broiling sun or in freezing cold. In heavy rainstorms standard equipment for each team was a large umbrella. The current transmitted was pushed up to 2 amperes and more. The batteries, designed for high voltages and low weight, polarized rapidly under this treatment: voltage was lower at each closure, the current had to be measured several times, averages had to be made, and so forth. Not until 1940, after adapting the photographic recorders for telluric surveys, would Marcel be able to improve the operation by recording the measurements on film. This subject will be discussed later in connection with tellurics (p. 90).

Toward increasing depths of investigation

By 1927 the resistivity measurement techniques had already been seasoned by years of field experimentation. Although these techniques remained basically unchanged, some new developments resulted from a broadening of their application to the study of deeper geological structures.

Until then, the location of salt ridges or domes had been determined by the red (conductive) areas of the resistivity map, which were supposed to indicate the zones where the upthrust of the salt had thinned out the superficial formations (alluvium) overlying the older and more conductive marls. For this purpose the use of a short spacing of the current electrodes (400 to 500 meters) was adequate, except that the image of the objective, namely, the top of the salt, was rather distorted. Furthermore, the red areas could be indicative of something other than the upthrust of the salt: as the overlying alluvium becomes more shaly, its conductivity increases. On the other hand, the alluvium even on top of the salt is sometimes so thick that such short spacings do not reveal anything. For these reasons it became important to increase the depth of investigation in order to obtain resistivity maps indicating the presence of the salt itself and to "see" it, at least at its structural highs. Since salt is electrically insulating, such highs would appear as blue (resistant) areas on the map. The tops of most salt domes lie at hundreds rather than tens of meters under the surface, necessitating a corresponding increase in the spacing of the current electrodes.

The same problems existed in regard to the reconnaissance of anticlinal structures without a salt core. In a stratigraphic series assumed to contain a petroleum deposit, at least one formation or system having a structure in conformity with that of the assumed reservoir had to be identified. This formation or system would then serve as a marker; it had to be rather thick and to have an average resistivity that differed significantly from that of the overlying formations. Yet the top of such a marker is often hundreds of meters deep, requiring systems that permit deep "seeing."

In theory there seemed to be no limit to the depth of investigation; the current electrode spacing could be increased indefinitely. The difficulty lay in the handling of very long cables and in the limited sensitivity of the measuring instruments. The greater the spacing between *A* and *B*, the smaller the potential difference between *M* and *N*. Efforts were made to use voltages as high as possible in the *AB* circuit, but safety precautions, as well as the bulk and weight of the batteries, restricted the *AB* circuit to 500 volts. Bulk and weight also imposed a limit on the possible reduction of the ohmic resistance of the circuit; further reduction would have required many long stakes and large-diameter cables.

Under these conditions the spacing could be increased to 2000, then 4000, and even 6000 meters in 1929, a record for that time. There was, however, one exception, a rather bold and entirely original experiment undertaken by Conrad and Marcel in the spring of 1928 near Vitré in Brittany. The experiment dealt with an electrical sounding with a current electrode spacing ranging from 2 to 200 kilometers, and a spacing between potential electrodes *M* and *N* of up to 20 kilometers. The sole purpose of the brothers' efforts was scientific research in exploring the thickness of the earth's crust on the assumption that beyond 10 or 20 kilometers the rocks become semifluid and much more conductive than the crust. Telegraph lines were made available by the Post and Telegraph Administration. The current was reversed for periods of up to 4 seconds for the long spacings to counteract the difficulties arising from telluric currents, mutual induction, and skin effect, and thereby obtain a dependable average reading. A high-capacity dry-cell battery could deliver a steady 2 amperes for several minutes. Since the subsurface in Brittany has a very high resistivity, the potential differences to be measured, even with a 200 kilometer spacing, were compatible with the sensitivity of the potentiometer.

The result was an electrical sounding curve. The remarkable regularity of the plotting attested to the dependability of the method. Resistivity increased sharply with depth without decreasing at any level; even with this 200 kilometer line, the conductive substratum was not felt. Another discovery the experimenters made was that it was possible to explore a huge slice of the earth's crust with only a few hundred watts of power—no more power than is needed to light a 200 square foot room.

Prospecting for metalliferous ores

No substantial change was made in the spontaneous polarization technique used by Conrad in his prewar tests, even though the findings of further experimentation proved valuable in fieldwork and in the interpretation of measurements as well. It was confirmed that formations made of conductive grains embedded in a resistant gangue do not produce spontaneous polarization. Experience showed that anthracite beds, as well as pyrites, can be the sources of spontaneous potentials, and that one had to beware of premature enthusiasm when observing the beautiful reactions of certain formations that happened to be devoid of any mining possibilities, for example, graphite-impregnated schists. Mineral reconnaissance was generally conducted by combining the plottings of spontaneous polarization and equipotential measurements, the latter being soon replaced by the resistivity map. The technique proved most effective in the search for all conductive ores: pyrites, copper sulfides, galena, pyrrothine, and so on. No conclusive result was achieved by various induced polarization tests.

In the early 1920's, the fledgling years, there was keen competition in prospecting for metalliferous ores, and Conrad and Marcel thought of extending their operations to include exploration for ores—mainly iron ores—which, in addition to being electrically conductive, had magnetic susceptibility. Good magnetometers, particularly the Schmidt balance, were on the market, but, as Conrad often said, "You don't develop an industry from what already exists." Therefore he endeavored to make a personal contribution to the advancement of the magnetic method.

Although of high quality, magnetometers were difficult to adjust and to handle. They were subject to drifting, and measurements were rather time consuming. Conrad thought of tackling the problem from the electromagnetic end and decided to build a prototype. Even though little came from the project, it is worthy of a short description. The instrument, conceived by Doll and engineered at "La Pros" by Chatelier and his team, was of the differential type and able to detect the gradients of the terrestial field. It was made of two circular frames 1 meter in diameter, each wired with the same number of turns and driven by a crank around a horizontal shaft. All the materials were nonmagnetic: wood, plywood, Bakelite, aluminum, and so on. The first "field" test took place in Paris, in front of the rue Fabert headquarters. The adjustment of the instrument was a delicate one that required an absolutely invariable magnetic field. A crowd soon gathered to examine the bulky rotating machine and to watch the white smocks scurrying around it. One cyclist in the crowd became quite fidgety, and each twitch of his bicycle displaced the needle of the galvanometer, which was already jumping wildly at the passing of every subway train. It was hopeless. The experiment ended with the arrival of a policeman, who dispersed the crowd. The tests were resumed at Marcel's estate at Cormeilles-en-Parisis (which for some time served as a testing ground for experiments in the open), but after further field studies the matter was dropped. As occurred often in the history of "La Pros," other, more urgent developments demanded attention.

First studies based on the anisotropy of formations

From the very beginning, Conrad had observed the effect of electric anisotropy of sedimentary formations. An electric current sent through the formations flows more easily in a direction parallel than in a direction perpendicular to the strata; the strata are thus said to be anisotropic with respect to electric conductivity.

What is the origin of this phenomenon? A sedimentary series often consists of a sequence of thin layers of limestones, shales, sandstones, and marls of uneven resistivity. The current is impeded in crossing the successive resistant layers frontally, and therefore finds it easier to flow alongside the more conductive layers parallel to their stratification. Moreover, the mineral grains which generally constitute the sediments are nearly flat or oblong and, in most cases, settle parallel to the strata.

This is clearly seen in clays, where the microscopic crystalline particles look like flattened leaves. This disposition of the grains adds to the tendency of the current to flow parallel rather than perpendicularly to the strata.

In an area where the underground is a steeply dipping sedimentary series like the May sandstones in Normandy, the equipotential curves plotted in the vicinity of one ground electrode (the other being at a distance sufficient to make its effect negligible) have an elongated, nearly elliptical shape. This phenomenon is explained by the mathematical theory which, when applied to geophysics, maintains that the equipotential surfaces are flattened ellipsoids with an axis which passes through the ground electrode and is perpendicular to the strata. Their intersection with the surface of the ground is an ellipse whose major axis is parallel to the stratification. Furthermore, experience shows that this effect can still be felt on the surface if the uplifted formations are hidden under an overburden which is rather uniform: not too thick or too conductive in relation to the substratum.

A systematic application of the phenomenon was made for the first time at May St. André, in conjunction with the survey in which the first resistivity profiles were plotted. However, without conclusive information regarding the direction of the dip and its value, the method was of little use. In his report of 1920 Conrad had noted that to obtain these data the source of the current would have to be placed at a certain depth within the sedimentary series; this would require drilling, thus considerably limiting the applicability of the method. One of the first operations involving a borehole took place at Velsen in the Saar coal basin during the winter of 1921. A few more were carried out in the spring of the same year at Molières-sur-Cèze in the Gard coal basin. There were not many others before the 1930's.

To detect the direction of the dip through surface observations, an electromagnetic method was devised. The basic idea is as follows: on the assumption that the surface is flat and horizontal, and the underground homogeneous and isotropic, a current is emitted between two ground electrodes, *A* and *B,* connected by a rectilinear wire. The electrical circuit is then the equivalent of a vertical frame, the surface wire being one of its sides. The result is that the average direction of the magnetic field created by the current is horizontal. If, however, the subsurface is made of dipping, anisotropic strata, and the wire *AB* runs parallel to the stratification, the frame tends to tilt in the direction of the dip, and the magnetic field straightens up accordingly. With alternating current, the direction of the dip can be detected through a receiver reacting to induction, provided that the dip is steep enough without being vertical.

A first version of the receiver was a coil about 1 meter in diameter, fitted with several turns of insulated cable and connected to an earphone. This device, however, was not employed beyond the experimental stage (Fig. 19). Another version, extensively used for several years, made it possible to employ the conventional equipment for resistivity measurements without any special addition. The receiver was a spire of cable of a simple, radially symmetrical shape: a square or an octagon of 50 to 200 meters with one diagonal placed alongside the emitting line. The ground electrodes were symmetrical about the center of the spire (Fig. 20). When a pulsated current was sent into *AB*, the electromotive force induced in the spire was rectified by the pulsator and observed on the potentiometer. When the device was above dipping, anisotropic layers and the *AB* line was running straight along the stratification, the magnetic field had a vertical component that induced a signal in the spire. Bending

Figure 19. Experiments with an electromagnetic system for the determination of formation dip (Jan Joachim Valley, California, 1928).

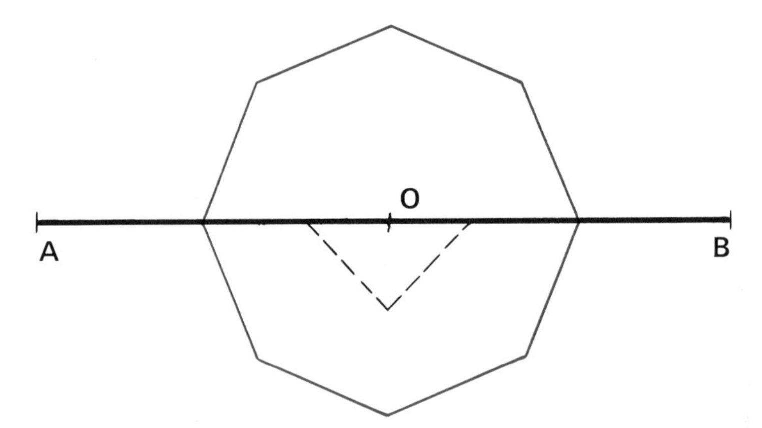

Figure 20. Schematic view of the spire.

the cable AB symmetrically about the center in such a way as to form an isosceles triangle modified the reading. It could be nulled by finding the proper size and orientation for the triangle. The side of AB where the tip of the triangle happened to be indicated the direction of the dip, and the area of the triangle was approximately proportional to the dip angle. These findings were shown on the topographic map by means of a vector. In the frequent case where the direction of the stratification was not well known, two sets of measurements were made at each station with two perpendicular AB directions, and the sum of the two vectors was plotted. In practice a reciprocal system was used in which the functions of AB and the spire (as transmitter and receiver) were exchanged. This gave the same results, but the sensitivity of the measurement was increased.

This simple and convenient method, requiring only lightweight equipment, aroused great hopes. In the San Joaquin Valley in California, where it was applied for many months, the dry climate and the ideally flat and bare terrain, over which vehicles moved unimpeded, together with a good operational organization, led to high performances: up to about ten 1000 meter stations per day. Unfortunately further experience showed limitations in the applicability of the method. It worked only on very flat ground, and the resistivity of the anisotropic layers had to be rather uniform throughout the tested field; the presence of thick enough and more resistant layers could upset the distribution of the electromagnetic field to the point of giving misleading measurements called "antidips."

These would result in lines on the map where the inversion of the direction of the vectors looked exactly like the traces of anticlinal axes, whereas they were merely resistant layers outcropping under the caprock. For these reasons the spire technique was dropped after a few years.

Laboratory work and a bit of mathematics

While methods and techniques were undergoing field tests, interpretation was being clarified and refined through laboratory observations and theoretical studies. In 1919 Conrad had resumed his model studies at the École des Mines in Paris. Perhaps it was an exaggeration to designate as a "laboratory" the collection of a few instruments and apparatus in a room of the rue Fabert building. At any rate, after a few years an expanded laboratory was transferred to a former barroom located at the corner of rue Saint-Dominique. The galvanized sink of the bar was kept for experimentation purposes, as well as the children's bathtub that had served for Conrad's research before the war. Later, larger and more complex tanks were used; but, as stated earlier, in the idiom of "La Pros" they continued to be called bathtubs.

Though of modest size, the bathtubs at that time proved to be valuable in experimental studies, and many models were built. By packing clay on one side and sand on the other, two formations with different resistivities and a variable dip were represented. With scaled-down *AMNB* configurations, measurements could be made alongside profiles perpendicular to the trace of the contact between the two formations. Thus, for each electrode spacing, an apparent resistivity profile could be plotted and the effect of all factors involved in the measurements assessed. By filling the bathtub with water, a homogeneous formation could be simulated and its resistivity adjusted by adding the necessary amount of salt. Plates of insulating material (usually ebonite) were placed across the tank to represent resistant layers of variable thickness and dip. This arrangement resembled sandstone layers embedded in shales. By dipping the upper edge of the plate below the surface of the water, an overlay concealing ancient formations was simulated.

The fact that the possibilities for experimentation were multiple and highly flexible led to partial clarification of a phenomenon that had been observed in the field. When the *AMNB* configuration is moved across the stratification and ground electrode *A* nears the trace of a high-resistance layer, the current emitted is impeded in its forward flow; the current density increases toward the rear, and so does the apparent resistivity recorded between *M* and *N*. On the contrary, once *A* has crossed the resistant layer, the current is impeded in its backward flow and the measured value decreases. The profiles show a peak followed by a trough, at a distance of the layer equal to *AB*/2; a similar effect occurs when ground electrode *B* crosses the resistant layer. The crossing of a conductive layer by *A* and *B* produces similar but inverted pictures. The result is two "electrode kicks" superimposed over the other peaks and troughs of the profiles, which appear when *MN* is on top of a geological feature. If the feature involves an alternation of several resistant and conductive layers, it becomes difficult to distinguish between the effects produced by the electrodes and those caused by geological features.

To increase the reliability of the interpretation, a variant of the "dragging" technique was worked out, involving two pairs of *MN* ground electrodes, adjoining the center of *AB*. At each station the measurements had to be repeated with two positions of *AB*, designated front and rear, and offset by a distance equal to *MN*. If either ground electrode *A* or *B* nears or crosses a bed of contrasting resistivity during the offset, the two measurements are different; otherwise they are the same. This technique, called "repetitive dragging," facilitated the unscrambling of the effect produced by the "electrode kicks." However, it also made the measurements more time consuming and more expensive.

Conrad had a definite gift for physics. His could have been the motto that Pavlov had engraved on the door of his laboratory: "Observe; always observe." Marcel, on the other hand, would say, "Take material and make it into something that works." Neither man was attracted by purely mathematical constructions because both saw them as dangerous oversimplifications of geophysical problems. Tongue in cheek, Conrad wrote:

> The calculations dealing with these problems are all taken directly from the theories of electrostatics . . . . There is plenty of room here for budding mathematicians to enjoy themselves. Either the questions address themselves to the high school level, or they are so unnecessarily complex that

they get lost in the clouds of still more complicated formulae; then the wisest thing to do is to turn the page.[15]

Doll, the best mathematician of the three, always maintained in his research a sensible perspective and a well-controlled appreciation of the value of mathematics.

In the early stages the results of the measurements made in the bathtubs were supplemented by calculations that, far from being "unnecessarily complex," were modeled on extreme simplicity: a vertical layer crossed by an electrode setup placed perpendicular to it.[16] Later, however, once his research required a more complete theoretical background, Conrad enlisted the assistance of Alfred Liénard, the deputy director of the École des Mines and a prominent mathematician. On September 15, 1925, with Liénard's help, he filed a patent claiming that, when the subsurface is made up of two homogeneous beds separated by a horizontal plane, the apparent resistivity can be calculated.[17] However, at that time he did not attempt applications involving the use of numerical data, and until about 1927 the interpretation of electrical sounding curves remained empirical, sometimes rather arbitrarily so. This was adequate for shallow investigations, but when operating with spacings of 4000 to 6000 meters, Conrad and Doll realized that interpretation had to rest on more dependable evidence, which only mathematics could provide. They knew that in a medium composed of homogeneous horizontal layers the potential created by the flow of a direct current emitted by a point source could be calculated by means of Maxwell's image theory.[18] A young and

[15] In *Le Puits Qui Parle,* No. 4, April 1921.

[16] The basic calculation is derived from Maxwell's image theory. At any point near an electrode, within a space divided by a plane in two media of different resistivities, the potential is calculated by introducing a fictitious electrode symmetrical to the real one about the plane. By analogy with optics, the fictitious electrode is called an image. In the case of a layer perpendicular to *AB* (two parallel planes of separation), this theory involves four infinities of images, yet the calculation is not too long because after a certain number of reflections the images are so remote that their effects become negligible.

[17] He was unaware of the study of O. H. Gish and W. J. Rooney on this problem, which was being published at the same time in the United States.

[18] The theory had already been used by N. Hummel in Germany in dealing with this kind of problem, where either one or two layers rested on top of a third one of infinite thickness. Without going into higher mathematics, the image theory led to lengthy and cumbersome calculations at a time when computers were not yet available. In the case of

brilliant mathematician, S. Stefanesco, worked out a more refined, flexible, and expeditious analytical solution, which for decades served as the basis for a number of theoretical curves calculated by a host of researchers and operators.

In early 1929 Stefanesco's method was limited to the calculation of a set of curves for two formations. This set was called the "chart of the overburden." At the same time the diagrammatic presentation was modified: the values of $AB/2$ were the abscissae, the apparent resistivities were the ordinates, and the logarithmic scale was adopted for the theoretical graphic presentation as well as the transfer of the field measurements. At that time the interpretation of a sounding curve consisted of superimposing over it a transparent overburden chart and finding the theoretical curve giving the best coincidence. However, the method was dependable only when the subsurface appeared as a series of two or, at most, three clearly distinct layers. In this case hypotheses were possible which, combined with data from other sources (geology, drilling), led to acceptable solutions. The treatment of a few typical configurations had shown that the interpretation of electric soundings was often obscured when a stratigraphic series could not be reduced to a very small number of layers.

During the years that followed, some researchers went beyond the overburden charts. In 1933 the Compagnie Générale de Géophysique (see p. 82), under the direction of Raymond Maillet, began the systematic calculation of curves for three formations. These were collected in a book from which it was possible to select the curves most suitable for interpreting specific measurements. The book contained more than 400 curves. Later, H. M. Mooney and W. W. Wetzel published a book with 2400 curves, of which 2000 are for four formations.[19]

There is at least one other method for the interpretation of electrical sounding curves. Upon first examination it is possible to imagine a model of horizontal layers, calculate their theoretical curve, compare it to the field measurements, and retouch the model step by step until a satisfactory coincidence is reached.[20] This approach with its cumbersome calculations became practicable only with the advent of the computer.

---

more than three layers the mathematics became inextricable. Hummel's calculations, as well as those of L. V. King in England, had no practical application.

[19] University of Minnesota Press, Minneapolis, 1956.

[20] Theory has shown that there are infinities of combinations in keeping with a measured

Until 1926 there was no dissemination of technical information among engineers in the form of service memoranda or bulletins. The prospectors—and there still were only a few of them—were trained by association. Sent to the field almost as soon as they were recruited, they were directed by a senior colleague from whom they received most of their knowledge. They also had the benefit of exchanges with Conrad and Marcel when they stopped over in Paris between missions. Additionally, they received letters from Conrad (his brother was not much of a letter writer) commenting, advising, or drawing conclusions from the surveys they were managing. It might be added that the head office was not markedly eager to disclose information that could be of interest to enterprising competitors, especially in the United States and Canada. This was one reason why written documents concerning methods as well as instruments and operations were sparingly distributed.

After 1927, however, technical memoranda became rather frequent. The sum of theoretical and experimental data serving as a basis for interpretation was extensively exposed in a letter of November 22, 1927, and was circulated to the the entire staff. It shed a new light on a number of things and helped to increase understanding of the meaning of the measurements. Nevertheless Conrad and Marcel maintained their belief that theoretical deductions led to oversimplification of hypotheses, and that field observations were the ultimate and safest elements of decision. This was the thinking behind the remark Conrad made to a prospector: "If you find something in the ground that does not show up on the measurements, it's better to believe the ground than the measurements." A quip indeed, but how realistic at a time when surveys dealt only with shallow structures, and careful examination of the terrain would help in the understanding and verification of the electrical measurements.

Operations

In 1920 electrical prospecting, as well as all other geophysical methods, was virtually unknown to the mining industry. Schlumberger's many explanations and demonstrations gradually introduced a new tech-

---

curve. To narrow the possible solutions of the problem, other data on the formations studied and their resistivities are required.

nique into the field of traditional mining operations: instead of digging to explore the underground, a small array of electrical equipment (wires, stakes, batteries, potentiometers) could be used to scan from the surface.

In addition to the traditional dowsers, quite a number of "inventors" were peddling their allegedly "scientific" wares, claiming that they had effected astounding discoveries. Some industrial concerns, unable to distinguish between such eccentricities and well-established methods, remained skeptical. The terms "black magic" and "geomancy" gained popularity.[21] The more realistic mine operators were less hesitant in making their choices. In France several mining concerns, already aware of Conrad's scientific authority, were quick to express a preference for his methods.

Other qualified researchers either became involved in, or continued operations in, geophysical prospecting: Lundberg, Mintrop, Ambronn, Koenigsberger, Slichter, and Mason were the principal ones. From the start, Schlumberger had to compete actively, not only in gravimetrics and seismics, but in the field of electrical prospecting as well. For the latter, certain operators used high-frequency electromagnetism, but it was all too easy to imitate the spontaneous polarization, the plotting of equipotentials, and resistivity measurements with short spacings.

From 1920 to 1929

In the early 1920's most of Schlumberger's activities consisted of short-term missions (generally a few weeks) scattered throughout a number of countries: France, North and South Africa, Spain, Italy, Yugoslavia, Bulgaria, Poland, the United States, Canada, and Japan. In France outstanding operations included the reconnaissance of Silurian synclines in Normandy, the location of faults in the Lorraine coal and the Briey iron ore basins, the use of spontaneous polarization in prospecting for anthracite at Saint-Michel de Maurienne, and the exploration for lignite in the Landes.

Mineral prospecting by electrical measurements had been introduced into the United States and Canada in the fall of 1921. The first experiment took place in the copper ore basin of Ducktown, Tennessee, where the

[21] In the French text of a 1919 patent, H. Lundberg, one of the best known geophysicists of that time, refers to the measuring ground electrodes as "divining rods." Perhaps commercial reasons led him to adopt this rather odd analogy with the dowser's tool.

shapes of the lodes at Mary Mine, Boyd, and Culchote were traced by spontaneous polarization. Other tests in Pennsylvania, Michigan, and Ontario led to commercial development, especially in Manitoba and Saskatchewan. Outstanding results were obtained at the Noranda mines in Canada, where an important deposit was discovered in the unexplored part of the lease.

Even though the activities of Conrad were far from widely known, he was contacted by companies wishing to license his processes. The first contract of this type was negotiated with a French concern prospecting in Ecuador. For 1 year and for the payment of 10,000 French francs, the Schlumberger Company guaranteed to supply all equipment and operating instructions. Under similar contracts drawn up during the following years with mining companies in Rhodesia, Katanga, and Japan (Yokei area and Nagamunu mine), extensions of already-known copper mines as well as new deposits were discovered.

In a country like France, which, according to most geologists, had no prospective petroleum wealth, electrical prospecting was limited to geological reconnaissance about the small oil field of Pechelbronn, north of Strasbourg. One of these explorations succeeded in locating the trace of the great Rhenish fault under the Quaternary terraces near Lobsann (1921). The news of these results spread to Steaua Romana, a French oil company operating in Rumania, where some geophysical work was in progress. A contract was signed in 1923, and except for an interruption from 1927 to 1931 Schlumberger crews worked there for more than 20 years.

Initially, Conrad's main thought was that the salinity of formations in the vicinity of petroleum deposits ought to be very high.[22] Consequently, areas where the readings were lowest were sought out on the resistivity maps. A few months after reconnaissance of the Prahova district west of Ploesti, Rumania, was begun, the map showed a red spot that stood out sharply near the village of Aricesti. Two boreholes had detected the flank of a structure to the north of the red spot; the electrical method confirmed its existence, and a contour map of equal-resistivity values showed its outline (Fig. 21). Other boreholes, sited on the basis of these data,

[22] In 1923 he tried to verify this assumption by asking Pechelbronn to make measurements on clays both near the producing zones and remote from them, but Pechelbronn was unable to comply.

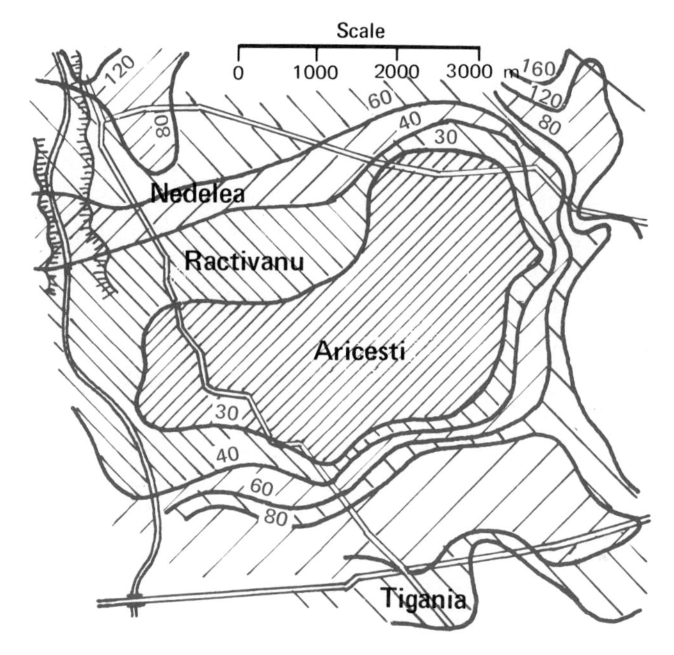

Figure 21. Resistivity map of the Aricesti, Rumania, salt dome (1923).

revealed a salt core and a huge reservoir of hydrocarbons. This was the first time in history that an oil-bearing structure had been determined by geophysics.

Although the conclusions drawn from the resistivity map were correct, the theory on which they rested was later disproved. Experience showed that saline formations were frequently present in oil-bearing basins, but because they extended far beyond the immediate vicinity of the producing zones, they could not be used as guides for oil exploration.[23] It took several years before a new and correct interpretation of the

[23] Another long-advocated idea was that salt domes diffused at the top and formed "salt aureoles" that served as the origin of the conductive zones on the maps. This assumption, too, was gradually disproved.

Aricesti map was made. The stratigraphic series contained a thick stratum of Tertiary marls overlain by a substantially more resistant Quaternary alluvium (sands and gravel). During the Quaternary the dome kept moving upward (and continues to do so to this day), causing the alluvium to thin out. The equal-resistivity curves reflected the topography of the top of the marls, while a conductive spot at the center located the crest of the structure.

In the United States petroleum exploration was beginning to apply other geophysical methods. In 1921, in Oklahoma, J. C. Karcher and W. P. Haseman ran the first successful tests of seismic reflection. The gravimetric method was being introduced in California and Texas, and seismic refraction on the Texas coast and in Mexico. In 1924 two spectacular discoveries of productive structures confirmed the value of these methods: the Nash Dome in Texas, located by the Rycade Oil Company with the help of the Eötvös balance, and the Orchard Dome, also in Texas, discovered through seismic refraction for the Gulf Oil Corporation by Seismos, a German firm.

The Royal Dutch Company's American affiliates had undertaken an intensive geophysical prospecting campaign. Informed of the success of the Schlumberger processes in Rumania, they decided to test the resistivity method in Texas. Accordingly, Marcel Schlumberger arrived in Houston in June 1925 with his nephew Marc, and proceeded with his first measurements (resistivity profiles with 400 meter spacing) on the Pierce Junction Dome, the southern flank of which was producing 500 barrels per day. Other tests followed in the same area (Blue Ridge, Humble, and Goose Creek domes) and continued toward Beaumont (Spindletop, Fannett, and Sour Lake). In spite of experimental difficulties with existing pipeline systems and leaks from cables lying in the saltwater puddles spilled by certain wells, the profiles showed a resistivity drop above the producing domes. Continuing with the notion that saline formations indicated the proximity of hydrocarbons, it was thought that electrical prospecting detected the pay zones, whereas the balance and seismic methods served strictly for locating salt domes, whether productive or not. The electrical method seemed to be more efficient; its equipment was easier to handle, and its operation much less costly. Thus the first contract with Royal Dutch was signed on September 23, 1925, on behalf of its affiliate, Roxana Petroleum, which was operating in Texas.

Marcel had returned to France before the end of the summer, leaving behind his nephew (later joined by Pierre Baron and Jacques Gallois) and a few junior engineers—Paul Charrin, Gilbert Deschatre, Julien Breusse, Marcel Jabiol, André Allégret, and Robert Roche—the first group to settle in the United States, in Freeport, Texas. In the early part of the following year, operations were extended into Louisiana (Fig. 22).

Few domes were known at that time, but geology led one to believe that many others could be discovered. The active competition among the various companies to acquire the most promising areas[24] gave geophysical

Figure 22. Electrical prospecting in Louisiana swamps (1925). (Reproduction from La revue pétroliére, *1935, p. 1608.)*

[24] In the United States the subsurface belongs to the owner of the land; therefore such areas had to be purchased or leased.

processes a strong boost. Roxana was interested in the work of "La Pros" but without great expectations of immediate results. At any rate, in May 1926 an office was opened in Dallas. Marc Schlumberger's visa had expired, and he was replaced by Charrin until October, when Léonardon arrived.

Conrad, on a visit in September, was greatly interested in a conductive area on the resistivity map of Bayou Serpent, Louisiana. Contrary to the opinion of Roxana's geologists, he suspected the possibility of a dome, and after thorough consideration he took it upon himself to suggest the location of a borehole. Roxana had other plans, however, and after several postponements finally shelved the proposal.

Having used electrical prospecting for 18 months with no conclusive results, Roxana figured that Schlumberger was merely experimenting, often blindly. Indeed there was some justification for this thinking: operations were still run with spacings which were too short, and no theory had evolved that gave a clear understanding of what the measurements represented. Moreover, their accuracy remained questionable. Giving preference to balance and seismic methods, Roxana finally canceled its contract in the spring of 1927.

Meanwhile, however, another contract had been signed on March 21, 1926, with the Shell Oil Company; this brought a Schlumberger crew (R. Nisse and R. Viry) to the San Joaquin Valley in California, in search of anticlinal folds as a possible extension of the Lost Hills and North Bellridge structures. No results had been obtained there in more than 2 years with the balance, and the only discoveries had come from boreholes near surface showings of asphalt.

The local working conditions were excellent: the region was a flat semidesert; travel was easy and the climate dry. No electrical leaks were expected. The spacing could be increased to 1000 meters. There were no salt domes or conductive spots showing on the maps, and the anticlines seemed to correspond to maximal resistivities. Conrad, who visited the mission after his trip to Texas in the fall of 1926, thought that this was indicative of rather hard sandstone beds marking the crests of the structures.[25] A second crew was stationed at Wasco, California, from the spring of 1927 until the early part of 1930, when the mission ended. Meanwhile,

[25] The same results could lead to different geological constructions, but they remained at the hypothetical stage.

the multielectrode resistivity method had been supplemented by the inductive method referred to as the spire, particularly for studies of dips.

However, in the wake of the disappointment caused by the suspension of work in Texas, Paris was inclined to give America second priority. Slow communications prevented careful monitoring of the fieldwork, and the market proved to be much tougher than in France. The trend was to repatriate the surplus engineers since they could be fully occupied in France. It required perseverance on the part of Léonardon to ensure that a half-dozen available prospectors were assigned to Canada, where mining exploration was in full swing. Changing from the Texas heat to the rigors of the Great North, the prospectors traveled from mine to mine between August and October; numerous vein and lode surveys by spontaneous polarization and resistivity were made under conditions of severe hardship. Backed by the unrelenting commercial efforts of Léonardon, who was now established in New York along with most of the mining companies, these activities continued, through thick and thin, until 1930.

Exploration for new deposits in the Alsatian potash basin of France began in 1926. There were two potash beds, intercalated in a thick formation of rock salt; therefore it was necessary to locate the areas where the rock salt came closest to the surface. It lay directly beneath a thick bed of uniform Oligocene marls, impregnated with highly mineralized water and overlain by Quaternary sands and gravels of much higher resistivity. Since this configuration was almost identical with that encountered in Rumania, a similar operating method was adopted: a resistivity system with rather short electrode spacing (AB = 350 meters). After a few weeks of prospecting, a conductive strip about 7 kilometers long, running along the Ill River, revealed a broad saline ridge: the Meyenheim anticline had been discovered (Fig. 23). In the following fall another red spot marked the Hettenschlag Dome, located north of Meyenheim (Fig. 24). Two boreholes confirmed that the salt was 100 meters deep, as indicated by vertical electrical soundings. These were momentous discoveries in the history of geophysics.

By late 1927 the number of engineers employed by "La Pros" had increased from 3 to 16. During the last 7 years some 50 missions had taken place, many outside France. Most had yielded positive results. The Schlumberger processes had become better and better known; some 10

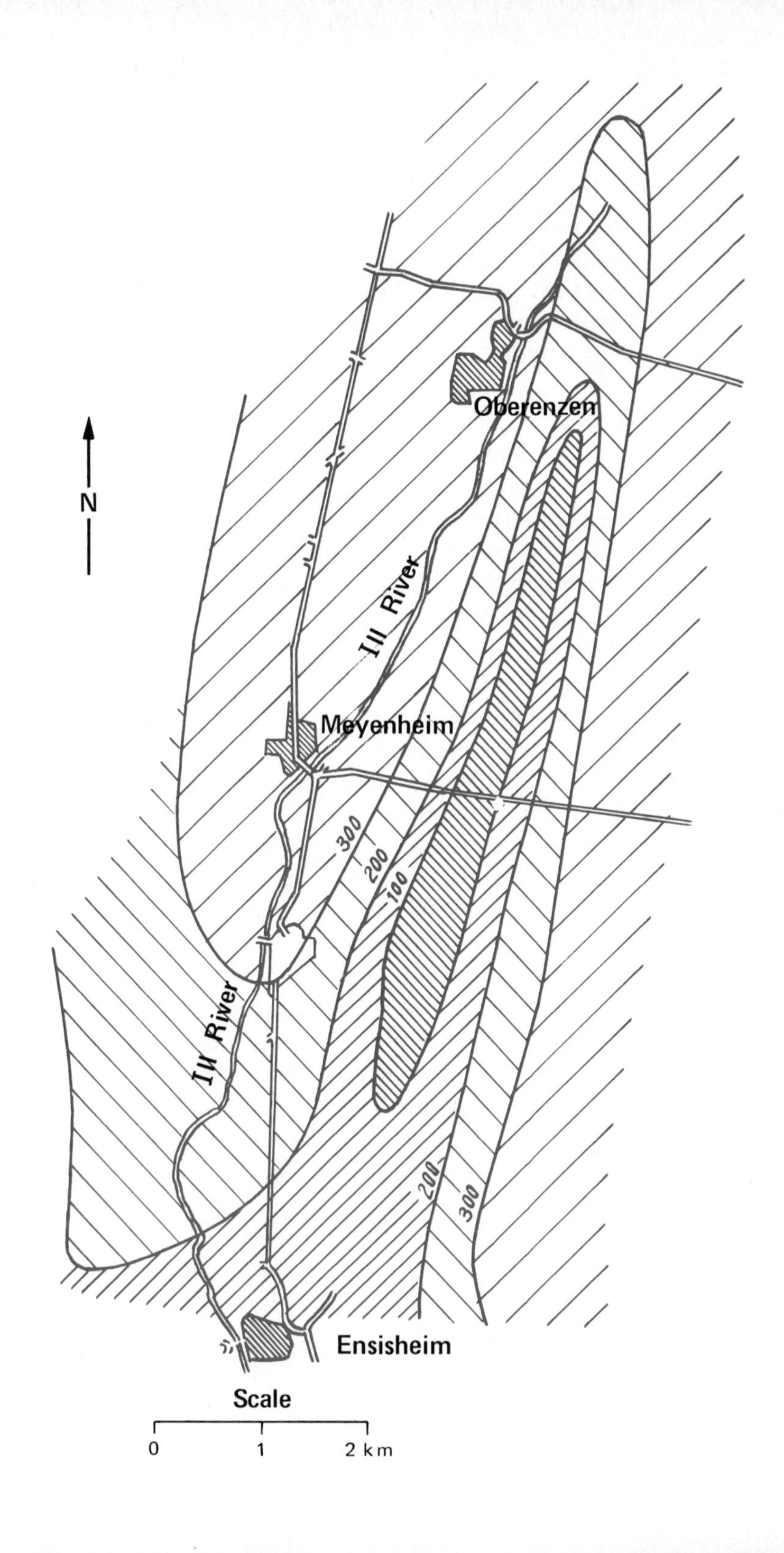
Oberenzen
N
Ill River
Meyenheim
300
200
100
Ill River
200
300
Ensisheim
Scale
0
1
2 km

publications (articles, brochures, and papers presented before various congresses) by Conrad Schlumberger, Léonardon, Charrin, and others had appeared in France and the United States. Mining companies in South Africa and Japan had disclosed the results of the processes in the professional journals of their countries. With regard to petroleum exploration, a field in which Schlumberger had aggressive competition, a number of decisive successes had proved the superiority of gravimetry and seismic refraction over electrical prospecting. Conrad thought it preferable, therefore, to focus the commercial effort on detailed geological studies (uplifted beds, faults, etc.), where the superiority of the resistivity method was well known and "La Pros" had a marked technical lead.

With the proposal of a dam on the Connecticut River, a new field of applicability opened up: civil engineering. For this project electrical soundings were used to determine the depth at which the dam should be seated (April–May 1928). Twenty surveys of this kind followed in the United States, Canada, and North Africa. In particular the preliminary reconnaissance of a dam site on the St. Lawrence River near Morrisburg, Ontario, in the spring of 1929 was among the outstanding achievements of electrical prospecting. This method was also used to locate competent rocks among sedimentary complexes and to study the depth of crystalline substrata, problems that arose during the digging of tunnels at Bridge River, British Columbia (summer of 1928), and in the Lièvre Valley near Masson, Québec (1929). Also in 1929 Jean-Paul Mathiez carried out a survey for several months on behalf of the American Zinc and Lead Smelting Company, Joplin, Missouri, to plot the subterranean topography of the roof of limestones covered by shales. The technical quality of the survey was such that it earned recognition at "La Pros" as a textbook example.

In the eastern hemisphere there was no need to pursue contracts and fight tough competition. Surface work proceeded smoothly. In addition to many small missions, long-term surveys included new studies of iron ore-bearing synclines in Normandy (Soumont, 1928–1929; Halouze, 1929; Sées, 1931); prospecting large areas for salt domes in the Landes in search of potash, a problem similar to that in Alsace (1929); discovery of a pyrite body at Chizeuil, Nièvre (1929); and a big success at Benissaf in

Figure 23. Meyenheim anticline (1926). The equal resistivity curves east of the Ill River draw the outline of the salt ridge.

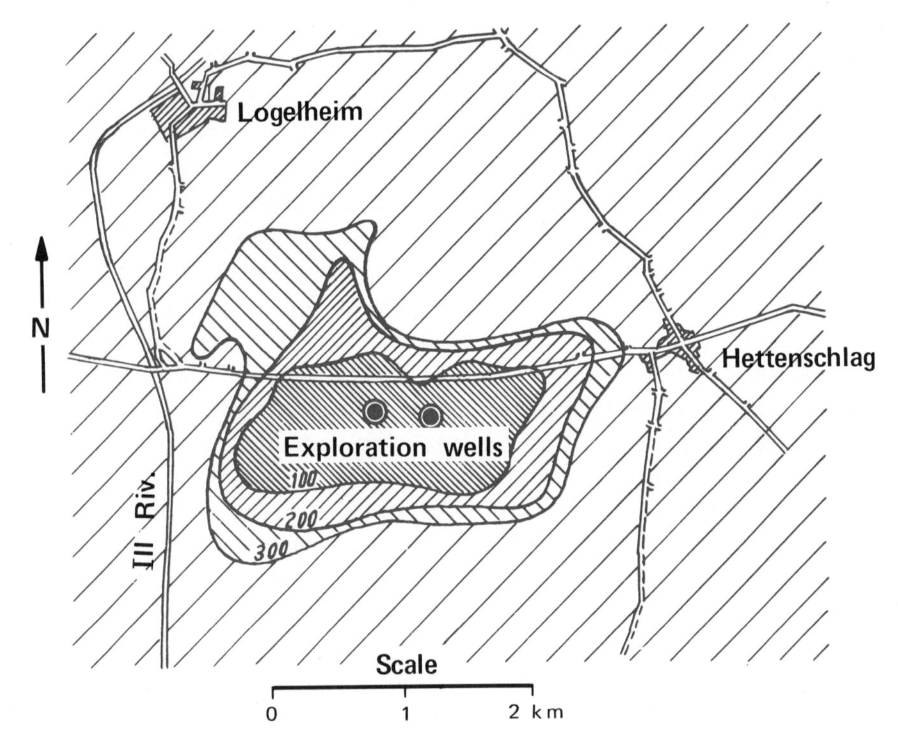

Figure 24. Hettenschlag Dome (1926) north of the Meyenheim ridge.

Algeria, where a previously unknown bed of iron ore 50 meters thick was uncovered by resistivity and magnetometry and confirmed by drilling (spring of 1929).

Abroad, the team of Paul Charrin and Jacques Castel was active in Spain between the fall of 1927 and the summer of 1928. The team had won out over the competition of seismic refraction and gravimetry in the coal basin of Villanueva-de-las-Minas in Andalusia, and then continued quite successfully with the search for metallic ores (copper, lead) in the Cordoba and Huelva regions, as well as in Spanish Morocco. The following year more missions operated in Africa: Ookiep (Union of South Africa), where a mission discovered an anomaly that, years later, led to the development of a copper mine; Upper Katanga, where the Union Minière depended on Schlumberger's technical assistance for the interpretation of surveys under license; and Minouli (Brazzaville area), where, presumably because of the lack of ore, the survey turned out to be disappointing.

The spring of 1929 marked the beginning of negotiations for an important contract with the U.S.S.R. At the same time electrical coring was introduced into Venezuela and was later tested in the United States. The central facilities in Paris were reorganized and expanded. New buildings were acquired at Nos. 40 and 42 rue Saint-Dominique; a new workshop was built; the operating, equipment, and administrative departments were restructured; and new staff was recruited. Such optimism was fully justified during 1929: by the end of that year, the staff had grown from 56 to 95, of whom 45 were field engineers.

The Wall Street crash of October 1929 hit the mining and petroleum industries hard. As early as 1930 exploration in the United States and Canada had slowed down or stopped altogether, engulfing geophysics in an acute crisis that was soon to extend to the whole world except the U.S.S.R. After years of what seemed to be unending growth, the bottom was falling out of everything. It would be several years before the oil and mining companies recovered from the shock and exploration resumed.

In the spring of 1930, after 2 years of almost totally unsuccessful prospecting, the Shell Oil Company terminated the California contract. The ensuing disappointment and the tendency to blame the economic depression could not alter the fact that the spire method, after prolonged large-scale operations, revealed false anticlinal axes along the trace of hards beds (see p. 61). This was also the end of operations in all other fields, electrical coring included, in the United States and in Canada. The last activities had been in Joplin, Missouri.

Electrical prospecting in the U.S.S.R.

In 1929, despite the fact that the U.S.S.R. had already established a large scientific organization, which had conducted extremely valuable theoretical research, the country was lagging in the development of geophysics. Petroleum exploration by geophysics had been limited to the torsion balance, used mainly in the Emba Basin north of the Caspian Sea, where a number of salt domes, some with small production had been located. Nothing had been done in seismic exploration at a time when first refraction and then reflection methods were already in use in the United States and elsewhere.

Geological considerations seemed to explain the lag in part. In the United States the main areas of production were the Mid-Continent[26] and the Gulf Coast. The former was characterized by smooth structures and stratigraphic series with compact beds (limestones) which offered good reflecting horizons; on the latter, the formations were mostly associated with salt domes, lending themselves to seismic refraction and gravimetry. By contrast, most of the U.S.S.R.'s production depended on the deposits of the northern Caucasus and Baku, located in generally soft and steeply dipping formations hardly suitable for the seismic prospecting methods of the time. Furthermore, since these deposits were found in anticlines without a salt core, gravimetry was practically useless. Such overall conditions were not unlike those in California, where seismic and gravimetric activities had been abandoned.

Yet there was a pressing need to undertake extensive geophysical exploration, especially for Grozneft, the petroleum trust of the Grozny region north of the Caucasus. The large prewar production of the "Old Fields" was decreasing after 20 years of exploitation. Although the drop in production was compensated for by recent discoveries of the same magnitude (the "New Fields"), additional reserves had to be found. With little success from geological prospecting and after several disappointing gravimetric surveys, Grozneft turned to electrical prospecting. On the basis of its experience in California, Schlumberger seemed to be the organization most qualified in this technique. A 2 year contract, signed in July 1929, provided for the utilization of resistivities and of the spire, together with the testing of electrical coring. An important provision was that Russian crews were to be gradually trained by French instructors.

The mission arrived in Grozny in early September, led by Vahé Melikian.[27] Operations started in the immediate vicinity of the Old and New Fields, marked by hills stretching in two long lines almost parallel to the Caucasus chain. By November a first resistivity map already showed

[26] The region extending from the Rockies on the west, the Mississippi River on the east, Iowa and South Dakota on the north, and northern Texas on the south.

[27] A young engineer of "La Pros," Melikian, a native of Baku, had had a leading role in the conclusion of the contract.

clearly the eastward extension and closure of the anticline. Moreover, the spires traced the axis of the anticline and confirmed the steady dipping of the formations. The Russian geologists, used to retiring to their offices as soon as fall came, were dumbfounded when they saw the French prospectors—Raymond Sauvage, André Poirault, Jean Lannuzel, Roger Jost, and Charles Scheibli—proceeding with their work throughout the winter in worn-out vehicles, handling kilometers of cables over snow-covered hills and steppes or through vast quagmires. Such tenacity, together with Melikian's skill and perseverance, was rewarded by an expanded contract, this time with Soyuzneft, the central petroleum organization. In April 1930 a second group of prospectors arrived in the U.S.S.R. Still others followed; some, like Castel, stayed for several years. In fact, until 1936 almost the whole of "La Pros" took turns going to the U.S.S.R.

The spring and summer campaign of 1930 involved mainly the prospecting of the Terek steppe, between the river and the Manich Basin to the north. From a borehole the thickness of the overburden had been estimated at several hundred meters, a depth that first excluded the use of the spires and then required resistivity equipment with a deep range of penetration. A series of north–south resistivity profiles with 4000 meter spacing was undertaken up to distances of 100 to 150 kilometers from the Terek. The term "dragging" previously used for this kind of operation was not appropriate here since, for the first time in the application of this technique, the cable was no longer dragged from station to station, but was laid and rewound in 400 meter lengths by means of truck-mounted reels. Most of the survey area was covered by dunes, and travel was very difficult; the sand gave high resistance to the ground electrodes, and the sensitivity of the readings was sharply reduced. Additionally, electrofiltration and telluric currents kept the needle of the potentiometer in continual agitation. Whereas in a field offering easier travel and better electrical grounding a team could produce 25 kilometers of profiling daily, only 3 to 4 kilometers could be completed, and that arduously, in the dunes.

By the fall of 1930, however, a vast territory had been prospected in the Grozny perimeter; the resistivity map showed a remarkable concordance with the geological map and had located all known anticlines. Therefore, in Grozny and in Moscow, Marcel believed he could safely state that, barring oil-bearing horizons beyond drilling range, the existence of other anticlines need not be considered.

In Baku, where operations started at the same time, geological conditions did not allow for reconnaissance surveys as extensive as those in Grozny. The huge Kura Plain south of Baku, the proposed site of a search for deep structures, covered a subsurface composed of loose sands and clays impregnated with salt water, where there seemed to be no layer of sufficient thickness and contrast to constitute a good electrical marker. Hence renewed attempts at electrical prospecting in this plain yielded results of dubious merit, and surface operations were confined to the Apsheron Peninsula, where limestone beds contrasted sharply with the surrounding clays.

One of these operations, offshore from the Bibi-Eibat field close to Baku, constituted the first example in history of underwater geophysics. The Bibi-Eibat structure, an anticline whose trace is marked by a limestone outcrop, was partially submerged in the sea. The sea had gradually been land filled as the field was developed. It was necessary to know how far this filling had to be extended—in other words, to determine the closure of the formation under the sea by plotting the extension of the limestone. The survey was made in a few weeks during the spring of 1931, using equipment borrowed from the electrical coring operation with spacings of $AM = 45$ meters and $MN = 10$ meters. The winch was on shore. For each profile the cable was pulled offshore perpendicular to the coastline, and the measurements were made, point for point, while it was being wound back. The plotting showed resistant bands outline the underwater trace of the limestone bed (Fig. 25). When the operation was resumed 3 years later, the Russians enlarged the resistivity map by a band several kilometers wide bordering the peninsula. This they did by dragging the cables on floats towed by a boat carrying batteries and potentiometer; in fact, this was the primitive precursor of the system that was to be used much later in underwater seismics.

In the Emba Basin some results were obtained by detailed study of half a dozen salt domes, but the measurements required exceeded the capabilities of the method because of the highly conductive formations and the hardships of working year round in a semidesert with extreme climatic conditions. It is no wonder that the accuracy, efficiency, and yield were far behind those obtained in Alsace.

When Conrad arrived in the U.S.S.R. at the end of 1931, the overall balance sheet of the operations, especially electrical coring,[28] showed

[28] Electrical coring in the U.S.S.R. is discussed in Part Two, "The Origins and Development of Logging."

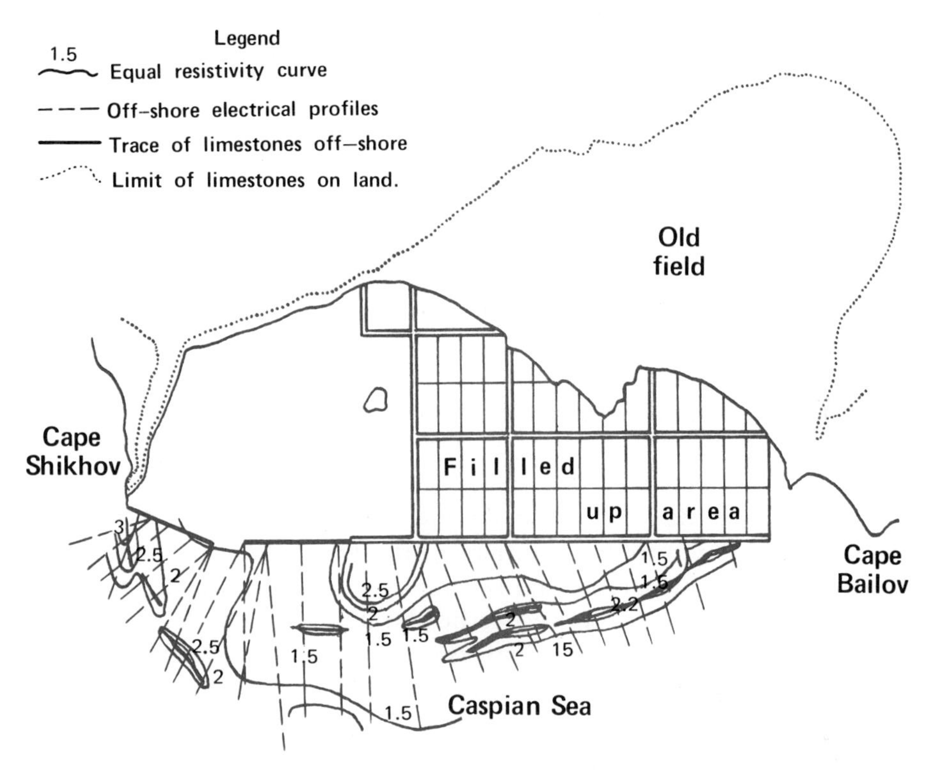

Figure 25. The first known example of underwater geophysical prospecting: tracing through resistivity measurements of the limestone beds demarcating the Bibi-Eibat structure under the sea (Baku area). (Reproduction from a reprint of Science et Industrie, *nonserial issue, "La technique des industries du petrole," 1932.*

brilliant success. One contract followed another until 1936. This collaboration enabled the Russians to expand the use of electrical methods both on the surface and in boreholes. In the 1930's operations were extended to Georgia, the Maikop region, and central Asia, and later to the West Ural and Volga basins, to the Donbas (coal), and finally to the region of Lake Baikal, Sakhalin, and Kamchatka. In addition, in 1935 Schlumberger played a decisive role in the launching of seismic reflection, which ever since has been applied widely in the U.S.S.R. Today the crews assigned to surface electrical prospecting and logging in the U.S.S.R. number in the hundreds, and many geophysical laboratories and research institutes can be found all over the country. In no other place in the world does there exist such a massive utilization of electrical processes, while elsewhere seismic reflection plays the leading role in petroleum exploration.

The success of Schlumberger in the U.S.S.R. was offset by a great tragedy; it was during his return from a trip to Moscow in May 1936 that Conrad suddenly died. His premature death deprived the Company of a prestigious chief whom everyone loved and the geophysical industry of a scientific mind that was one of the most original and creative of the era.

Marcel then took over the management of the Company.

After the establishment of the Compagnie Générale de Géophysique

In 1929 the Banque Mirabaud and French mining companies had become the licensees of the German Ambronn and Selfeld processes of electromagnetic prospecting and seismic refraction, and had established the Société Géophysique de Recherches Minières (S.G.R.M.). Hit almost immediately by the economic depression, this company sought to make an arrangement with Schlumberger, which led to the creation of the Compagnie Générale de Géophysique (C.G.G.) in March 1931. The new corporation combined the management of both partners' operations except for the Russian contracts and electrical coring. Raymond Maillet, the general manager of S.G.R.M., brought his highly scientific expertise to the group.

The structure of C.G.G. was to change several times. In 1935 the Société de Prospection Geóphysique, a joint venture of Compagnie Française des Pétroles and the Banque de Paris et des Pays-Bas, merged with the C.G.G. group, contributing gravimetric prospecting with the Eötvös balance. With the acquisition of the license of American seismic reflection processes in 1934, C.G.G. was equipped to handle any geophysical method. Later, having given up its managing function, it became an operating company. After World War II, under the leadership of Léon Migaux, assisted by J. Cunin, it became the second largest geophysical company in the world. Seismic reflection was the key to its expansion, first under French programs and then on an international scale. Schlumberger withdrew from C.G.G. in the 1950's.

At a time when electricity still accounted for the bulk of C.G.G.'s activities, some outstanding results were achieved, among them the discovery of the Bucsani anticline in Rumania (1933) and the geological study of the Digboi anticline in the Assam jungle (1936). This period also marked the dawn of the application of the electrical method to groundwater exploration. Several important surveys were conducted under the personal supervision of Schlumberger: continuation of the survey in the Alsatian Plain, underwater prospecting in the Algiers roadstead, corrosion studies of buried pipes, and measurement of telluric currents.

In 1931–1932 prospecting in Alsace involved resistivity profiles with 6000 meter spacings and stations 500 meters apart. This was a region crisscrossed by roads, railroads, and canals and covered by farmland and forests, where it was difficult if not impossible to stretch the cables in a straight line as could be done in the semidesert plains of California or the U.S.S.R. With consideration for the topographic constraints, the location of the ground electrodes was determined on the map. Once the electrodes had been planted, a truck-mounted automatic spooler connected them, along the most practicable route, to each other, to the batteries, and to the potentiometer. The stray currents created by the trolley lines often required nighttime operations.

This prospecting was to elucidate the salt structure of the Alsatian subsurface. Resistant bands on the map revealed four saline ridges parallel to the Rhine River and a transversal ridge. With an accuracy that was confirmed by drilling, electrical soundings assessed the depth of the salt at about 500 meters. Simultaneously, Schlumberger undertook balance surveys, especially around the Hettenschlag Dome, but the results, although generally in accord with those of the electrical measurements, appeared less complete with regard to depth assessment. Also, the balance proved slower than electricity. These resistivity measurements were remarkable with respect to salt tectonics and seemed to be superior to results achieved by seismics on the Gulf Coast. One might ask whether, under the circumstances, an opportunity to successfully resume operations in the United States had not been missed. In all likelihood, with gravimetry and seismics by then firmly rooted in that country, the chance of developing a substantial resistivity market would have been remote.

In 1932 a new underwater survey of about the same nature as that in

Bibi-Eibat was made. This survey was more difficult, however, because it had to determine the thickness of the silt deposited on the bottom of the Algiers roadstead, which lay 30 meters under water of a much higher salinity than that of the Caspian Sea. Consequently, there was a risk that this would reduce the sensitivity and the accuracy of the measurements. However, the survey was facilitated by the fact that the sea and the silt were homogeneous media of known resistivity, while that of the underlying rock was practically infinite. A special kind of electrical sounding was required for this operation, whereby the cables were dragged on top of the silt. Thanks to a chart of theoretical curves calculated for this configuration, the measurements could be correctly interpreted.

During the same year Conrad thought of using his methods to solve the problems arising from the corrosion of buried pipes (e.g., liquid or gas pipelines, or the sheathing of telegraph or telephone lines), which were inadequately protected by the insulating coatings of the time. Thorough studies, among them those of the U.S. Bureau of Standards, had revealed the crucial relationship between the nature of the soil and corrosion. Because of the mineralization of the impregnating waters, the metal was chemically etched and eaten away in places. Even worse was autogalvanic corrosion. Pipelines buried in soils more or less aerated, or containing different salts, are the source of spontaneous currents. At the spot where the current leaves the pipe, metallic ions are diffused in the surrounding water, thus progressively eroding the pipe. A still more destructive action is caused by stray currents. This action usually occurs in pipelines located in urban areas near industrial installations which generate the currents. The study of these phenomena had entailed many potential measurements with electrodes placed very close to the pipes, requiring the digging of holes or trenches at a cost that would have been prohibitive on a larger scale. Conrad tackled the problem with surface measurements. From laboratory experiments he was able to refine and complete the data already published, and to define the methods suited for the various phases of the problem, a description of which follows.

The layout of a pipeline could be determined with the help of a resistivity map plotted with very short electrode spacing (a few meters) along a strip straddling the proposed route. Where possible, the pipe had to skirt the most conductive, or ionically active, zones, and to avoid crossing the boundaries between zones of different conductivities. The process was used in the U.S.S.R. for laying an oil pipeline 1000 kilometers long between Guriev, north of the Caspian, and Omsk in southwest Siberia (June–November 1932). For urban piping systems the electrical field of the stray currents was measured throughout the industrialized region with an apparatus similar to that used for the tellurics (see p. 88). Since the resistivity of the ground was determined at the same time, the value of the current density at each point of observation permitted a more effective planning of the pipe-laying operation.

Delineating the zones that were corroded or corrosion threatened by the autogalvanic effect required that potential profiles be run straight above the pipes between one nonpolarizing electrode and another fixed-reference electrode. The potential peaks pointing to the zones of outflowing current indicated the need for adequate protection. Since this measurement of potentials could not apply to pipes exposed to erratic stray currents, the current outflow zones were located through a specially designed differential device. These methods were applied to a few urban centers in the early 1930's.

In geophysical language the term "tellurics" refers to irregular currents that flow hundreds and thousands of kilometers through the crust of the earth, usually in the form of nearly horizontal sheets, and reflect electrical phenomena of the upper atmosphere. They are distinct from all other currents flowing through the ground: the strays, often stronger but spatially more circumscribed; the currents of spontaneous polarization confined to the vicinity of the bodies where they originate; and localized currents produced by electrofiltration in zones where, for example, mineralized water has seeped upward toward the surface of the ground by evaporation.

Until about 1920 the scientific work published had consisted of measuring the potential differences created by the tellurics between widely spaced (several tens of kilometers) ground electrodes, generally by using out-of-service telegraph lines. With such long spacings the potential differences reached several volts, and the measurements did not require especially sensitive instruments. However, because they encompassed vast areas, these measurements did not permit an adequate analysis of localized soil resistivity effects. Conrad tried to determine these effects by using his potentiometer and rather short electrode spacings.

In 1921 he instructed the crew operating near Lobsann in Alsace to observe the telluric currents on both sides of the great Rhenish fault, which brought conductive shales into contact with the resistant Vosges sandstones. On each side of the fault and perpendicular to it, two lines of the same length were connected to two galvanometers whose operators made simultaneous readings. These very brief observations confirmed that the tellurics varied constantly and that their average amplitude was much higher for sandstones than for shales. However crude, this was the very first indication of the effect produced by a geological feature on the telluric field.

Other observations were recorded by Conrad in several unpublished papers written in 1922–1923. One of these describes the observation he made in the course of an experiment in the Cherbourg roadstead (May 1922) in order to locate metallic wrecks through the spontaneous polarization phenomena they produce.[29] His conclusion had been that the conductivity of the seas, high in relation to that of the continents or of submerged rocks, could be assumed to play an important role in the overall distribution of the telluric currents over the surface of the earth. "We hardly can be wrong," he wrote, "in stating that, up until this time, all available documentation on this new aspect of the telluric currents (which we shall call pelagic currents, since the term 'marine currents' has quite a different meaning) does not go beyond the few observations we made offshore the Cherbourg roadstead in May of the past year." He was considering measurements made at sea, from which oceanographers could draw valuable data on submarine topography: shoals, continental shelves, abyssal troughs, and so on.

The first systematic study of the phenomenon, as reported in another paper, took place at Paray, Haute-Marne (June 3–9, 1922). The electrode configuration consisted of two lines, each 150 meters long, connected at each end with nonpolarizing electrodes and oriented west–east and south–north, respectively. As in the Lobsann experiments, the readings were made on two galvanometers. The plotting of the many curves showed rather irregular pulsations with a period of 30 seconds to 1 minute, simultaneous on both lines, and superimposed over movements of much longer duration (several hours). Vectors proportional to the amplitude of the pulsations recorded between two instants on the two lines were marked on rectangular coordinates, and the resultant gave the

[29] A submerged metallic hull forms a huge pile because it is made of diverse metals.

average direction of the tellurics, as well as a value proportional to their intensity over the same time interval. Conrad had observed that this direction had a tendency to line up with certain azimuths, varying according to the time of day. Furthermore, simultaneous measurements using two identical cross-shaped configurations separated by about 2 kilometers had shown concordance of the pulsations at both stations.[30]

Still another unpublished paper set forth the basic ideas of prospecting by tellurics. These currents, if they spread through formations of uniform resistivity, would take the form of sheets of parallel lines. It was as if they had been created by artificial A and B sources several hundred kilometers apart, with the AB alignment and the intensity of the current varying constantly. Since the soil is not homogeneous, its configuration and consequently its average resistivity affect the magnitude and the direction of the telluric vector at any moment and at any point of the surface. However, since neither the length of the imaginary current emission line nor the intensity of the current is known, no absolute value of the resistivity can be obtained from telluric measurements; the study is limited to its spatial variations. To this end the measurements made at a mobile base moved from station to station across the area being prospected are compared with those recorded within the same time intervals at a fixed reference base. The technique amounted to a resistivity map made with very long spacings and lightweight equipment, like the one used for the spontaneous polarization surveys. Although easy to carry out, the method did not match the current emission techniques where the spacings and the direction of profiles could be selected to adapt to the characteristics of a particular structure. Since only operations with shallow depth of investigation and spacings of a few hundred meters were then under consideration, experimentation with tellurics did not go any further at the time.

After several years of experience had demonstrated how difficult it

[30] These were the manifestations of the tellurics proper. Variations would occur from time to time, reflecting the meteoric origin of these currents: lightning (abrupt induction effect on the lines), wind, rain, sunshine (change in the evaporation rate of subsoil waters, and hence in the amplitude of the electrocapillary potentials). Conrad's reflections on the latter phenomena, their mechanisms and their behavior according to topography, temperature, sunshine, vegetation, and so forth, actually touched the field of physics of the globe; they supported the original assumption that electrofiltration might be the cause of the negative charge of the terrestrial globe.

was to operate with long current spacings, the measurement of tellurics was given fresh consideration in 1934. Marcel, always in favor of lightweight and convenient equipment, furthermore perceived these measurements as an expeditious way of conducting large-scale reconnaissance. The much costlier seismic reflection method would then be used to detail zones demarcated by the anomalies of the tellurics map.

As Conrad had indicated in his notes, the technique involved one fixed and one mobile base. The hand recorders used were adapted from the electrical coring. The readings on each mobile base took 10 to 15 minutes. Even over long distances (i.e., 10 kilometers), there was excellent correlation of the pulsations observed at the two bases. Since experimentation had shown a linear vectorial relationship between these recordings, it was convenient to scale the vectors plotted at each station, taking the synchronous vector of the fixed base as a unit: the ends of these vectors would define ellipses whose areas were proportional to the average resistivity of the ground.

One of the most significant surveys of that time took place near Hettenschlag in Alsace, the site of many previous resistivity measurements (Fig. 26). At each measuring station the vectors indicated the average direction of the tellurics: they were channeled between the Vosges in the west and the Black Forest in the east, flowing between the surface and the electric insulator constituted by the salt. The orientation of these currents was nearly parallel to the Rhine River except in the vicinity of the north–south saline ridges, which they crossed frontally.

This study and others showed that recording by hand was both inconvenient and costly (four permanent operators), and that the future success of the method required automatic equipment of the type already devised in connection with the first photographic recorders used in logging. The underlying principles, the housings, and the optical systems of the two recorders were the same, except that for tellurics the galvanometers were more sensitive and slower and the film was unwound by an automatic timing device.[31] These instruments were ready for field use when World War II broke out.

[31] This category of measurements was not based on the null method as with the potentiometer, but on the deviation of the galvanometers: the total resistance of the circuit, ground electrodes included, was adjusted for each base to a common known value, which

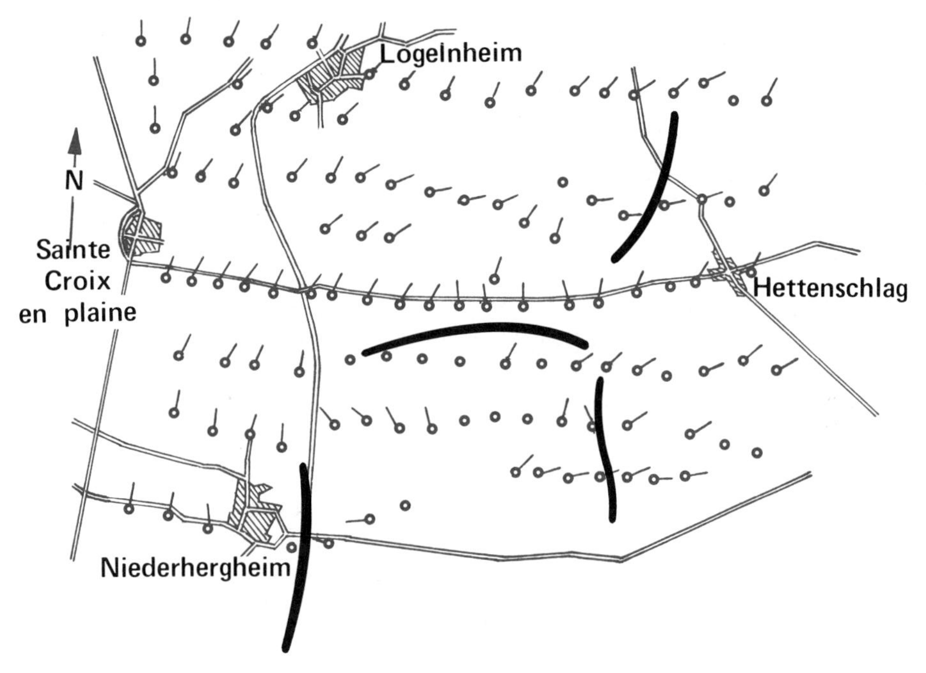

Figure 26. Telluric current survey of the Hettenschlag area. The solid black lines show the outline of the saline ridges.

After the German invasion in the summer of 1940, part of the staff, with equipment from S.P.E. and C.G.G., gathered at Saint-Gaudens, Haute-Garonne. Only a few months earlier, a large gas-bearing formation had been discovered in a well located nearby on the Small Pyrénées chain, a line of hills parallel to the main chain and reflecting a series of closed anticlines. As a result a new field was being developed near the village of Saint-Marcet. A discovery of such importance led one to believe that the region contained other deposits. Although a few structures were visible from the surface, geophysics was required on the search for those that were not. On the basis of the available geological data, Marcel assumed that the region was favorable for electrical prospecting and proposed its

made it possible to calibrate the deviations in potential-difference values. The adjustment of the circuit was possible thanks to an accurate electromotive force calibration standard equal to 1 volt.

application, mainly by tellurics. Thus, as early as 1940, a large part of the Aquitaine Basin was prospected by this method, which clarified several structures and traced their outlines.[32]

Part of the region was also prospected by resistivities with long spacings. The technique (basically like that used in Vitré; see p. 56) was to reverse the current at regular intervals of a few seconds and to record on film the potential differences between M and N with the instruments used for tellurics. The measurements, each a few minutes long, were plotted in diagrams against a background of fluctuations created by noise, where sharp deviation appeared at each reversal of the current. The average amplitude of these deviations gave the value of the potential difference generated by the current emission. The "kicks" resulting from induction and skin effect were easily eliminated by visual examination. This was a safer and more convenient process than the one provided by traditional measurement with the old potentiometer.

During the same years prospecting by tellurics was extended to the Languedoc, Bresse, and Bas-Dauphiné basins, as well as to part of the Rharb Plain in Morocco. It proved especially valuable in the detection of uplifted structures hidden by overlaps: outcrops, faults, steeply dipping anticlines, piercement folds, and the like, where seismic reflection often fails.

In charge of surface prospecting and logging in southern France between 1943 and 1946 was Eric Boissonnas, whose role in the development of tellurics was outstanding. Later he sought to apply his experience with the method in the United States, particularly on the Haynesville Dome in Texas, which had already been thoroughly explored by drilling. Its shape, as outlined by the telluric survey, was in keeping with the data provided by earlier surveys. In spite of these results, however, tellurics did not succeed in gaining recognition in the United States as a tool for petroleum exploration.

After World War II C.G.G. continued to improve the resistivity method. Theoretical studies undertaken at the highest level included the calculation of charts for multiple horizontal layers, the treatment of cer-

[32] In the southern part of the basin, the measurements were perturbed by intense stray currents from the Bayonne–Toulouse railroad line, to the point of making the values geologically meaningless.

tain cases of dipping formations, and the use of computers for the interpretation of electrical soundings with electrode spacings reaching 10 to 12 kilometers. From 1957 to 1958 the investigation of saline ridges was again resumed in Alsace and was extended to Baden and Wurtemberg. In addition, C.G.G. undertook several telluric surveys, particularly in the Parisian Basin (a detailed exploration of its eastern part was conducted in 1949) and in Algeria (Hodna Basin, 1948–1950). These surveys contributed to the study of certain aspects of the phenomenon by mathematical treatment, on laboratory models, and under specific field conditions. One survey was conducted with bases thousands of kilometers apart (France–Madagascar–Gabon), between which the recordings showed correlations (1945).

From the 1950's on, surface electrical prospecting was used chiefly outside the field of petroleum exploration. The resistivity method was applied in studying the configuration of captive aquifers, in locating contacts between fresh and brackish waters, in assessing the thickness of alluvial deposits containing groundwater tables, and so on (France, Italy, North Africa, and certain arid countries). The electrofiltration phenomena were put to use for the application of spontaneous polarization measurements in the reconnaissance of privileged water circulation zones. Electrical methods entered the geothermal exploration field in Larderello, Italy, and later in Kizilder, Turkey. One of the most important spontaneous polarization surveys for metallic ores took place in Mauritania (copper sulfide and magnetite).

Induced polarization, which had caught Conrad's attention as early as 1912, was the subject of important theoretical research work and was used in prospecting for pyrite, blende, and galena deposits. Finally, surface electrical prospecting was applied in civil engineering (dam foundations) and public works surveys.

PART
TWO

The Origins and Development of Logging

INTRODUCTION

Petroleum and natural gas occur within the pores of certain permeable formations (sands, sandstones, and limestones) of sedimentary origin. Also contained in a sedimentary series are other formations such as shales which, though porous, are practically impermeable, or rocks altogether compact like hard sandstones or limestones, gypsum, rock salt, and flintstone, as well as all the intermediate types: shaly sandstones, sandy clays, marly limestones, silicified clays, and so on. Vertically there is a very great variety among all these sediments, whereas laterally, at least within the same geologic unit, they may retain almost constant lithological characteristics over wide areas. Moreover, should these characteristics change laterally—say a limestone bed becomes marly—it often happens that the fossils remain the same.

Petroleum is produced by the transformation of certain live organisms buried, together with very fine sediments, in (generally) marine waters, and gradually compressed under the weight of further deposits accumulating on top. Formed between mineral grains and then in large part expelled by compression, petroleum travels through the openings available—permeable layers, faults, fractures—until it reaches the surface, where its traces soon disappear unless stopped inside a permeable layer by a barrier of impervious rock. If, for example, the permeable layer were folded, all the petroleum that could penetrate it would rise above the water toward the upper part of the fold or anticline. Such a configuration is called a structural trap. If there were also gas, it would move to the top of the petroleum, in which case the reservoir rock would, from bottom to top, enclose oil and gas. The real situation is more complex; a certain proportion of water remains in the pores by capillarity even at the very top of the structure. Structural traps are provided by other configurations: the top of a dipping permeable layer may be sealed off by clays brought about by a fault, or it may butt against the sheer flank of a salt dome.

On the other hand, a permeable layer may contain an increasing amount of argillaceous material and become altogether impermeable at the top, or it may taper off and disappear. Many petroleum deposits are also found in sand lenses or other reservoirs completely surrounded by shale. These are the so-called stratigraphic traps.

After surface geological and geophysical surveys have established the presence of potential oil- or gas-bearing structures, the only way to ascertain that these structures actually contain such deposits is to drill one or several exploratory boreholes.[1] The purpose of these boreholes is to identify the productive zones, if any, and to provide overall geological data on the formations penetrated. From the fact that the lithological and paleontological characteristics of the formations remain laterally constant it is possible to establish correlations among boreholes and to determine the structural features of a deposit or a basin.[2] Furthermore, exploration may sometimes be guided toward the discovery of stratigraphic traps by comparing borehole data from several wells, including rock lithology, dip, thickness, and fossils.

Once reconnaissance has located a commercial deposit, a production program is drawn up involving the drilling of a certain number of additional wells, called development wells. Exploration holes in the discovery area are usually converted to production.

In the 1920's numerous wells, particularly in shallow areas of the United States, were still drilled by the old cable tool method, in which a

[1] Very recent developments in seismic reflection make it possible to detect, in certain favorable cases, the presence of a gas deposit.

[2] Data gathered from the observation of outcrops can be rapidly complemented and refined by drilling small-diameter boreholes no deeper than a few hundred meters.

heavy steel ram was projected at the lower end of a cable. However, for the past 20 years or so, the cable tool process had been largely supplanted by the rotary method with its much higher drilling speed and ability to drill much deeper.[3]

The drilling tool employed in the rotary method is called a bit. It is rotated like a twist drill at the end of hollow drill pipes and is driven by surface machinery. To begin with, the bit is screwed to the bottom of the first drill pipe; as it penetrates the soil, a second length is screwed to the top of the first, then a third, and so on to the depth desired. The derrick —formerly a wooden structure, now steel—allows vertical handling through a crown block and elevators of pipe lengths from 60 to 90 feet. The drill pipe is rotated through a huge gear called the rotary table, which is driven by steam, electrical, or diesel power. In the course of drilling, water, charged with various ingredients and called mud, is pumped through the drill pipe down to the bit; it then returns upward through the annular space between the pipe and the wall of the borehole. This mud is composed mainly of suspended clay, but also contains a number of mineral and organic additives which, dispersed or dissolved, give to the mud its desired qualities: density, viscosity, and colloidal and chemical properties, according to the specific conditions of the borehole (depth, temperature, diameter, nature of formations drilled through, etc.).

By strengthening the walls and applying hydrostatic pressure on the formations, the mud makes it possible to drill over hundreds of feet without the need of lowering a protective casing. It also lubricates the bit and the drill pipe and carries the fragments cut out by the bit to the surface. The density of the mud is adjusted in such a way that, barring the unforeseen, its pressure at the level of each permeable horizon remains higher than that of the formation fluids—water, oil, or gas. Thus the mud prevents uncontrolled blowouts with their disastrous consequences, such as destruction of installations and fire.

After setting a short section of large-diameter conductor pipe, drilling begins with a borehole about 12 to 16 inches in diameter. At a depth of 800 to 1200 feet, a first string of casing (surface pipe) is lowered into the hole and sealed against the wall by the injection of cement. This casing is usually set below all freshwater sands to protect them from borehole fluid contamination. Drilling is then continued with a bit diameter of

[3] In the 1920's the depth record was 7500 feet; it is now 30,000 feet, drilled in 2 years.

about nine inches to the depth determined from the geological conditions or the location of the pay zones drilled through or hoped for, after which a new string of casing is lowered and cemented. According to the circumstances, drilling ends at this point or is continued to the final objective.

Before electrical logging was known, the only expeditious way to identify formations was to monitor the rate of drilling, thus obtaining some idea of the relative hardness of the rocks, to observe the mud (for level drops, changing salinity, gas bubbles, or oil droplets), and to examine the cuttings as they reached the surface. With the techniques then available these procedures were vague and unreliable. The most accurate one was mechanical coring. In this operation special crown-shaped bits cut cylindrical samples a few feet long called cores, which are drawn to the surface for examination. Although valuable, the data gathered by mechanical coring have shortcomings. The frequent retrieving, lowering, and disassembling of the tool, as well as the whole string of drill pipe, make drilling slower, more complicated, and more costly. Furthermore, the cores are often crushed by the tool, and in very soft formations core recovery remains incomplete. Also, cores coming from permeable rocks are contaminated by the mud and, while the tool is being retrieved, lose most of the oil or gas they may have contained, thus giving only a distorted picture of the actual formation. Neither is it always easy to obtain from the cores stratigraphic data which can serve as a basis for well correlations.

These shortcomings were more acute 50 years ago, when the techniques available for the handling and observation of cores had not reached their current degree of refinement and accuracy. It may be added that, to limit costs and risks, operators endeavored to core only where the presence of key horizons or pay zones could be anticipated from geological data and drilling incidents. Such a procedure was often disappointing; the cored lengths could be devoid of interest, or the cores could have been taken too late, after the promising formations had already been drilled. There remained room for much improvement.

When electrical coring was invented, most of the oil produced came from fields in Tertiary basins made up of sands, sandstones, and shales that were generally unconsolidated, with here and there a few compact beds. Oil fields in Venezuela, California, on the Gulf Coast of the United States, in the Caucasian provinces of the U.S.S.R., in Rumania, and in

Indonesia were of that type. These were also fields where electrical coring rapidly became an efficient and economical auxiliary to exploration and production (Fig. 27, left).

Another type of field, peculiar to the Mid-Continent of the United States, Mexico, and the Middle East, has since represented an increasing share of world production as new discoveries have taken place in these countries and such others as the U.S.S.R. and Venezuela. What is found there are thick and hard beds, mostly limestone, where in places the porous, permeable, and usually fissured rock yields petroleum or gas, often with spectacular outputs (Fig. 27, right). Whereas scant use had

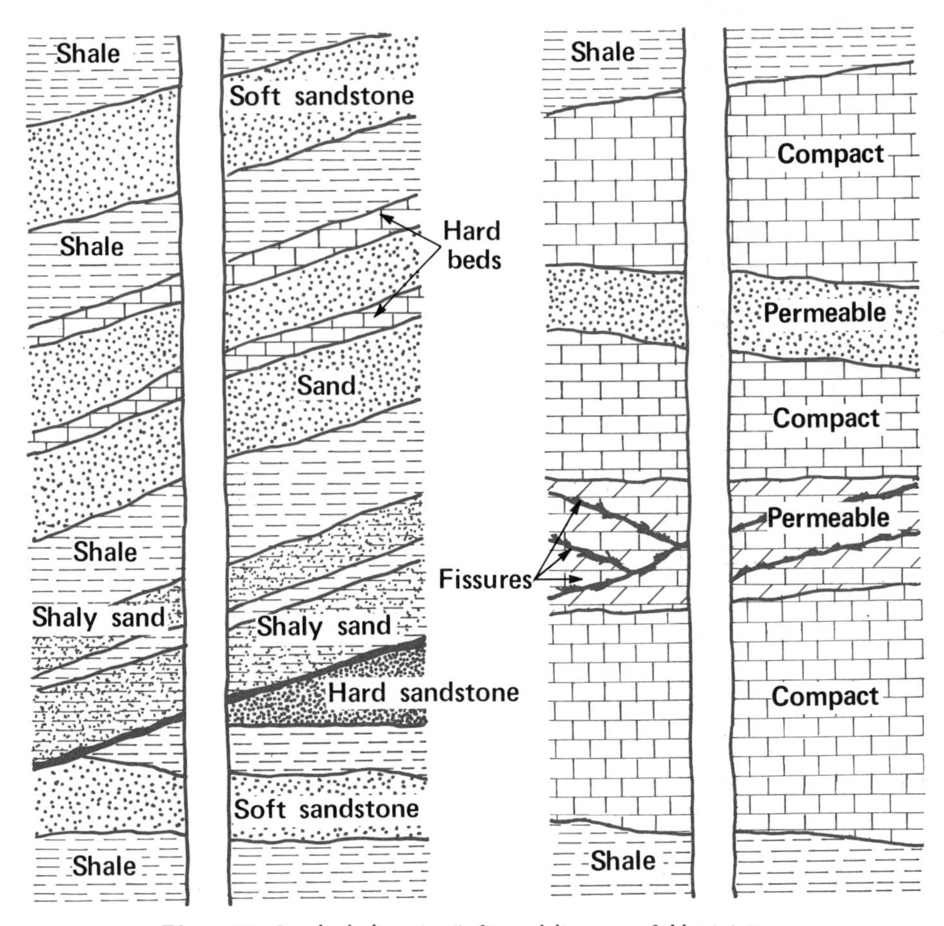

Figure 27. Sand–shale series (left) *and limestone field* (right).

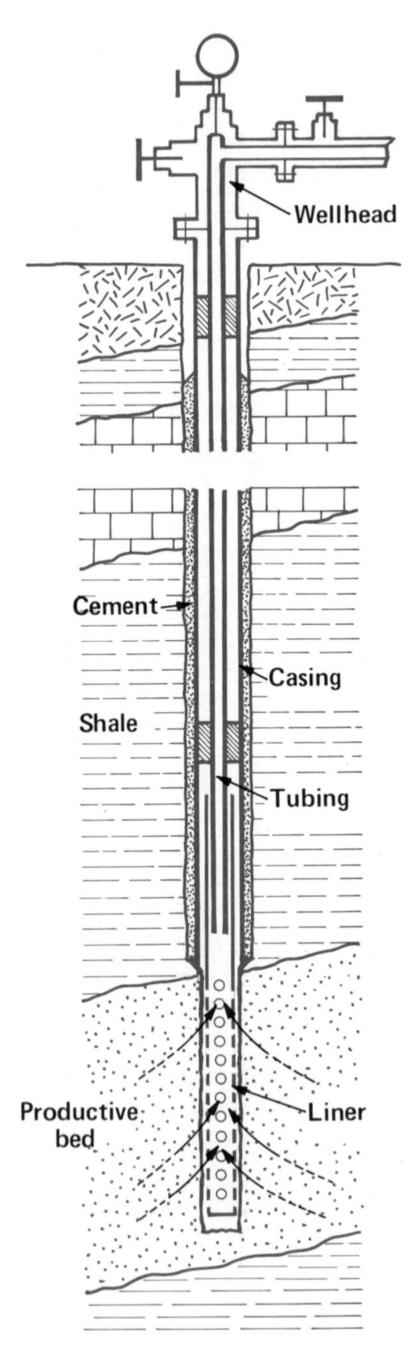

Figure 28. Schematic cross section of a completed rotary oil well. Such a layout was in worldwide use up to the early 1930's.

originally been found for electrical coring in this category of formation, this situation changed much later when new techniques had improved its capability and scope.[4]

When one or several horizons have been identified as pay zones, the oil well is completed (Fig. 28). In the 1920's only one horizon at a time could be produced, usually beginning with the shallowest. The sequence in this operation was, first, to lower and cement a string of casing reaching to the top of the pay zone, so as to seal off the upper aquifers (water shutoff), and, second, to position a perforated liner in the pay zone and lower a small-diameter tubing. The mud column was then lightened by swabbing the tubing or by injecting gas or water into it. If the formation pressure was high enough, the oil reached the surface on its own; otherwise it had to be pumped. Most of the same procedures are still employed today. Only later, presumably in the wake of electrical coring, did perforation techniques appear. Perforated liners have been largely replaced by casing perforations. As a result multiple pay horizons may now be produced at the same time; if horizons are produced one at a time, the usual order is from the bottom to the top.

[4] According to prevailing terminology, these two types of fields will hereafter be called sand–shale series and limestone fields.

Phase One: 1927–1932

Pechelbronn: A first resistivity curve

It will be helpful here to look back a few years. In March 1921 Marcel Schlumberger, Eugène Léonardon, Pierre Baron, and Jacques Gallois were conducting a survey by plotting equipotentials in the Bessèges coal basin near Molières-sur-Cèze, Gard, where a reconnaissance borehole was being drilled. Marcel recognized there a unique opportunity to determine *in situ* the resistivity of the subsoil formations for the purpose of enhancing the interpretation of the surface measurements. The same borehole was to serve a little later, in July, for a dip study.

Some 2500 feet deep and cased to about 1500 feet, the hole was full of water. There are no detailed records of the operation except for the fact that between March 20 and 22 several resistivity measurements (over a few feet at the bottom of the hole) did reflect the variations in the nature of the formations. This was no doubt the first operation ever of this kind in the history of mineral exploration; it was not followed by such another until 1927.[1] The time lag can be explained by the fact that every effort was then being devoted to surface prospecting.

The Pechelbronn company, which for years had been contracting with Schlumberger for surface work, was eager to know whether resistivity measurements in boreholes could be of assistance to its geologists. The

[1] Marcel had nonpolarizing electrodes lowered into an available ventilation shaft; one of their components was . . . flower pots. How he managed their contact with the wall of the shaft has not been recorded. This episode is mentioned only as an example of the improvisations often required for the success of an experiment.

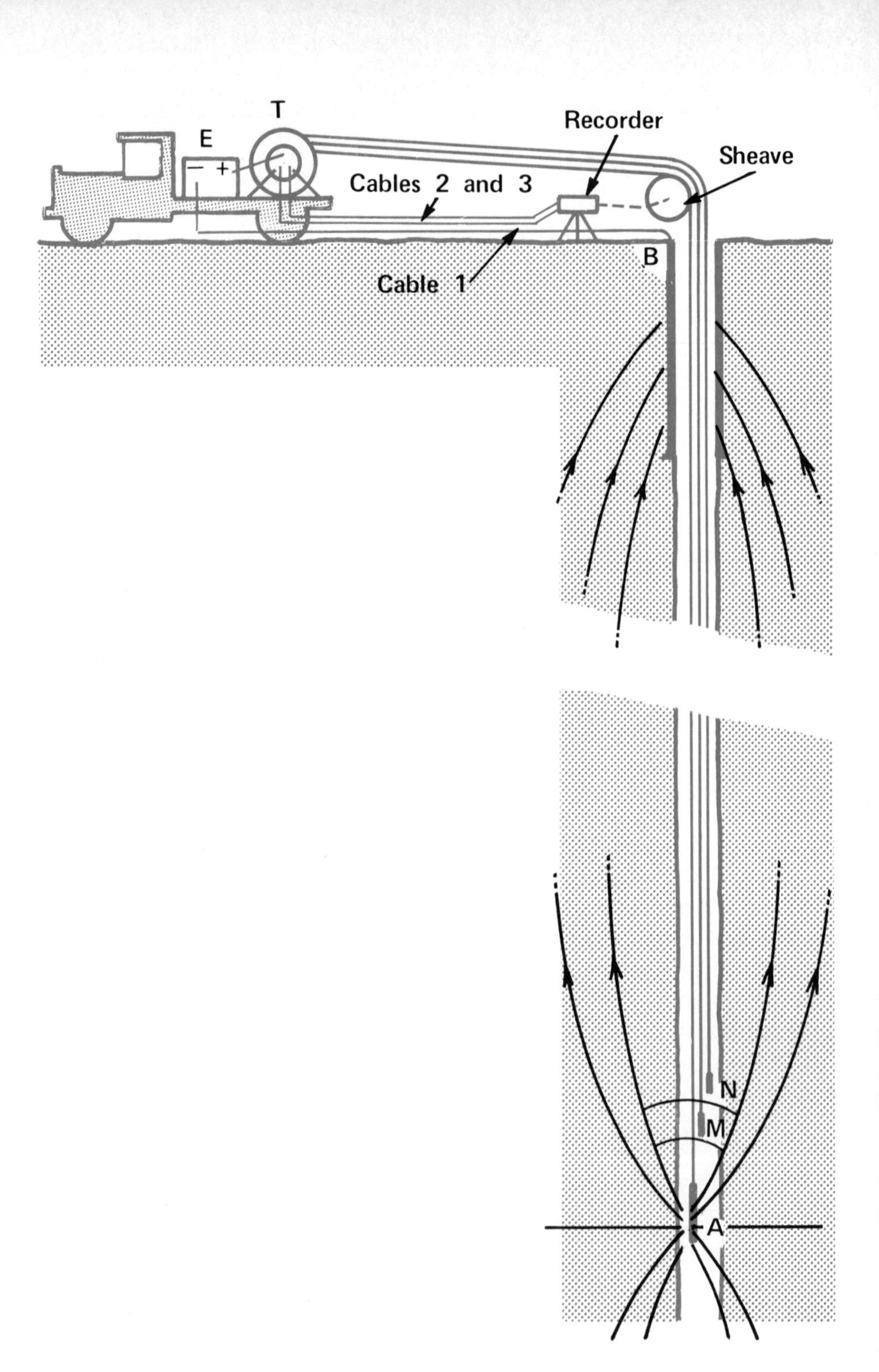
T
E
− +
Recorder
Sheave
Cables 2 and 3
Cable 1
B
N
M
A

company was especially interested in the location of a key horizon (the top of a bed of Hydrobiae marls), which drilling sometimes failed to reveal. This suggestion was welcomed by Conrad, who for a long time had had a similar idea in the back of his mind.[2] This was the starting point of a project from which electrical coring was to be born. If such measurements proved feasible, he thought, they would contribute primarily to refining the geological cross section rather than actually detecting the petroleum horizons, as scarce and thin as they were in that basin; from these measurements, furthermore, data could be gathered on the resistivities of deep formations that would shed light on surface work.

In a note dated April 28, 1927, and entitled, *Recherches Électriques dans les Sondages* ("Electrical Research in Boreholes"), Conrad outlined the principle of the new method, which from then on was called electrical coring (Fig. 29). Doll was given the responsibility for equipment design and testing.

The actual invention consisted of proving by facts that open-hole resistivity measurements were feasible and significant—something which, at the outset, was far from obvious. Some thought that the current would flow through the mud without penetrating the far less conductive formations. Although this claim had seemed to be invalidated by the few tests made at Molières-sur-Cèze, these tests had been too few and had been undertaken under conditions too special to allow the drawing of a conclusion. Now Conrad, aside from any experimentation, had demonstrated through simple reasoning that the current emitted by the elec-

[2] In fact, it seems that the idea of electrical coring originated in Conrad's mind before any suggestion was made by Pechelbronn. During a conversation at lunch, in early spring 1927, a mining industry executive emphasized the difficulties encountered in recognizing bottom-hole formations. Thereupon Conrad disclosed to P. Charrin the idea of using to that end the recording of resistivities in boreholes, and gave him a sketchy description of the measuring system. Charrin was entrusted with the preparation of the equipment, but, because of an urgent mission abroad, he had to turn the project over to Doll. (Communication from P. Charrin subsequent to publication of the French edition of this book.)

Figure 29. Schematic view of electrical coring. Three electrodes, A, M, *and* N, *are lowered into the borehole, each at the end of an insulated conductor. The current emitted by* A *flows through the mud and spreads across the formations. The voltage created between* M *and* N *is transmitted to the surface and measured there. From this measurement and that of the current intensity it is possible to deduce the value of the apparent resistivity. As in practice no borehole ever is strictly vertical, the three electrodes are in contact with the wall.*

trode ought to flow through the mud and then spread inside the formations.[3] Others, among whom were men of repute such as Professor R. Ambronn, maintained that below a certain depth all geological formations became altogether compact and hence infinitely resistant. Nothing of the kind had been shown by surface measurements, but since they did not penetrate very deeply, positive proof was lacking.

The first electrical operation was undertaken on September 5, 1927, at Diefenbach in well No. 2905, rig No. 7, by Doll, assisted by Scheibli and Jost. Nothing would be more eloquent than to quote an excerpt from the description by Doll 32 years later of this memorable experiment.[4]

> *AM* was three meters long and *MN* was one meter long. We made a sonde by connecting four meter-long sections of Bakelite tubing by means of short lengths of brass tubing, fastening them to each other with brass screws. The electrodes were wired to the Bakelite tubes. We contrived a weight, or plummet, for the bottom of the sonde, making it of one meter of brass tubing, four centimeters in diameter, and filling it with lead pellets like those used in duck shooting. It was plugged at both ends and weighed about 25 pounds. The whole assembly looked like a long black snake with five joints.
>
> The cable, if you could call it that, was three lengths of rubber-insulated copper wire, of the kind used on spark plugs in cars.[5] It had a tensile strength of about 80 pounds per wire. The wires weren't spliced together, as was done in later surveys, but were allowed to wind onto the winch drum loose from each other.
>
> The winch had an X-shaped wooden frame; the drum was made with wooden flanges and the core of a large Bakelite tube. It was assembled by long brass bars and nuts. To turn the drum, we had a big pinion connected to a smaller pinion by a motorcycle chain. The moving axle was steel, with a bicycle pedal mounted at either end. One of us would get on one side and one on the other, and turn the pedals. There was a ratchet to keep the drum from unwinding.
>
> We had no collector. Instead, we had a plug, much like a common wall plug, at the side of the winch flange. When the winch had to be turned, the cable connection to the potentiometer was unplugged so the turn could be made. Then the cable was plugged back in so that we could make the readings.

[3] Several years later highly advanced calculations confirmed the correctness of his demonstration, which in the meantime had been verified by thousands of measurements.

[4] See *Sonde Off*, September 1959, p. 22.

[5] This was actually the cable then used in surface prospecting.

The sheave[6] was made of wood with an eccentric axle. It had a long tail as a counterbalance. This served as our strain gauge.[7] We were very worried about the wires breaking; and by watching the rise and fall of the tail, you could tell what kind of pull was being exerted on the wires. For depth measurement, we had a counter on the sheave wheel like the mileage indicator on a car. We planned to take readings at intervals of one meter.

We made our measurements with a standard potentiometer mounted on a tripod like those we used in our surface exploration work.

It was a nice fall, a decent day. We drove out to the well—it was Diefenbach 2905, Tower No. 7—in an old station wagon that had been used in surface prospecting and was completely worn out.

The well was about 500 meters deep which is about 1500 feet. We couldn't have gone much deeper, for we only had about 1800 feet of wire.

We began making our measurements. Someone had to unplug the connector, someone else turned the winch, someone had to run up on the rig floor to look at the counter on the sheave . . . there was a lot of running back and forth. I wrote down the measurements on a pad, together with the depth reading. Then it was unplug, roll up one meter to the next station, and plug back in. Make the next reading. And so on, one meter at a time.

At first, we jumped around a good bit, but soon we got the swing of things and before long, we were able to take about 50 stations per hour. At one meter per station, that's around 150 feet per hour surveying time.

The whole arrangement worked well. Everything went off as we had planned, except for one incident. When we came out of the hole, and had the sonde hanging in the derrick, we unfastened the weight to remove it from the bottom of the sonde. We forgot that the upper tube, being hollow, had filled with mud — which showered all over us — and we got thoroughly messy. We ended the day by going into the nearby village to take a bath.

In preparing the experiment, Doll had thought of how to control the electrical leaks that would occur, mainly in the cables after they had been dipped in hot saline mud under a pressure of several tens of atmospheres. To this end he had provided the sonde with a relay resistor not unlike the ones used in surface work. The control consisted in making sure that, when the current was reversed to actuate the relay and to introduce the resistor into the circuit in series, the values measured did not vary; the

[6] A grooved pulley at the wellhead redirecting the cable toward the winch.

[7] This instrument has been named the Roman balance; the sheave was mounted on the short arm, where the pull of the cable was balanced by the weight of the long arm.

operation showed that in this first experiment the leaks had not played a significant role.

Back in Paris, Doll plotted his measurements on a strip of graph paper and drew the first of the typical diagrams that were to become familiar to the petroleum industry. A part of this venerable document is reproduced in Figure 30. It shows a zone of rather uniform resistivities corresponding to the Hydrobiae marls whose top served as a marker and, above, a sequence of peaks and troughs where the geological cross section indicated hard marls, conglomerates, and sandstones. As could be foreseen, the resistivity dropped to zero at the lower end of the casing:[8] the depth at which the drop occurred was within 6 feet of that indicated by the drillers. In view of the rather primitive metering of the cable, more could hardly have been expected.

The tests were continued at Pechelbronn under generally difficult operating conditions, mostly at night when drilling was usually suspended (Fig. 31). Cave-ins or exceedingly heavy or viscous mud frequently prevented the lowering of the sonde; and since boreholes are rarely vertical, the dragging of the sonde and the cable against the wall when being pulled out resulted in frictional stress close to tensile strength. Contrary to the prevailing concepts of Marcel, heavier weights and stronger cables were required. As access to drilling sites was difficult in rainy weather, an odd suggestion came from Paris: to unload the truck when stuck in the mud and push the winch like a wheelbarrow, with the reel serving as the wheel. Needless to say, the first attempt converted the whole mechanism into a huge mud ball.

In spite of these difficulties and, in particular, the inadequate lengths of open hole available—only 60 to 150 feet—positive conclusions could be drawn in the very first weeks of operations. Hard layers appeared on the diagrams as peaks contrasting clearly with the soft and conductive marls. From the similarities in the log features, accurate correlations could be established between boreholes from which the configuration of the formations could be determined throughout the field. Comparing the measurements with the cores in a sufficient number of holes allowed the identification on the logs of most peaks and troughs. Thus electrical coring gradually came to replace most mechanical coring.

[8] Compared to the resistivity of the formations, that of the metallic casing is practically zero.

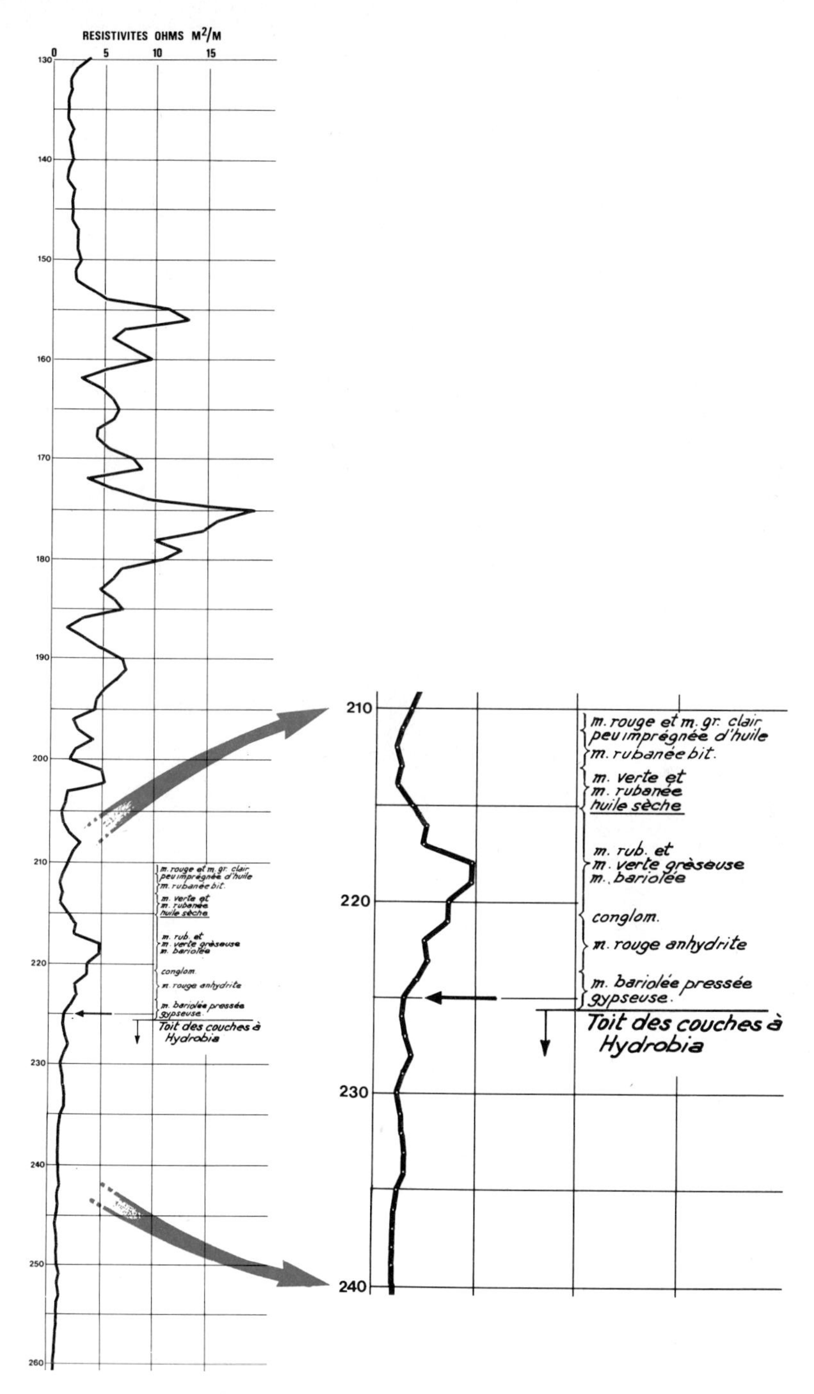

Figure 30. The first electrical coring diagram (Pechelbronn, Alsace, September 1927).

Figure 31. The winch, cable, and pickup truck used in the beginnings of electrical coring at Pechelbronn.

Developments in basic equipment

Initial results obtained at Pechelbronn warranted an effort to provide more advanced equipment for electrical coring. Later, when the method had proved its value in the large petroleum basins of Venezuela and the U.S.S.R., it became the main objective of engineering and manufacturing activities, with surface operations taking second place. This led the management to a reorientation of the Company's technical plans and goals.

This new method opened for Conrad an attractive field of activity; yet, since it was merely a technical service performed at a day-to-day operational level in support of the oil companies' drilling activities, it could not provide him with the same stimulus as geophysics proper, with its much broader objectives at the very forefront of exploration. The intellectual interest of electrical coring lay mainly in its multiple technological and operational aspects, whereas Conrad's preference was for research work relevant to physics and geology. This explains why (although often intervening in crucial decisions involving the orientation of electrical coring research and techniques), he took personal initiative only in a few special projects where he felt most at ease, such as the design of the "guarded monoelectrode sonde"[9] and laboratory work on spontaneous potentials. However, in the rare moments not devoted to the overall management of the Company, he followed surface work with a keen interest, hoping that he could return to this favored activity once electrical coring was well on its way and required less of his personal attention.

The contributions of Marcel and Doll to surface prospecting were important indeed; but to the former, with his mechanical genius, and the latter, with his mastery of electricity, electrical coring presented a wider and more varied range of problems awaiting solution through application of their respective talents. Most of the technical developments of the Company were theirs, and they knew not only how to recruit a highly qualified staff, but also how to give it inspiration and guidance in the design and manufacture of highly specialized equipment. Only its principal components will now be reviewed.

Cables

It will be recalled that the initial operations in Pechelbronn utilized the cables employed in surface work, tied every few meters with insulating tape. The three wires became entangled when being rewound on the winch, however, and it took great effort to unravel them when again lowering the sonde into the borehole. Moreover, as it was impossible to equalize the stress among the three wires, breakages were frequent. Sometimes only the copper conductor would be ruptured under the insulation, and then it was necessary to lay the cable flat on the ground to allow a step-by-step inspection to pinpoint the leaks. To this end the

[9] Also called "current output sonde"; see p. 123 ff.

operator had to slip his hand under the electrically charged cable until a slight pricking located the point of discharge. All this was time consuming, clumsy, and inaccurate, and the ensuing frequent delays were not appreciated by the Pechelbronn company. The prospectors considered that a cable better suited to the task was indispensable for the successful marketing of electrical coring. What they were demanding was a real, industrially made tricable with good tensile strength and adequate insulation. Doll was promoting the same idea, and the first cable made by French industry to Schlumberger's specifications was put into operation in June 1928. It consisted of three conductors, each made of about 20 stranded steel wires and coated by rubber insulation and tape. In turn, these three conductors, with an appropriate packing, were stranded together and enclosed in a sheath protected by a tarred cotton braid. The tensile strength of this cable was 1 ton. With it field service improved immediately as lost time decreased. Thereupon a greater tensile strength was sought; in 1930 it reached 2 tons for the French-made, and 1.3 tons for the first American-made, cable. Both had a textile-braid protection (jute in France, cotton in the United States).

Nevertheless, in spite of tight specifications and detailed inspections, after a few operations the electrical insulation of the cable was often reduced to the point where errors in measurements became unacceptable. Insulation control in the field and localization and repair of leaks therefore became the focus of continuing studies, as witnessed by numerous articles on this subject published in the early 1930's in *Proselec.*[10]

To measure the insulation of a cable was a rather simple thing, but to locate the leaks was much more difficult. Improvements in the latter respect are worth reporting. In 1930 the Paris engineering office had designed a rather complicated system based on the Wheatstone bridge principle; its shortcoming was that it functioned properly only when a conductor developed a single, relatively minor leak. Another process was tried a little later. After the ends of the conductors had been insulated, the electrically loaded cable was lowered into a borehole; as soon as a leak reached the top of the mud, the circuit was closed, and the potentiometer

[10] *Proselec* appeared first in mimeographed form as a technical publication for the use of the Schlumberger staff only. Initially, it covered the whole range of activities of the Company. From April 1931 on, it comprised two parts: *Proselec Carottage (Proselec Coring)* and *Proselec Surface*. The latter part was replaced in early 1934 by *Cégégec*, published by C. G. G. Both publications ended with World War II.

deviated. Unfortunately the effective functioning of this system required sizable leaks, and it proved of little use. In Baku in early 1931, R. Sauvage and G. Delamotte resorted to the extreme remedy of "arcing" the leaks. The cable was dipped into a tank filled with water and then connected to a 2000 volt transformer; this ruptured the insulation at its weak points, as shown by sparks, bubbles, and smoke. This was, without doubt, a radical process, and not without risk to the operators and the overstrained cable itself. Furthermore, few crews could find an appropriate installation locally.

Finally, the so-called guard ring method was devised in Paris by Henri Doll and Gilbert Deschatre. The operation took place when the cable came out of the hole. While it was passing between sheave and winch, its braid well soaked with water, three wire coils (called rings), wound around a wet sponge, were held by hand against the braid. Each of these coils was properly grounded, the middle one across the potentiometer. The cable was connected to a battery; whenever there was a leak outside the coil system, the current flowed directly through the braid into the ground without affecting the potentiometer. On the contrary, when a leak came within the two outer coils, part of the current flowed through the central coil, and the needle of the potentiometer deviated. This system was simple, reliable, and accurate and made it possible to pinpoint a leak within a few inches. It was tested by Doll in the U.S.S.R. during the spring of 1932 and was utilized by all crews as long as textile-braided cables remained in use.

Winches

The new winches were practically a personal achievement of Marcel. The first improvement was to fasten a collector with four copper rings on one of the flanges of the wooden, crank-driven reel used in the Pechelbronn tests. Three of the rings were connected to the ends of the conductors wound on the reel. One brush was connected with the battery, and two others with the potentiometer. The fourth ring was grounded as an additional precaution against leaks that could occur between the ring of the emission circuit and those of the measuring circuit.

When operating in boreholes deeper than those at Pechelbronn, that is, when the cable was subjected to increasing stress, the need arose for winches of greater capacity and durability. The trend was to build all-metal reels with steel-plated flanges and an iron core. However, to reduce to acceptable limits the induction phenomena occurring in the conductors wound on the winch at each switching on and off of the circuits, a better solution was adopted: metal flanges and a Bakelite core. In 1930 and 1931 there were winches of two tricable capacities: 4000 and 7000 feet. They were electrically driven, with a gearshift and a clutch. When electric power was not available at the drilling site, the winch was driven by the engine of the truck. The arrangement was simple and rugged: jack up one rear wheel of the truck, prop up the chassis, and connect through a flat belt the tire of the lifted wheel with a pulley at the end of the winch drive shaft. This was the first model of what Americans, with a smile, were to call the "French power takeoff." Someone even had the idea that the spare wheel of the truck could be used as a pulley! Indeed, as early as 1929, prospectors arriving in Venezuela had tinkered with a chain drive for the winch made with a sprocket on the winch shaft and another fixed on a wheelhub. This system was later to be improved with a powerful clutch and special props to give firm support to the rear of the truck. The year 1930 also witnessed the first winding device to provide for the regular reeling of the cable on the winch.

The winch problem illustrates the diversity of equipment then produced in the Paris workshop. Such diversity was inherent in the variety of operating conditions from one country to another (average well depth, density and viscosity of muds, etc.); it also reflected Marcel's ceaseless drive to devise improvements. Thus, although by the end of 1932 there were only half a dozen logging crews operating in the world (the U.S.S.R. excepted), nine models of winches were already in operation. (Fig. 32).

Recorders

Doll has described (see p. 104) his first electrical coring operation, at a time when the measurements involved the same reading techniques as surface prospecting, that is, point by point, interrupting the upward movement of the sonde at regular 1 meter intervals (Fig. 33). The variations in resistance of electrode *A* were negligible as compared to those of the emission line (as for electrode *B*, since it stayed at a fixed point on the ground, its resistance obviously remained unchanged throughout the operation). It was sufficient, therefore, to measure the current just once at the beginning of the operation and then check at the end to see that it had not changed: only the voltage, Δ_v (see Fig. 29), had to be

Figure 32. Truck-mounted winch (1932).

measured at each station. As it became more and more urgent to seek continuous measurements without interrupting the upward motion of the sonde, Marcel designed the first hand-operated recorder, which was to serve for many years with only minor changes.

In this system the potentiometer and the null method served as before, but to avoid a situation in which the measurements were affected by the polarization of electrodes *MN* and other parasitic effects, the operation resorted to pulsated-alternating current (see p. 52 ff). The potentiometer was enclosed in an aluminum casing, on the top of which a circular plate with a handle could be rotated around a vertical axis. Through a rather sophisticated transmission gear (a groove in the shape of an Archimedes spiral milled in the plate was its main component), the operation of the handle drove the unit and tens knobs of the potentiometer together with a pencil-holder, whose linear displacements were proportional to the measured values. Once the current was plugged in and the pulsator rotating, the operator maintained the potentiometer needle at zero as the sonde moved upward. The pencil drew a curve on a paper strip which unwound mechanically in relation to the motion of the cable. The curve was a continuous resistivity diagram drawn as a function of depth. The scales selected for resistivity were obtained by adjusting the

Figure 33. Electrical coring measurement with potentiometer (1929).

intensity of the current and the sensitivity of the potentiometer; for the desired depth scales the proper set of pinions had to be inserted.

The first recorder was fixed on a console containing the switches, rheostats, and other accessories for the adjustment of the emission current and the control of the electric motor driving the winch. During a survey this console rested on a trunk used to store and transport the various tools (Fig. 34). The whole arrangement was set up between the winch and the drilling rig. Under the console the cable passed between two pulleys, one of which drove the cylinders with the paper strip. This type of recorder was sent in 1929 to the crews in Venezuela, the United States, and the Dutch Indies and somewhat later to Grozny in the U.S.S.R.

With such a setup, it could happen that the cable would suddenly become so taut that the whole apparatus would go flying into the air before the astounded gaze of the operator. In 1931 an improved version was introduced, in which the arrangement of the accessories was more compact and convenient, and the tripod-mounted recorder was located on the drilling floor; the motion of the paper was controlled by the rotation of the sheave through a flexible drive shaft. In 1932 the recorder (or, rather, recorders, since at that time the spontaneous potential was being

Figure 34. The first hand-operated recorder (1929). The cable, passing between two pulleys on the side of the box, controlled the motion of the paper.

measured simultaneously) was placed next to the winch, the flexible shaft being driven by a set of pulleys built into the winding device.

Formation resistivity being a widely variable parameter (from a fraction of an ohm-meter for soft shales or saltwater-saturated sands, to tens of ohm-meters for oil-bearing sands or sandstones, and practically to infinity for compact rocks such as rock salt or gypsum), the prospector had to adjust the scale during the recording lest the curve be truncated when the sonde passed high-resistivity formations, and a blank be left instead of what were perhaps the most significant values. Such an adjustment was possible whenever correlation with diagrams from neighboring wells provided some foreknowledge of how the curve would look in the well currently being surveyed. Otherwise, the sonde had to be lowered again and the measurements resumed on a smaller scale. This had a negative effect on the duration and cost of the operation; moreover, when the well was immobilized too long, there was a risk that the sonde could not be lowered again, or that the sonde and the cable would become stuck when coming out.

One elegant way out of the difficulty seemed to be the use of a logarithmic resistivity scale.[11] Some potentiometers were adapted to this kind of scale, but not all geologists liked it because it weakened the character of the diagrams. It was abandoned after a few years, only to be successfully reintroduced 30 years later.

At the end of 1931 the various improvements in electrical coring had increased the reliability of the method to the extent that Conrad could state in a communication to the U.S.S.R. Academy of Sciences:

> With a good electrical cable, a good winch and good measuring instruments that are both sensitive and sturdy, it is possible to operate at the bottom of a 2000 meter well and make highly delicate measurements of the phenomena generating potential differences of only a few millivolts, and involving practically negligible amounts of energy. Such a mating of the finest of potentiometers and a powerful and rugged rotary rig certainly makes for an odd marriage, but so far everything seems to be going well.

The spontaneous potential curve

In many operations—at Pechelbronn, and even more clearly in Venezuela and the U.S.S.R.—it had been observed that, even when no current was being emitted, the potential in the *MN* circuit varied with the depth. Various causes were assumed: instability in the polarization of the electrodes; a somewhat heterogeneous mud column; currents generated by the casing, since the latter's oxidization may vary over its whole length and therefore behave like one or several batteries; the action of stray or telluric currents; etc. Any of these effects could generate electromotive forces as a function of time and depth. Conrad and Doll thought that, rather than compensating for these potential differences or eliminating

[11] The displacement of the pencil then becomes proportional to the *ratio* between resistivity values. For instance, when the measurement increases from 1 to 10 ohms, the deviation of the pencil is 3 centimeters; when it increases from 10 to 100 ohms, the deviation is the same; and so on. A diagram with a 1 : 1000 ratio can be drawn without changing scale; this is usually sufficient.

their effects, as in the case of resistivity recording, it might be worthwhile to measure them.

The study began in Pechelbronn in November 1930. Despite the rather unfavorable geological conditions of this minor oil field (most of the permeable layers were very thin), it had the great advantage of being close to Paris; therefore Doll could direct the tests and analyze their results personally without having to spend all his time in the field and neglect his many other tasks. Paul Chabas, then head of the Pechelbronn mission, was in charge of carrying out the measurements.

Rather crude devices rendered the electrodes nonpolarizing.[12] The potential differences, measured every 20 inches between two electrodes that distance apart, constituted "gradients." To facilitate interpretation, Doll patiently reconstituted the potential curve as it would have been recorded between a mobile electrode in the borehole and a reference electrode at the surface. The problem was to make the sum of the potential differences or, more specifically, to integrate the gradient curve.

The first tests showed that repeated measurements in each borehole gave a true reproduction of the values, that disturbances (stray currents, etc.) were intermittent and infrequent, and that, except in the vicinity of its shoe, the presence of the casing did not manifest itself. On the other hand, the fact that in these first boreholes the layers subjected to measurements were mainly shales and hard beds resulted in curves without relief. Finally, on February 5, 1931, a bed of conglomerate produced a "kick," stable with time, which, after gradient integration reached a minimum of some 10 millivolts, and stood in sharp contrast to the rather flat line recorded through the surrounding formation (Fig. 35). A conglomerate being a rock made of more or less cemented gravels and sands and generally permeable, the invasion of the bed by the mud could have generated an electromotive filtration force; the lines of the current produced circulated in the conglomerate and through the adjacent shales and closed through the drilling mud. The value measured was the ohmic potential drop due to the passage of the return current along the hole. Furthermore, according to the mechanism of electrofiltration, the poten-

[12] Each electrode was made of a lead wire winding and enclosed in a sleeve of heavy, fire-hose-like cloth, tied on both sides of the electrode and filled with a saturated solution of lead acetate with an excess of crystals. When it eventually appeared that polarization was weak and stable, such precautions proved superfluous.

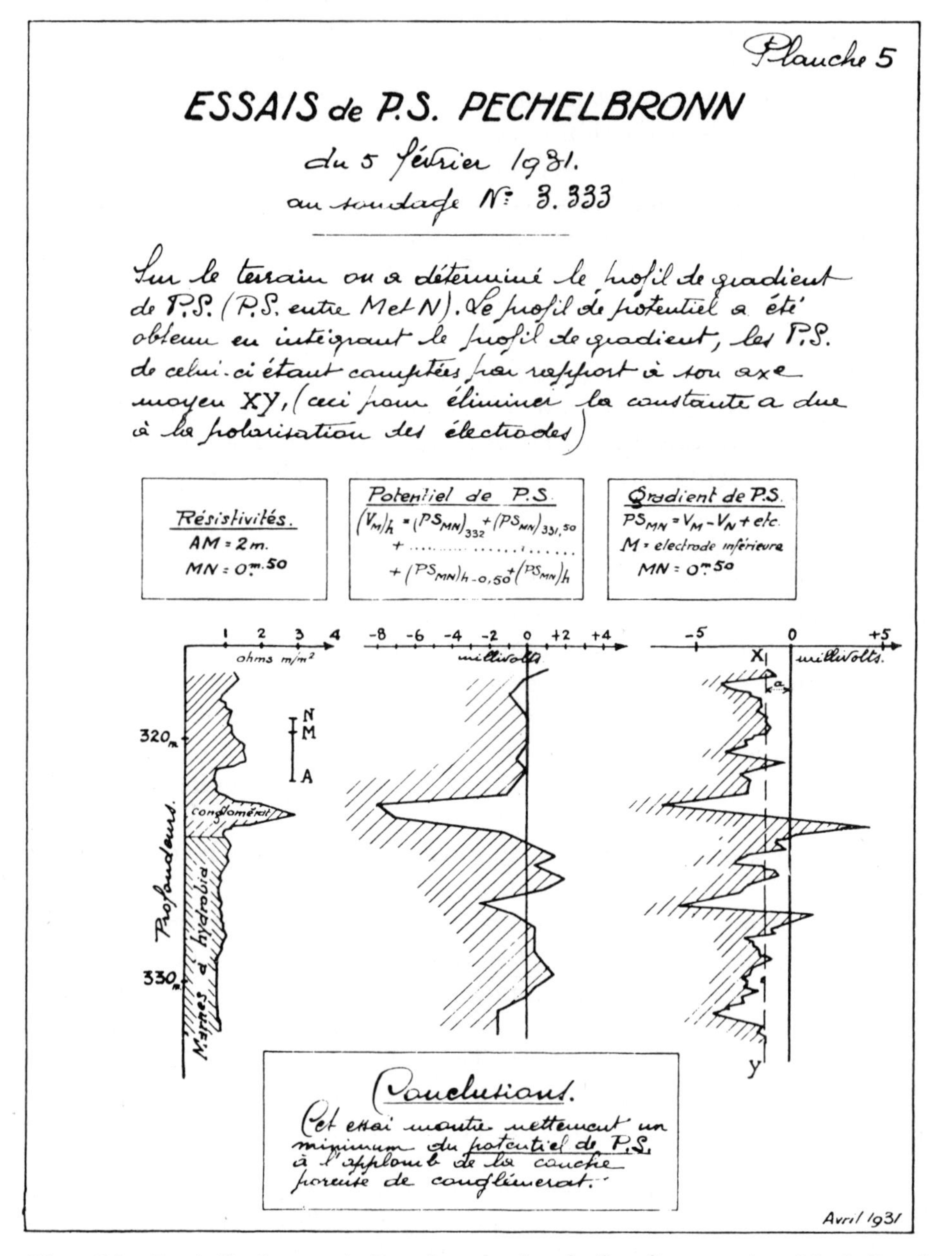

Figure 35. Facsimile of a page in Proselec, *showing the first diagram where S.P. indicated distinctly a bed of porous conglomerate (Pechelbronn, February 1931).*

tial inside the permeable layer increases in the direction of the flow; therefore the potential in the mud facing the layer was expected to be negative (Fig. 36). Observations confirmed this theory precisely. Since the phenomenon had been induced by the contact of a hole full of mud with the formations cut by the drill, it was, strictly speaking, not a natural one. Nevertheless, it took place without any artificial source of current and so was called "spontaneous potential" (S.P.).[13] Renewed tests in various boreholes during ensuing weeks confirmed the consistency of the measurements and showed, on other diagrams, the existence of similar kicks at the level of permeable layers.

These results were immediately communicated to the crews operating abroad; in fact, the entire May 1931 issue of *Proselec* dealt with them. The engineers were asked to measure this new parameter whenever feasible by using one electrode in the borehole and one at the surface for the direct production of potential diagrams. At that time electrical coring was in operation in basins generally made up of shale, sand, and sandstone series, where the many reservoirs reached thicknesses of 15 to 30 feet or more. Such conditions were much more favorable than those at Pechelbronn and produced spectacular results: whereas on the S.P. diagrams the shales appeared as almost straight lines, the top and bottom of the sands were marked by sharp deflections, and the departures[14] were no longer of a few units but of several tens of millivolts.

Here was a new and far-reaching discovery. The definition by the S.P. curve of every permeable layer provided an invaluable complement to the resistivity curves, which were much less reliable and accurate in the location of oil-bearing strata than in the definition of correlations among boreholes. Indeed, although oil and gas contained in a horizon generally produce peaks in the resistivity curve, compact beds intercalated in softer formations likewise give peaks. In most cases the ambiguity could be removed by the S.P.; a resistivity peak without an S.P. anomaly was most likely to mean a nonpermeable and hence dry horizon. Moreover water-bearing horizons and their limits, a subject of concern to the drillers, were

[13] It has become customary to use the abbreviation "S.P." for "spontaneous potential curve."

[14] The departures are also called "S.P. anomalies," an extension of geophysical terminology.

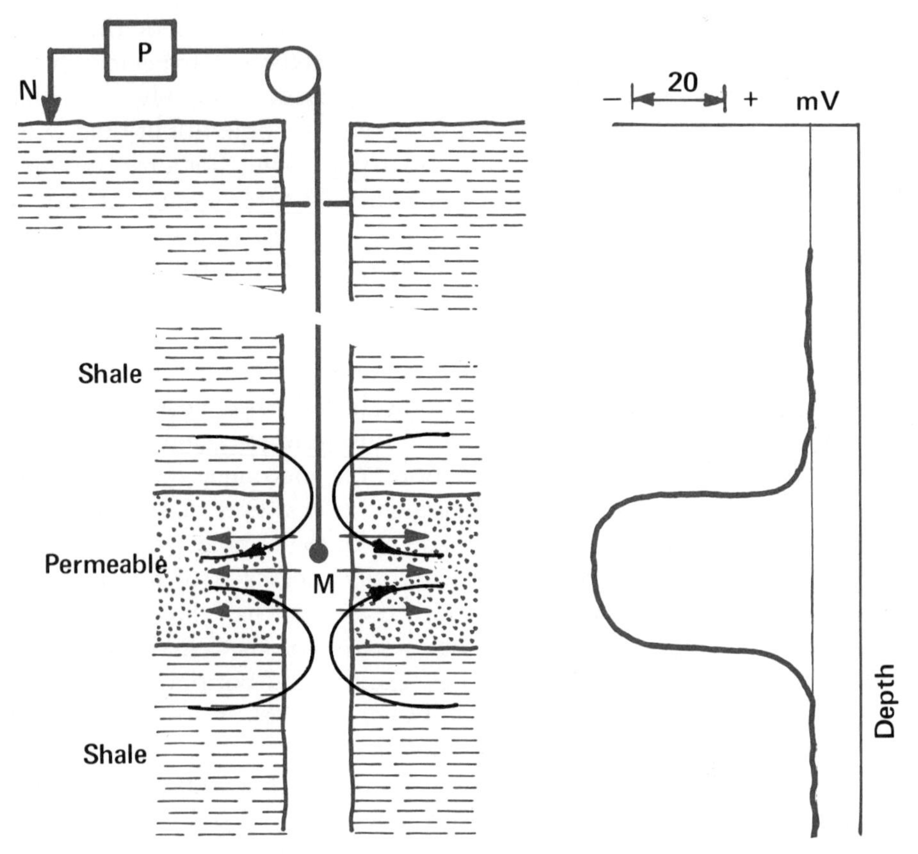

Figure 36. Left: *Schematic cross section showing mud filtration in a permeable layer and circulation of currents due to spontaneous potentials.* Right: *Recording of ohmic drops created by the current along the borehole.*

now marked, an achievement hardly possible with resistivity alone. In short, the efficiency of electrical coring was substantially enhanced.[15]

Once the usefulness and potentialities of the S.P. measurements

[15] When the S.P. curve was introduced into the oil basins (the U.S.S.R. excluded), it was presented under the name "porosity curve," a physical term more meaningful for industry than "spontaneous potentials." Indeed, for most professionals, porosity connoted permeability; it referred to the ability of a rock both to contain fluids and to allow their circulation, thus excluding the shales, with pores so small that any flow is practically impossible. The S.P. curve marked precisely the limit between the shales and the porous layers as understood by professionals.

were confirmed, the technique was soon perfected. Adjustments were made in the circuits so that S.P. and resistivity could be measured simultaneously. To this end, advantage was taken of a characteristic of the *AMNB* quadripole expressed in the reciprocity theorem, according to which apparent resistivity remains the same when the roles of *AB* and *MN* are inverted.[16] The electrical coring configuration *AMN,* with *B* at the surface, was thus replaced by an equivalent *MAB* configuration with *N* at the surface: the bottom electrode *M* transmitted through the same conductor both S.P. and resistivity signals (the voltage corresponding to the current emitted between *A* and *B*). For a few months the plotting took place point by point; the signals passed through two potentiometers in series, and at each station the S.P. and resistivity measurements were made, one after another. This very primitive form of telemetering involving a chronological system was soon replaced by the continuous and simultaneous recording of both parameters through an appropriate connection in the circuit of the recorders and the pulsator. The principle of this wiring remained unchanged for several decades, as long as the pulsated system was used.

In accordance with the terms of the contract between Schlumberger and the U.S.S.R. petroleum organizations, Doll arrived in Grozny in February 1932, to conduct tests. The S.P. measurements had become routine procedure by that time, and hundreds of diagrams had been plotted. They were considered only as accurate qualitative indicators of the limits of permeable layers, and interpretation of the magnitude of the low values produced by these layers had not been the subject of any study. Nevertheless, according to the theory of electrofiltration, the electromotive force generating the S.P. had to be proportional to the resistivity of the mud and to the pressure differential between the mud and the fluids inside the layer, whether water, oil, or gas. A simple calculation showed that, if the mud pressure varied by a known quantity, everything else being equal, the formation pressure should, in principle, be obtained by comparing the S.P. amplitudes.

[16] This theorem, long known in the case of linear circuits, had been set forth by Wenner in 1912. Its demonstration is immediate for a homogeneous indefinite medium. In the 1930's it was mathematically demonstrated by Lienard for any heterogeneous medium, and experimentally verified in surface work and in the Pechelbronn oil wells. The reciprocity of the quadripole has been so frequently applied that today it is taken for granted.

One or two rather promising experiments to this effect had been made at Pechelbronn, and part of Doll's program was to pursue them in the U.S.S.R. In the Grozny area the muds had low salinities and hence rather high resistivities, and thick sandstones frequently gave S.P. values 80 to 100 millivolts in magnitude. Two series of measurements were made within a span straddling such a sandstone. In the first series the mud level was lowered to some 120 feet below the wellhead: in the second the well was filled to the top, making for a pressure differential of several atmospheres between the two recordings. The S.P. value of the sandstone showed substantial variation, and the value computed from the measurements for the pressure appeared reasonable to local geologists.

The test was repeated a few weeks later in Baku. As the muds were considerably more saline, the S.P., although providing a good definition of the boundaries of sands and sandstones, was of a lower average value. Contrary to expectations, when the mud level changed, the curves showed no discernible difference, opposite formations with S.P. anomalies of 20 to 30 millivolts, this with a pressure differential about the same as in Grozny. Calculation immediately showed that to explain S.P. values of such magnitude by the electrofiltration hypothesis amounted to assuming that the formation pressure was essentially nonexistent—a practical impossibility. The question arose: if the pressure change effect on the S.P. was perhaps not instantaneous, might it not have had time to occur within the experiment? Indeed, plans were made to resume the test at a more leisurely pace as soon as a borehole became available, but experience already suggested that S.P. might have other causes than electrofiltration.

Thereupon Louis Bordat, who had a crew in Surakhany (one of the best producing fields of Baku), reported to Doll the existence of a sand horizon, saturated with highly saline water, where the S.P. value reached 80 millivolts and more, that is, above what had been observed anywhere else. Doll assumed from Bordat's observation that the salinity of the formation waters could play a part in the S.P., that an electromotive force of the electrochemical type could develop at the contact of these waters with the less saline mud,[17] and that the currents generated flowed and closed through the borehole in the same manner and direction as the electrofiltration currents. A few hasty experiments by Conrad in Paris showed that such an assumption was not altogether unfounded. Thus in

[17] The two waters constituted what physicists call a concentration cell.

April 1932 came the discovery that at the depth of formations containing salt water the S.P. has an electrochemical component. It was established somewhat later that in formations containing hydrocarbons water in capillary form occurs everywhere, even if production is practically water free, and that here, too, the electrochemical S.P. can be observed.[18]

Current output sonde and normal and lateral sondes

The problem of the span of the electrode configuration (or sonde) had arisen with the first Pechelbronn experiments. After a rather haphazard beginning with $AM = 3$ meters and $MN = 1$ meter, the spacings underwent various adjustments in order to obtain diagrams as sharp and differentiated as possible. It was known that the quantities being measured were apparent resistivities[19] whose values were affected by the mud-filled borehole. In 1927 no calculation had yet been made to define this point more accurately. However, on the assumption that the medium in which the sonde is placed is homogeneous (as is the case when the mud and formation resistivities are the same), the distribution of the equipotentials had become standard knowledge. On this basis it was assumed as a first approximation that the volume of ground comprised in the measurement could be represented by a horizontal disk, with a thickness equal to MN and a diameter equal to AM. However rough, this interpretation was convenient, not unlike the $AB/4$ value long adopted in surface prospecting to represent the depth of investigation.

The fact that the volume of ground was of the order of several cubic meters offered an advantage over the situation prevailing when the information provided was confined to the diameter of the mechanical coring

[18] For the record, Conrad had already envisaged the possibility of electrochemical reactions occurring between the various formations encountered in drilling. Reference to this had been made in his communication of December 10, 1931, to the U.S.S.R. Academy of Sciences, but he considered such reactions to be very weak, and negligible compared to electrofiltration potentials.

[19] When an electrode configuration AMN (with B remote) or its equivalent MAB (with N remote) is placed in an indefinite homogeneous medium, the value measured is, because of the calibration of the system, equal to the resistivity of that medium. However, what is practically measured by electrical coring (as well as by surface prospecting) is an apparent resistivity (see p. 38), in other words, an average value that is a function of the geometry and the resistivities of the media within range of investigation of the system.

tool. However, to the extent that cores are recovered, geologists can examine them almost inch by inch (in reality, they do this only in special cases and usually limit themselves to one or two observations per foot). To reach such a fine degree of resolution seemed very difficult with the sondes in use; their size could not be unduly reduced lest the mud effect become overwhelming and blur the peaks of the diagrams. To distinguish between even the thinnest layers, Conrad turned to quite a different measuring device.

In a very large medium the resistance of an electrode is defined as that between the surface of the electrode and infinity. In fact, beyond a certain distance (which is itself a function of the size and shape of the electrode), the equipotential surfaces become very large, and the resistance practically zero. For an electrode, say 2 inches in length and diameter, 90 percent of the resistance will be confined within a 10 inch radius, and 98 percent within a radius of 4.5 feet.

The left-hand side of Figure 37 is a cross section of a short electrode (say 2 inches long) suspended at the lower end of an insulated conductor at the level of a very thin horizontal layer (e.g., 8 inches), which is more resistant than the adjacent formations, for example, a hard sandstone bed between salty clays. The current lines flowing out of the electrode endeavor to avoid the resistant formation. Therefore the resistance encountered by the current is hardly affected by the hard bed; and if the sonde moves in the borehole, the hard bed will show on the diagram, if at all, only as a weak, flattened-out peak.

Let us now move to the right-hand side of the figure and assume that electrode A is flanked by two elongated electrodes, A' and A'', and that all three are maintained at the same potential. All the current flowing out of A is now forced to pass through the hard bed, and on the diagram the resistance of A will reach a much higher value when passing in front of this bed than it will for the adjacent formations; the result will be a well-contrasted peak. "Guard electrodes" was the name Conrad gave to the two added electrodes which emitted current for the purposes of preventing that flowing out of A from dispersing, thus forcing it to squeeze within a nearly horizontal sheet.

The same design, resumed years later, was realized through electronic circuits. The wiring of Conrad was utterly simple. The two guard electrodes, A' and A'', were short-circuited and connected by one of the conductors of the tricable to a source of current at the surface. The casing

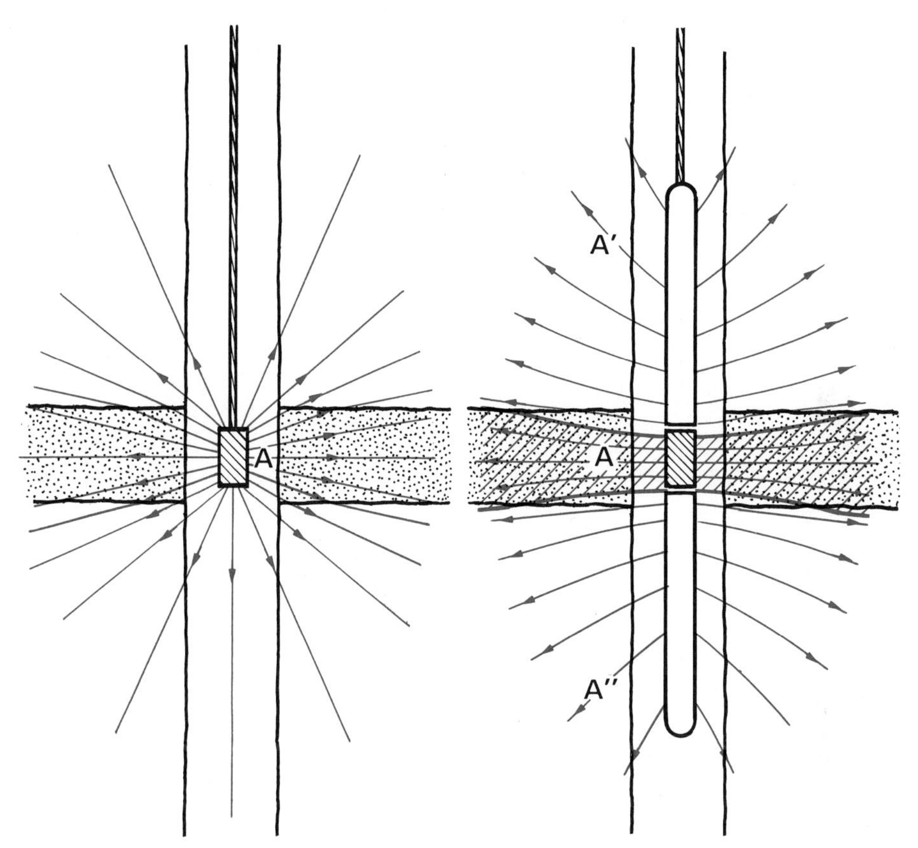

Figure 37. Distribution of the current emitted by an electrode. Left: *Without guard electrodes.* Right: *With guard electrodes.*

constituted the return electrode. *A* and *A'* were connected by a shunt whose resistance was sufficiently low for the guard and *A* electrodes to be at about the same potential, yet high enough for the potential difference at the terminals of the shunt (connected with the two other wires) to be measurable. At each station the potential difference was measured between *A* and a surface electrode, as well as the difference between shunt terminals, from which measurement the value of the current emitted by *A* was computed (Fig. 38). The ratio between the two values was the resistance of electrode *A*. Calibration of the system in a homogeneous medium of known resistivity (a miniature sonde in a tank full of water) led to the conversion of the resistance diagram into a diagram of apparent resistivities.

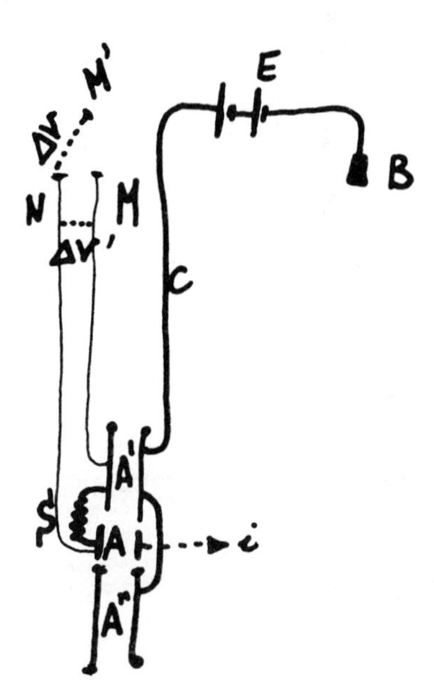

Figure 38. Facsimile of a sketch by Conrad Schlumberger of the current output sonde (October 13, 1927).

The device was called a "current output sonde" because Conrad wanted to measure the current emitted by the central electrode maintained at a constant potential. The first test took place at Pechelbronn in December 1927; although others followed, the system never found practical application. The measurements took more time, and it was impossible to reconcile the concern for detail with the depth of investigation, as can be done with the *AMN* configuration. After a series of experiments from which the Pechelbronn mission adopted electrode configuration values of $AM = 4.5$ feet and $MN = 20$ inches, the current output sonde was put aside.

However, the idea of using guard electrodes was not altogether abandoned. In a long handwritten note dated May 1, 1932, Conrad no longer envisaged elongated guards but visualized electrodes which, as in standard sondes, would be small enough in respect to their spacing to be assimilated to points. In 1950 Doll reverted to the concept of a sonde with guard electrodes restraining the central flow. He designed and perfected point electrode systems that offered much greater flexibility than the current output sonde and that made possible the simultaneous recording of the S.P. as well as other logging methods. Under the name

"Laterolog," they have today become basic tools of electrical logging (see p. 254).

By mid-1930 it had been amply demonstrated that electrical coring was able to delineate bed boundaries and define correlations. On the other hand, in Grozny (New Fields) and in Venezuela (Concepcion and Cabimas) the sole resistivity curve had already made it possible to distinguish between certain water- and oil-bearing formations, and even to forecast productivity. The results were empiric, obtained as they were by extrapolating the data resulting from the comparison between the production and the amplitude of the peaks on the diagrams. It was recognized that only very favorable geological circumstances had provided these results and that, in general, the correspondence between production and apparent resistivity is not so simple. In 1930, however, not all the factors involved in this relationship had been detected or analyzed. For the sake of clarity, it seems appropriate at this point to anticipate the sequence of events and to summarize the present degree of knowledge on this question.

Production is contingent on various factors, particularly the amount of hydrocarbons contained in a bed. This amount can be calculated from the thickness, the porosity, and the saturation of the horizon. Saturation (note 10, p. 10) in turn is reflected in the value of the true resistivity of the bed; true resistivity is also a function of several other formation variables: its porosity, the resistivity of the water occupying part of the pores, the amount of shale contained in the rock,[20] and, to a lesser extent, such other characteristics as grain size and shape.

Figure 39, which shows a cross section of a borehole drilled through a permeable formation—sandstone, for example—explains the relationship between the *true* and the *apparent* resistivities of a reservoir. Such a formation is always more or less deeply invaded by the water contained in the mud, which repels the fluids from their original place. On the drawing the slightly undulating line marks the limit of the invasion. It is known today that there is no such sharp limit and that the transition to the virgin

[20] The shale may occur in the form of very fine beds (laminations) or of minute grains disseminated in the pores of the reservoir. The shale content of the formation is often low enough to have a negligible effect on the resistivity of the formation.

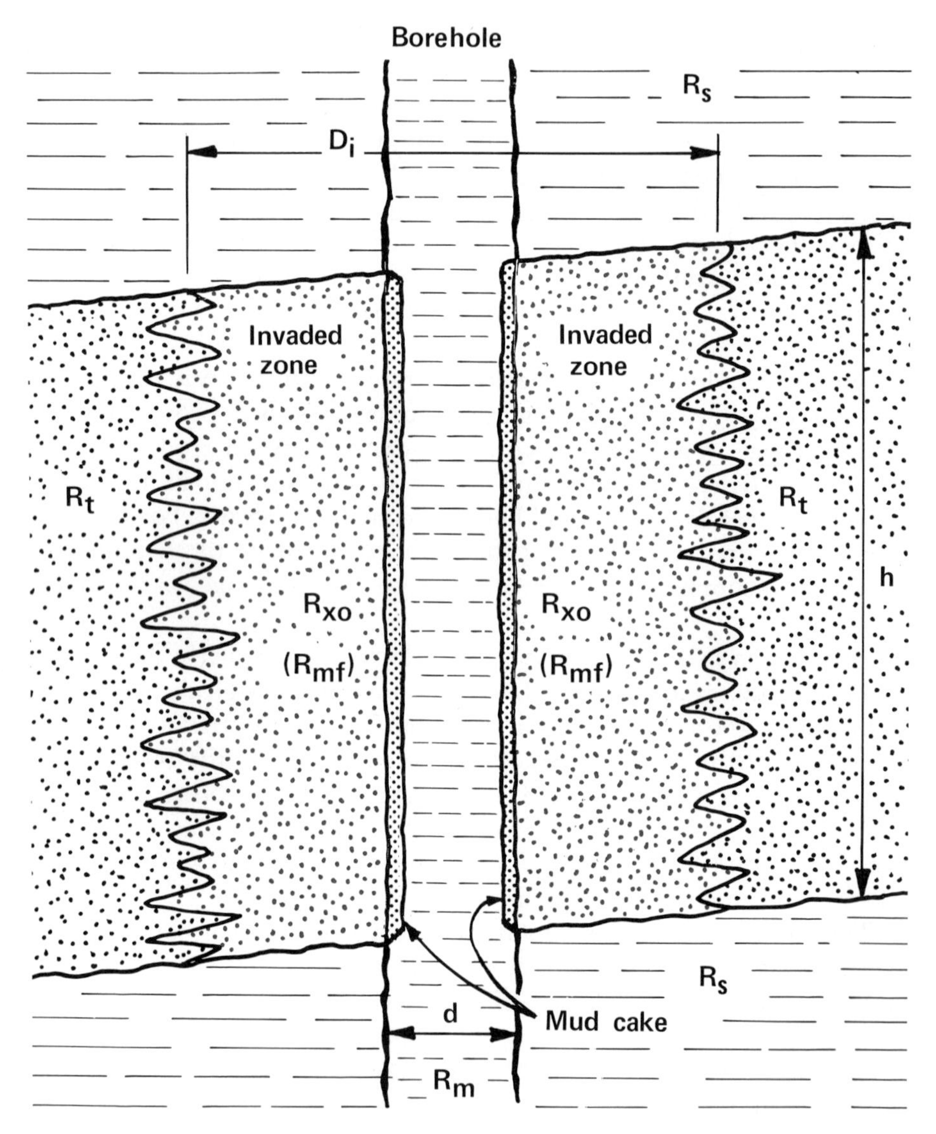

Figure 39. Vertical cross section of a formation invaded by mud filtrate.

zone is gradual, but both calculations and experience show that the intermediate zone may be disregarded with little error. It is further known that, when the mud deposits its clay particles along the wall of the hole, a kind of crust (mud cake) is formed and that, beyond this crust of a fraction of one inch, only clear water—the mud filtrate—penetrates the rock. The apparent resistivity (R_a) measured by the sonde depends not only on the true resistivity (R_t), but also on the borehole diameter (d), the resistivity of the mud (R_m), the diameter (D_i), and the resistivity (R_{x_0}) of the invaded zone. Furthermore, unless the formation is very thick compared to the spacing of the sonde's electrodes, R_a will be influenced by the shoulders of the formation, and this will involve their resistivity (R_s) and the thickness (h) of the formation.[21]

The above description provides an idea of the number of factors standing between a formation's production and its apparent resistivity. The outstanding results obtained in the U.S.S.R. and Venezuela were due to the fact that, saturation excepted, the multiple factors influencing productions and true resistivities remained rather constant in each horizon and there was little difference between apparent and true resistivities (rather thick beds and shallow invasions).

In most cases the logging tools used today give the values of porosity and true resistivity. On the other hand, it is generally possible to know the resistivity of the formation water and to assess the effect of the interstitial shale. From these parameters experimental relations lead to the almost exact value of the saturation, and from there to the volume of hydrocarbons in place. In 1930, however, when electrical coring could give only the formation's thickness and apparent resistivity, hardly anything was known of the relation between the latter and the true resistivity. Moreover, although the porosity could be obtained from core analysis (with serious limitations), the above-mentioned experimental relations were still unknown. The history of logging is marked by the stages of progress made toward gathering, under the most diverse geological conditions, the fullest data on the fluid contents of the formations and their ability to produce.

[21] Usually D_i varies from an inch or so to as much as 6 feet. All the notation here is taken from recent English terminology: *m* for mud, *s* for shale, *t* for true, *i* for invaded. Laboratory and field experience shows that the apparent resistivities are practically the same for a dipping layer as for a horizontal one.

In the effort to reach what is called the petroleum diagnosis of a horizon, an initial idea in 1930 was to lengthen *AM* to about 30 feet, thus increasing the range of investigation to the point of "seeing" beyond the invaded zone. One or two tests of this kind were conducted in Indonesia, but it quickly became apparent that mere extension of the sonde was no answer to the problem. Invasion was not the sole factor producing the discrepancy between true and apparent resistivities: the thickness of the formations also played an important role. When the thickness of a resistant layer was near or below *AM,* nothing much showed on the diagrams, whereas very thick layers could exhibit very high apparent resistivities.

There was also a practical problem requiring urgent solution. Anxious as they were to have information on the potentially productive horizons as soon as the bit penetrated them, the oil companies demanded that measurements start as close as possible to the well bottom. Unfortunately, because of the length of even the shortest devices (*AM* between 7.5 and 9 feet) and of the weight extension below the sonde, the lowest possible measurement described mainly the formations some 15 feet above bottom, which meant that often the most significant part of the operation might be missing on the diagrams. A solution to the problem was seen in inverting the configuration, that is, placing *MN* below *A* (the result was called the "*MN*-below" sonde). The proposal appeared to be sound, since everything happens as if the surface electrode were infinitely remote; and consequently, in an infinite homogeneous medium, the value measured will be the same whether *MN* is above *A* or below. There was, however, a suspicion that in a real medium composed of layers with various resistivities the diagrams obtained with the two positions would not be identical because the *AMN* configuration was asymmetrical. This was confirmed by an operation in the northern coal basin of France.

In February 1931 a coal mining company had asked for the electrical coring of a reconnaissance borehole near Courrières. Drilling had revealed a rather thick coal seam, about 2000 feet deep, but extremely fragmented cores had not permitted an accurate location of its limits. A first diagram, recorded with a 7.5 foot *AMN* sonde and *MN* above *A,* showed very sharply a peak approximately 4.5 feet thick. A second record, obtained after inverting the device, revealed a peak of the same width and height, but shifted by 7.5 feet toward the bottom in relation to the first diagram. The prospector was intrigued by this difference; he

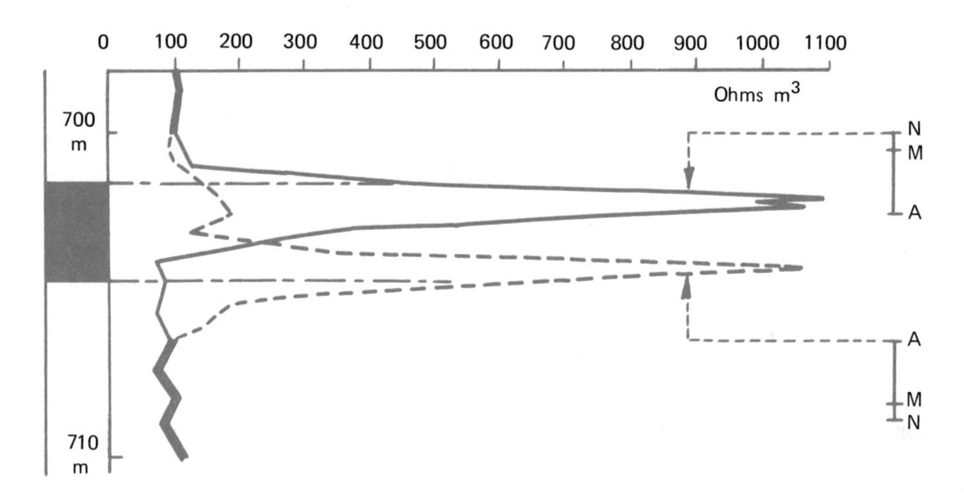

Figure 40. Two resistivity diagrams through a coal seam, proving that the AMN *configuration was asymmetrical.*

suspected some mistake and repeated the recordings, taking the utmost care with the depth measurements. The offset was confirmed (Fig. 40).

A thorough and systematic study followed in the wake of this experimental illustration of the effect produced by the asymmetry of the *AMN* system. Little light had been shed on the problem thus far by the few calculations made on the basis of highly simplified assumptions, the principal one being to disregard the borehole altogether. Doll improved these calculations by an approximate mathematical treatment relating to the case of drilling through infinitely resistant layers[22]—a more representative model of the real configurations. Tank measurements, where limestone slabs simulated formations, expanded and confirmed the calculated results. These studies were recorded in March 1932 in a house publication; they marked a crucial stage in the development of the Schlumberger techniques. They were the first to give a more accurate definition of the curves recorded with the *AMN* sonde. When *MN* is below, a resistant

[22] This simplification was based on the working assumption that oil- or gas-bearing horizons had an infinite resistivity, whereas, according to the theories on petroleum migration, a fraction of the pore volume was always occupied by more or less mineralized water, retained by capillarity. Many diagrams from different regions rapidly proved that this resistivity is always finite, that is, that the capillary water forms a continuous, electrically conductive system.

layer thicker than *AM* shows as a peak near the bottom, and a low-resistivity zone underneath its top, approximately equal to *AM*. The top of the layer is then said to be "eaten." With *MN* above, the curve is inverted. This explained the offset observed at Courrières; furthermore, the true thickness of the coal seam had to be 12 feet (7.5 + 4.5), a value that indeed corresponded with the information gathered from the cuttings. Another conclusion was that it was hopeless to use the *MN*-below arrangement to study the formations opened up by the last 12 or 15 feet of the well.

The design of a new electrode system, described by Conrad on May 1, 1932, was yet another result of these studies. In this system *N* was placed at a considerable distance from *AM*. What was measured was, in fact, the potential of *M* with respect to infinity (this was called the "*AM* ∞ sonde").

In September of the same year the "1932 sonde" marked new progress over the *AMN* device (Fig. 41). On the left-hand side of the figure the current is emitted between *A* and *B* (24 feet apart), while the potential is measured between *M* (1.5 feet below *A*) and the surface electrode. The system is equivalent to the AM ∞ sonde, since the distance from *B* to *A* is 16 times greater than the distance *AM,* and hence the effect of *B* on the measurement is negligible. This system, where the measurement was recorded at the center *0* of *AM,* was called the "normal sonde." Its advantage was that it required only a very weak current. The diagrams recorded were symmetrical about the center of a layer, whose top and bottom were accurately marked. The *AM* spacing being very short, measurements could be made very close to the bottom of the hole. On the other hand, because the system had only a small radius of investigation, the measurements were generally affected by the invasion. Finally, it responded very weakly to the thin layers.

On the right-hand side of Figure 41, the current is emitted between the closely spaced *A* and *B* electrodes, while the potential is measured between the remote electrode, *M,* and the surface: this is now the *MAB* configuration, which, according to the reciprocity theorem, is the equivalent of *AMN.* Called the "inverse sonde," it was not unlike the one used since the beginning of electrical coring. With *AM* 24 feet long, the radius of investigation became substantial—hence the other name, "lateral sonde." This sonde was well suited for detecting thin layers, even though it could not measure their true resistivity. Its shortcomings were that the

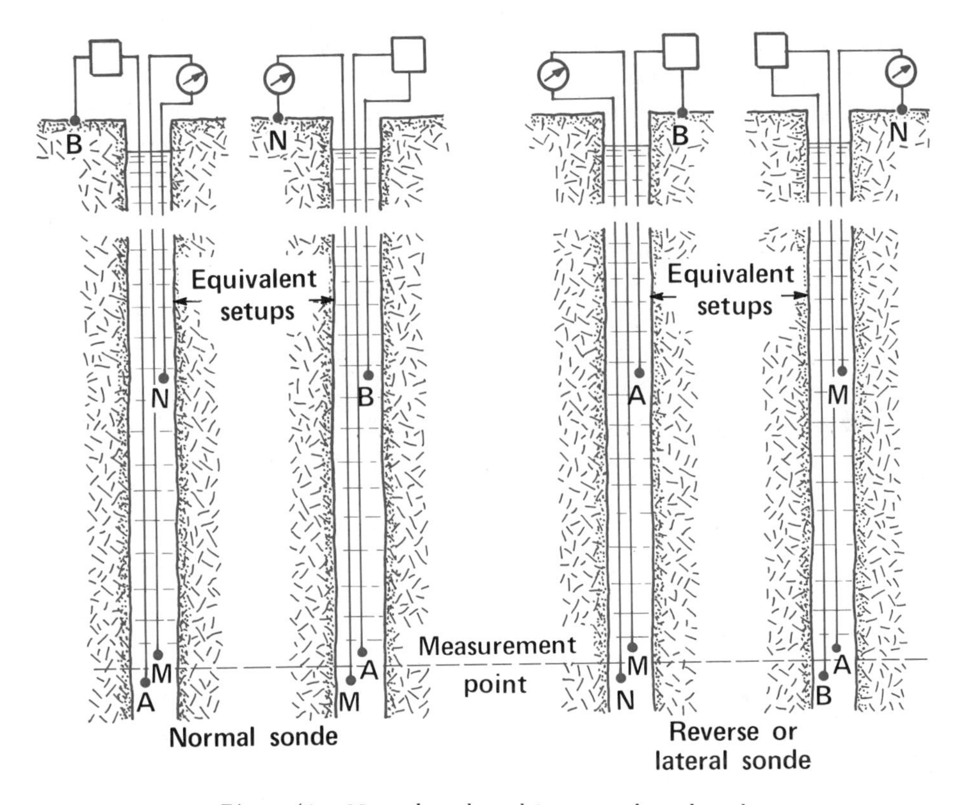

Figure 41. Normal sonde and inverse or lateral sonde.

top of a thick, resistant layer was deeply "eaten," and that a very strong current was required. With either type of sonde, normal or lateral, the *M* electrode simultaneously recorded the S.P. curve. Moreover, when both lower electrodes were eventually placed on a rigid cylindrical body doubling as a weight, the measurements could practically reach the bottom of the hole.

The question was raised as to whether, with the normal sonde, one conductor could not be made available by bringing *B* to the surface: this would have been strictly an *AM* sonde. However, experience soon showed the proposal to be impracticable with pulsated-alternating current recording. As a current wire and a measure wire wound on the winch amounted to an induction coil, the pulsing on and off of the current circuit generated a dangerous overvoltage, and the electromotive force induced in the measure circuit considerably altered the reading. The use of the third wire for the current returning to electrode B near the bottom of the hole reduced these effects to acceptable limits, although it was impossible to prevent them altogether because of the asymmetry between the two wires. However, the width of the dead segments on the measure collector of the pulsator was such that at 6 revolutions per second the induced millivolts had disappeared by the time the measure circuit was closed (see p. 53).

From 1932 on, the normal sonde generally became the basic tool (the U.S.S.R. excepted), whereas for several years the lateral sonde was used only in specific cases. The dimensions of both sondes underwent various adjustments to suit local geological conditions, and they remained in use until they were replaced, at least in oil wells, by induction logging and the Laterolog. They continue to be a part of the simplified and portable equipment employed in hydrological surveys.

Other measurements in boreholes

Together with the developments described above, technical studies were oriented as early as 1930 toward other projects, the principal ones being a device to measure the dip of the strata, an electrical resistance thermometer, and a technique for locating water inflows in boreholes. Inspired by Conrad, these various projects were essentially realized by Doll. Simultaneously, Marcel was tackling the problems of sidewall formation sampling (lateral coring) and casing perforation (see p. 182 ff). He had already considered pressure measurements at the bottom of boreholes, a problem that was to be taken up again 25 years later.

When a first borehole is located on the basis of geological and geophysical surveys, it is usually very difficult to foresee what, at the presumed depth of the pay zones, its precise position will be with respect to the overall configuration of the field under exploration. In the case of an anticline, for example, little is generally known as to whether the borehole will reach the interesting formations close to, or far away from, the axis, or even on which flank this will occur. There are several reasons for this uncertainty: geophysical interpretations allow a certain leeway; various features—faults, transgressions, unconformities—may result in a deep structure being quite different from its representation on the basis of surface geological observations; since no borehole is ever exactly vertical, the difference between bottom and wellhead coordinates may amount to as much as 50 feet or more.

The azimuth and the angle of dip of the layers encountered in drilling, and the drift of the hole itself, contribute substantially to defining the configuration of the deep objectives and guiding their exploration. To be specific, when a first well reaches an oil-bearing horizon, it is important to know whether a second well will intersect the same horizon at a point nearer the axis of the structure or more remote from it; in other words, whether, as geologists say, the second well will be located updip or downdip in relation to the first one. This is the reason why geologists have always tried to determine the magnitude of the dip from examination of the cores, especially when the cores come from finely layered formations like alternating shales and sandstones. Yet all the method yields is an angle with a plane perpendicular to the axis of the hole for the depth of each core. The values observed are merely apparent dips, and the difference between them and the true dip can be important if the hole is sufficiently slanted. Moreover, little is known of the orientation of the core, and hence the azimuth of the dip.

It was therefore important for the petroleum industry to find ways of determining the dips. A proposal to use electrical measurements for this purpose was made by Conrad at the time of the first Pechelbronn experiments in October 1927. The principle was nothing new: as in surface prospecting, it consisted of utilizing the electrical anisotropy of the sedimentary formations. All the rest, then, was mere technology—but how complex! Let us recall that, if a current is emitted by an electrode inside a borehole at the depth of an anisotropic, inclined, sufficiently homogeneous and thick layer, the equipotential surfaces are not spheres, but are flattened revolution ellipsoids whose axes are perpendicular to the plane of the strata (Fig. 42).[23] For the sake of simplification the figure has been drawn in the vertical plane passing through the steepest gradient of the strata. Placed symmetrically about the axis of the drillhole and at a certain distance (say 3 feet) above the power electrode, the two electrodes R and Q are on two equipotentials, and the potential difference between R and Q is a function of the slope of the ellipses at the point where they are

[23] These ellipsoids are actually slightly deformed at their intersection with the borehole, but this is only a secondary and generally negligible effect.

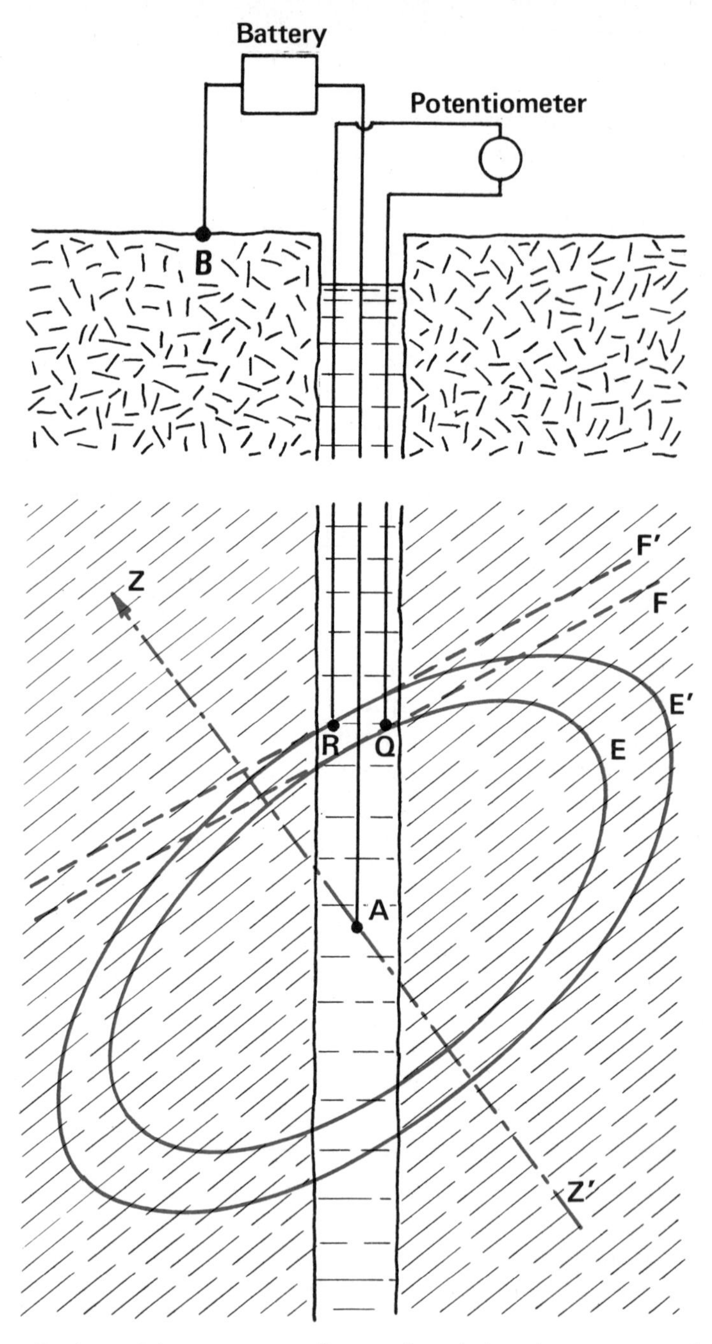

Figure 42. Equipotential curves surrounding an electrode in a homogeneous and anisotropic dipping formation.

intersected by the borehole. From the measurement of this difference, it is possible to determine the angle of slope of the ellipses, this angle being less than the true angle of dip. In practice the cable, when lowered into the borehole, has a tendency to untwist under the increasing stress of its weight; the sonde revolves several times around its axis, so that the orientation of electrodes *R* and *Q* varies with the depth. For this reason the system comprised two identical and perpendicular pairs of electrodes; by composing the values measured on each pair, a vector was obtained giving the azimuth of the dip with respect to these electrodes, that is, to the orientation of the sonde around its axis. This orientation then had to be determined with respect to the north. Furthermore, the determination of the true dip angle required both the value of the apparent dip as determined from the cores, and the angle of slope and the azimuth of the sonde's axis.[24] From there on, a rather simple spherical trigonometrical computation gave the true dip. This led to the design of a sonde in two parts. The first one, initially called the "inclinometer" and later the "teleclinometer," indicated at each measurement level the inclination of the sonde and its orientation with regard to its axis; the second one, termed the "dipmeter extension," gave the azimuth and angle of the dip with regard to the orientation of the sonde.

To design the circuitry, Doll had to manage with the tricable as the only electrical and mechanical connection between surface and drillhole. It was out of the question to manufacture a cable with a greater number of conductors since this would have required major and exceedingly costly changes in equipment design, especially for the winches. As one conductor of the tricable was mobilized for the emission of the current actuating the system, only two remained available for the return of the four signals of the inclinometer and the two of the dipmeter, each of the order of a few millivolts. The borehole apparatus required a distribution relay controlled from the surface. For purely technical reasons the project was achieved in two stages: the inclinometer first, and the dipmeter extension some years later.

The inclinometer was the first Schlumberger instrument lowered into a borehole that was no longer limited to three sections of lead wire serving as electrodes, but involved a delicate assembly of electromechani-

[24] Centering tools on the sonde ensured the actual coincidence of its axis with that of the borehole.

cal components enclosed in a cylindrical housing only a few inches in diameter that protected the instruments from being contaminated and crushed by the mud. In spite of their small size and of temperatures reaching 120 to 180 degrees F, the components were expected to work without failure. The watertight bronze housing contained an induction compass[25] made of a billiard ball driven by an electric motor around a diameter coinciding with the axis of the sonde. A coil of insulated wire was wound around a groove machined into the plane of this diameter and was connected with a collector contacted by two cross-shaped pairs of brushes. The same current fed the motor and excited an electromagnet maintained by a universal joint in a vertical position above the compass (Fig. 43). The two electromotive forces simultaneously induced in the compass by the terrestrial magnetic field and the electromagnet were rectified through the two diametrically opposed dead segments in the collector ring. To measure them separately, the current was reversed; since at the instant of reversal the direction of the motor's rotation did not change, the electromotive force induced by the terrestrial field did not vary, whereas that induced by the electromagnet changed sign. The former defined the orientation of the sonde with respect to the magnetic north; the latter served to compute the angle and the azimuth of its inclination.

The whole was dipped into an insulating liquid—a mixture of oil and kerosene—filling the housing. A metallic bellows at the bottom of the sonde transmitted the mud pressure to the liquid and prevented the thin bronze housing from collapsing. The electrical connection between the inside wiring of the sonde and the three conductors of the cable was achieved by elements designed to ensure, despite the pull of the cable, a positive seal against mud intrusion and good electrical insulation. Such devices, reminiscent of those used in motor engineering, were called spark plugs. The distribution relay included a ratchet cleverly driven by the rotation of the billiard ball, which, drilled through its center and working like a centrifugal pump, created in the oil an overpressure that pushed a piston driving, in turn, the ratchet, with the result that the latter moved by one notch at each switching on or off of the current. In this

[25] An instrument designed to define the direction of the magnetic field of the earth by utilizing the electromotive force induced by this field in a coil revolving around a vertical axis.

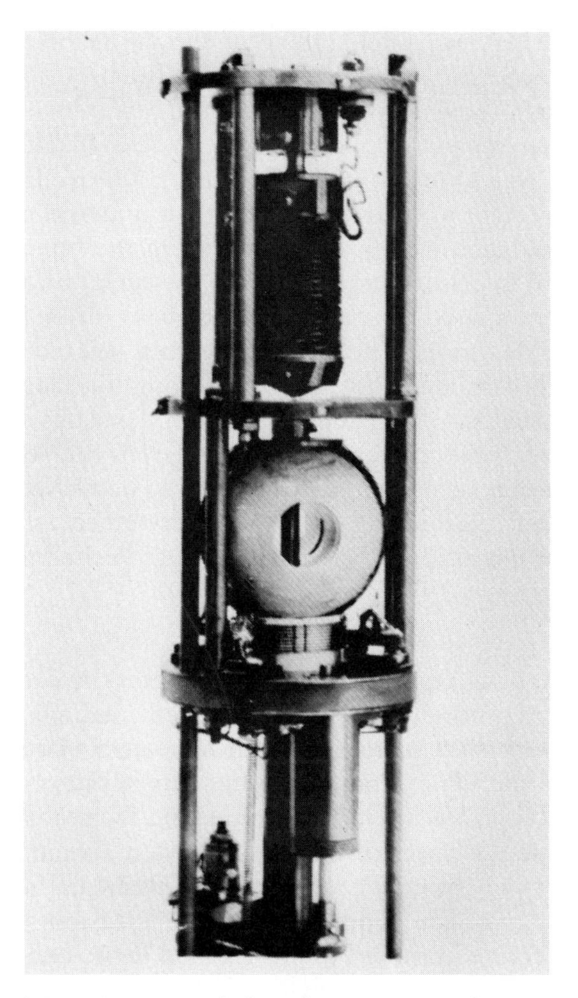

Figure 43. Part of the mechanism inside the inclinometer. Top: *Electromagnet suspended on its dial.* Center: *Hollow billiard ball used as an induction compass.* Bottom: *Piston controlling the ratchet of the relay.*

manner, at each station—every 30 or 50 feet—the four average values of the voltages between the brushes of the compass were obtained, from which the inclinometry data could be computed. Two additional positions had been provided for the relay to serve for measurements between the electrodes of the dipmeter extension. The results of the computation were interpreted through charts, especially drawn for each region (the local value of the terrestrial field intervened in the calibration), and it was possible to make the measurements and draw the profile of the drillhole for the entire open-hole section (Fig. 44).

The inclinometer (or teleclinometer), made available to industry in 1932, answered an important need. Because all oil wells drift from the vertical, it is important to know accurately their underground positions at reservoir and key geological marker levels. On the other hand, slanted holes gave rise to many drilling mishaps (e.g., stuck drill pipe). Methods had already been devised to measure the drift of drillholes,[26] but none had industrial application. The various devices utilized in Baku were inaccurate and inconvenient, and none (including the Shakhnazaroff device) gave the azimuth of the inclination. This explains the enthusiasm with which the teleclinometer was welcomed by the oil industry; its success and widespread utilization were almost immediate.

It took longer to achieve the complete instrument: the teleclinometer and dipmeter extension. After many tests at Pechelbronn, it actually reached the market in 1935.

Another important problem for oil operators was the location of water intrusions in boreholes. This had been studied from the very earliest days of electrical coring. It often happened that petroleum was accompanied not merely by traces of water driven out of the pay zones, but also by unacceptable discharges originating from active aquifers located either within the producing zone or above it behind the casing (see Fig. 28) and flowing through faulty cement. Remedial action required accurate knowledge of the depth of the water intrusion. The method in use began by injecting into the borehole a mud dense enough to prevent the intrusion of any fluid—in oil field terms, to "kill' the well. The mud was then circulated to ensure homogeneity throughout the zone of study, after which the mud level was gradually lowered to the point where the

[26] Particular mention should be made of a U.S. patent granted in 1915 to H. M. Smitt, in which an instrument is described whereby a photograph is taken of a circular bubble gauge, maintained by a gyroscope in a fixed and known azimuth. The principle of the design was to be developed and marketed much later by Sperry–Sun in the United States and O. Martiensen in Germany. These instruments were less convenient than the Schlumberger teleclinometer, but since they were not based on the terrestrial magnetic field, they worked inside the casing.

Figure 44. Doll calibrating a teleclinometer in Baku j(1932).

water could begin to flow. The intrusion level was located through an appropriate device lowered at the end of a cable.

The Schlumberger process was based on measuring the resistivity of the liquid filling the hole from top to bottom. A special sonde was designed, including three small lead electrodes about one inch apart, and a cylindrical shell that kept them sufficiently remote from the borehole wall (whether casing or open hole) so that the measurement would not be affected. The system was calibrated above ground in a tank filled with water of a known resistivity. In general, this resistivimeter was rather easy to handle in the borehole; all that was needed was to circulate ordinary mud, which normally is much less salty than the water to be detected.

Over the years the determination of water flows with the resistivimeter had become routine, but the method was used less and less frequently as the techniques of deep exploration and production were perfected. In particular, the completion of wells by perforation (see p. 188 ff) provided a much more effective way of sealing off the pay zones against water intrusion.

The increase of temperature with depth—a phenomenon characterized by the geothermal gradient—was also utilized to locate water intrusions. When the mud had been circulated in a borehole and the temperature then measured from top to bottom, the measured gradient is much less than the true gradient. The reason is that below 50 or 100 feet the mud is colder than the underground. Since formation water flowing into the borehole is warmer than the mud at the same depth, a new temperature diagram recorded at a later time will show an anomaly locating the water intrusions.

The idea of utilizing temperature measurements led to the study of a borehole thermometer that could be adapted to the electrical coring equipment. The only instruments employed in the industry around 1930 were maximum–minimum thermometers that required delicate handling and, moreover, produced only one measurement for each run at the cost of an hour to reach temperature equilibrium. Around 1928 an electrical thermometer had been used in Virginia, but in addition to being very expensive it did not allow for continuous recording. The principle of the Schlumberger thermometer was based on a Wheatstone bridge, one leg of which was a ferronickel resistor very sensitive to temperature; the other three were of invar alloy. The instrument not only had to be both sturdy and highly sensitive (to a fraction of a degree), but also had to have a low time constant and good electrical insulation. Several models were tested in succession with dubious results. Finally, in 1933 the required performance was achieved by a model in which the ferronickel wire, insulated by a thin layer of varnish, was wound in a helicoid groove machined in a steel core; the whole was encased in a copper sleeve forced under pressure against the wire. There was a close contact between copper and wire, without any damage to the insulation, and hence a very rapid temperature response.

The thermometer had various applications. It was used not only for the location of water intrusions but also for the detection of gas or oil inflows, the study of fluids circulating between two permeable layers with different pressures behind a defectively cemented casing, and the control of cementing (see p. 179). Thought had also been given at some time to utilizing the thermometric diagram for the differentiation of formations according to their thermal conductivity, but this parameter proved much less sensitive than electrical conductivity and hence without practical interest. Today the drillhole thermometer, greatly improved over the model of the 1930's, is a standard tool of production logging.

Field operations

In a letter to Léonardon dated May 31, 1927, Conrad and Marcel had suggested that a drilling company, preferably an American one, might be associated with the industrial development of electrical coring. The proposal remained vague and was promptly abandoned. The initial operations at Pechelbronn seemed too promising to warrant association with outsiders. By establishing correlations among boreholes over several miles, electrical coring reduced and often replaced mechanical coring, thus effecting economies in oil exploration. These results, together with improved equipment, led the Pechelbronn company to enter into a service contract with "La Pros." The contract, signed on July 12, 1928, marked the date that the Schlumberger process became commercial. The contract provided that in return for a monthly fee of 12,000 francs a crew was to be made permanently available to Pechelbronn; it further provided Schlumberger with the right of access to boreholes for the purpose of continued testing and development of new techniques. The Pechelbronn oil wells thus became a kind of extension of the Paris engineering office. This cooperation lasted until World War II.

However encouraging this first commercial success, nobody could foresee in 1927–1928 that what was involved was an invention with a commercial future far beyond that of surface electrical prospecting. At that time electrical coring was regarded as just another process, applicable to the location of key geological horizons, ore bodies, coal seams, and water-bearing formations, as well as to petroleum exploration; in other words, as an additional volume of business. Indeed, the operating conditions at Pechelbronn—percussion drilling, very short open-hole sections, and mainly thin, low-pressure, oil-bearing strata shown on the diagrams as blunted peaks—were too special to allow any foresight of what the method could yield in large oil fields outside France.

Electrical coring was sometimes used to determine the location and thickness of coal seams more accurately than was possible with cuttings or cores. The first operation of this kind was conducted by Sauvage in August 1928 in a borehole of the Lorraine coal basin at Saint-Avold. More followed; Courrières in February 1931 was particularly significant because that diagram became the source of very important studies. Other surveys to assist in the interpretation of surface measurements took place in exploration boreholes for iron ore in Normandy. Still, such operations were few, and it soon became evident that efforts had to be concentrated on the oil industry, where the number of boreholes exceeds by far that of all other drilling activities combined. Proposals were submitted to Royal Dutch–Shell; the use this company had already made of surface electrical prospecting had given it an appreciation of Schlumberger standards. The merits of electrical coring were stressed by Marcel in a letter of October 8, 1928, emphasizing in particular the low cost of the process as compared to mechanical coring. Royal Dutch was interested and sent Dr. Mekel, director of its geophysical department, to observe an operation in the field. The utmost care was taken in preparing for his visit. A well was selected where the top of the Hydrobiae marls was known to stand out sharply on the diagram. Deschâtre and Sauvage tidied everything up; the access track to the well was sanded, and the wheels of the old pickup truck received a new coat of paint. Mekel arrived in the morning with Conrad. He was shown the equipment, and the operation was explained to him. Sauvage made point-by-point measurements, immediately followed by the plotting of the diagram on a strip of paper. Mekel wished to repeat the measurements by himself. The sonde was lowered again; he found exactly the same values, confirming the authenticity of Sauvage's diagram. This apparently conclusive visit resulted in the signing of a contract on November 8, according to which Royal Dutch was to test electrical coring in its fields in Venezuela and the Dutch Indies.

A few months later the merits of electrical coring were obvious enough to justify its public introduction to the petroleum industry. The occasion was the Second International Drilling Congress in Paris in September 1929; the paper, written in English, was captioned *The Electrical Coring* (Fig. 45). It set forth the principles of resistivity measurements with several examples of correlations provided by Pechelbronn and one by Venezuela. This was the first of a long series of papers to be written on logging.

THE ELECTRICAL CORING

PAPER SUBMITTED BY

C. et M. SCHLUMBERGER

AT THE SECOND INTERNATIONAL DRILLING CONGRESS

Paris (September 1929)

SOCIÉTÉ DE PROSPECTION ÉLECTRIQUE

(PROCÉDÉS SCHLUMBERGER)

30, RUE FABERT – PARIS - VIIe

AND

SCHLUMBERGER ELECTRICAL PROSPECTING METHODS

25, BROADWAY – NEW-YORK CITY

Figure 45. Facsimile of the cover page of the first article on electrical coring (1929).

The Royal Dutch contract for Venezuela provided for an electrical coring crew of two engineers to be made available for a monthly sum of $2500. It was agreed that the crew, when available, could be assigned to surface prospecting. However, between 1929 and 1931 there were only a few short surface surveys of minor local problems.

Responding to a telegram from The Hague, Gilbert Deschâtre and Pierre Bayle presented themselves in February 1929 at the Shell office in Maracaibo, unaware of the nature of the problems they would be asked to handle. Here was a new situation: the problem was no longer, as in surface operations, to conduct measurements in an exploration area where at least some preliminary plan had been drawn up from the outset, and where the prospectors were granted some leeway in their work by the client. The local geologists, somewhat uninformed about electrical coring, told Deschâtre and Bayle that their primary concern would be not with exploration drilling but with day-to-day production, and especially with the accurate definition of the depth to which the casing was to be lowered for cementing and water shutoff (see p. 99).

At that time Venezuelan production, almost in its entirety, was provided by the western part of the country: Lake Maracaibo and its surroundings, Ambrosia, La Rosa (Cabimas), Lagunillas, Mene Grande, Concepcion, La Paz. The production of these fields seemed adequate for a shrinking world market, and the three main operators—affiliates of Shell, Gulf, and Standard of Indiana—were doing little exploration. In eastern Venezuela, 20 years of exploration had been unsuccessful except for the newly discovered Quiriquire field.

Oil and gas occurred in sand or sandstone horizons separated by generally rather soft shales. Tectonics were often complicated, for example, anticlines fragmented by many faults. In spite of the many cores required by the geologists, there were frequent errors in forecasting the depths of the pay zones and of the water shutoff. The prospectors had no doubt that electrical coring would be of substantial help in solving these problems. One asset was the great number of wells being drilled—especially in La Rosa, where electrical coring was to begin—and hence the possibility of rapidly establishing correlations from well to well to demonstrate the validity of the method. The local geologists were rather reticent. However receptive to new techniques, these seasoned professionals who thought they knew everything about their fields could only regard skeptically the youngsters who, under an agreement concluded between remote head offices, were supposed to help them with paper strips covered with diagrams. After all, what kind of experience did they have outside of some negligible French oil field?

The prospectors were required not only to conduct the measurements but also to retrace the diagrams by hand and interpret them, that is, to identify the electrical markers, to establish correlations from which to deduce the deep configurations (especially the existence and throw of faults), and to forecast the depth of the pay zones. It was not that the geologists were uncooperative: their experience and data were available; but, at least in the beginning, they left to the Schlumberger staff the full task and responsibility of practical diagram interpretation.

The first operation took place on March 6, 1929, in the La Rosa field (Fig. 46). Dozens of diagrams followed, and after 10 months the number of crews had grown from one to three. For the Shell field staff, Schlumberger had become a regular auxiliary with a recognized contribution. In 1930 operations were extended to the oil wells of Gulf, Standard of Indiana, and Richmond.

Addressing a meeting of geologists of the Maracaibo Basin, Bayle reviewed the results of 18 months of electrical coring. In certain areas, multiple markers on the diagrams provided clear correlations over tens of miles, eliminating almost completely the need for mechanical coring; in other, tectonically more complex areas, the data gathered from the diagrams complemented, refined, and corrected the geological data. The results were less outstanding with respect to oil-bearing horizons. They showed beautiful peaks whose heights, in certain zones, provided a clue for production; but in general it was difficult to distinguish between oil-bearing and water-bearing strata, or simply between permeable and impermeable layers. It will be recalled that at the time electrical coring involved only a single curve recorded with a 9 foot lateral sonde.

After their initial reticence toward electrical coring, many geologists and operators became inclined to expect more from it than it could give, such as the differentiation of aquifers, the determination of the water contents in oil-bearing horizons, and the detection of gas occurrences. These problems would not be solved until much later in most oil fields.

In any case the results were spectacular: 800 electrical coring operations in 300 wells. These totals had not been achieved without the strenuous efforts of the crews. Monthly performance reached 50 operations under conditions of extreme physical hardship: difficult transport;

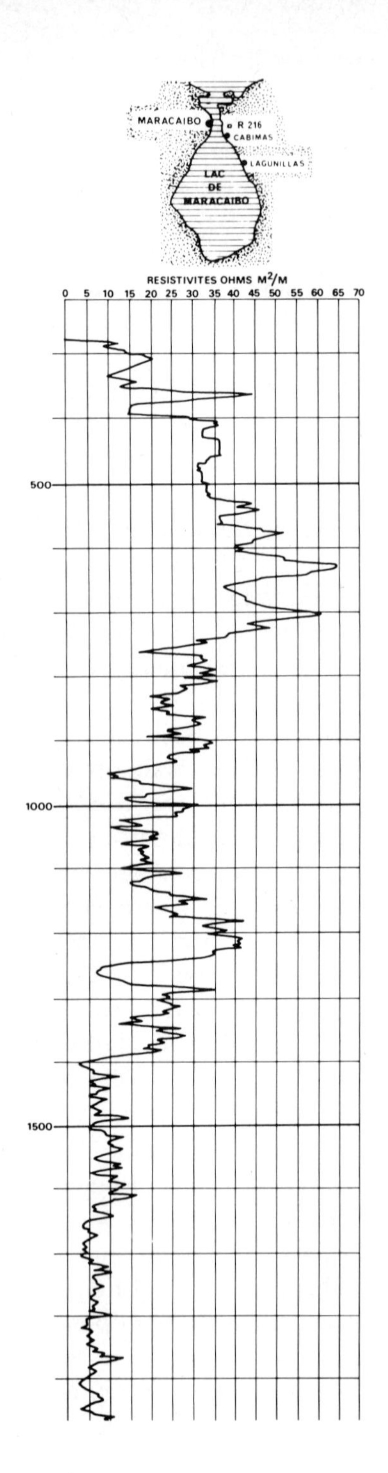

Figure 46. First electrical coring diagram outside France (La Rosa, Venezuela, March 1929).

debilitating climate; round-the-clock, short-notice calls; long, uninterrupted night operations on the rigs—all of which, by the way, continue to be the lot of well-logging operators in many parts of the world. What has changed a great deal is the ease of communications and the quality of equipment. In 1930 there was no transatlantic airline to South America. In Venezuela the oil companies already flew their own aircraft, but for prospectors travel generally meant almost impassable trails or slow, uncomfortable, and crowded lake boats. Often a trek of several days, interrupted by all sorts of obstacles—mud holes, sand dunes, fords, rickety bridges—was required to reach a few wells located in the heart of the jungle. Also, communications with Paris were very slow. For reasons of economy, cables were used in emergencies only. In 1929 mail took 1 full month to reach France; this time was reduced to 2 to 3 weeks with the advent of airmail between Venezuela and the United States. It took almost 2 months after customs clearance for equipment, shipped mostly from Paris, to reach the field. Of course, this gave the prospectors a great measure of autonomy.

In addition to everything else, the equipment was a constant source of headache. The original set had been adapted from the Pechelbronn experience, whereas the Venezuelan oil wells were much deeper—3000 to 4500 feet—and drilled with thicker and more viscous mud. Heavier and heavier weights were needed before the drillers could be convinced to circulate a lighter mud, thus making the walls of the hole as smooth and "clean" as possible. In cases where an obstacle blocked the descent of the tool, an expedient, often used elsewhere, was to lower the drill pipe without the bit to a point just below the obstruction, and then run the sonde through the pipe to the bottom of the hole. Cables wore out quickly; since replacements from Paris were slow to arrive, it happened more than once that sections of old cables had to be pieced together. The power of the electric motor was so inadequate that the prospectors preferred to drive the winch with a wheel of the truck. In addition, several months passed before Shell complied with the contract to provide roadworthy trucks and helpers who were of some use. As to the prospectors' board and lodging, these necessities were provided by the oil companies in their more or less comfortable field camps. It was a life without privacy, and not always in the choicest environment.

In 1931 all this activity collapsed. World economic depression led the companies first to curtail and then to almost completely cease drilling operations. By the end of the year, electrical coring operations were reduced to one or two a month. Only one Schlumberger engineer, Jean Mathieu, remained on the spot. By mid-1932, however, things took a turn for the better.

Operations in Trinidad began in July 1932, after 3 years of negotiations with British Controlled Oilfields, one of the English companies operating on the island. The company had experimented with electrical coring in the state of Falcon, Venezuela.

Trinidad's oil reservoirs were lenticular: very thick sand formations often enclosed in impervious shales. The lenses were many but small, and the fields were so fragmented that production required a considerable number of wells and intensive mechanical coring. Furthermore, exceptionally high formation pressures necessitated very heavy muds (density near 2). Consequently, production costs were very high, and in the middle of the world depression in 1931 the Trinidadian companies questioned the usefulness of pursuing their operations. It was under these circumstances that they turned to electrical coring.

The first diagrams, already including resistivity and S.P. curves, were a success; they outlined accurately the permeable zones and differentiated clearly between oil- and water-bearing horizons. Within a few weeks the use of electrical coring led to a substantial reduction in production costs, particularly through the elimination of mechanical coring. This, together with the general economic recovery, gave the Trinidad petroleum industry the opportunity not only to survive but also to expand.

In the United States the first tests were conducted on August 27, 1929, in Shell's B well between Hanford and Kettleman Hills, California. With a hand-driven winch, a 1 ton cable, and no recorder (in short, the equipment of the earliest Pechelbronn tests), Deschâtre and Roche successfully made (with the loss of a weight) a series of measurements from 2700 feet to the casing shoe at 54 feet (Fig. 47). The ensuing operations were slow and painful, and on September 19, one of them, at the 6000 foot deep Ruthenfort wildcat, proved a disaster: pressure and temperature made the cable so leaky that no dependable measurement was possible. All these operations pointed to the inadequacy of the equipment; and with wells spaced so far apart that correlations were impossible,

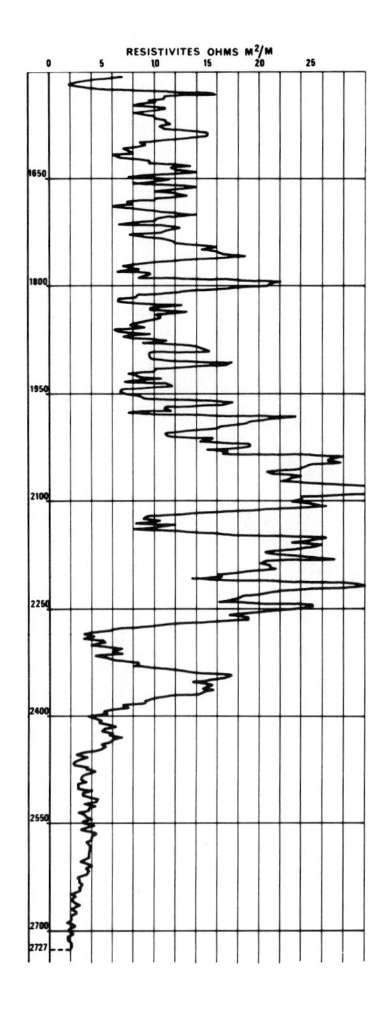

Figure 47. First electrical coring diagram in the United States.

the local geologists, regarding the method as useless, terminated the tests after a few weeks (Fig. 48).

A contract with Gipsy, an affiliate of Gulf Oil, took the operations to Oklahoma. The wells were less deep (2500 to 3600 feet); the formations were harder, and lowering the sonde into the hole was easier. Operations ran rather smoothly, though at the cost of endless troubleshooting and repairs. The 23 jobs carried out at Seminole between October 15 and November 15 (13,600 point-by-point measurements) kept the two engineers at a daily work average of 15 hours; moreover, 8000 feet of

Figure 48. One among the earliest electrical logging jobs in the United States (California, 1929).

ruptured cable had to be abandoned in the wells. At this point the crew received a hand recorder and a winch with electric drive.

Operations spread to Kansas, where Gipsy had undertaken the drilling of a great number of 750 foot deep structural coreholes. In both areas electrical coring revealed several markers from which accurate correlations could be drawn. This achievement afforded the geologists little satisfaction, however, for either the correlations affected series of no special interest or electrical markers could have been just as easily located by examining the cuttings,[27] at less cost and without any interruption in drilling. Finally, electrical coring failed to differentiate adequately the oil- and water-bearing strata and the hard beds. For these reasons and because of the fact that Schlumberger could not accept the 50 percent reduction in

[27] Even in deep exploration the routine practice at Gipsy was to rely on examination of the cuttings rather than on mechanical coring.

its already modest rates demanded by the client, the contract was canceled at the end of the year.

Nevertheless, interest in the Schlumberger methods continued. In January 1930 the Hugush Group (Humble, Gulf, and Shell) signed a 6 month, lump-sum contract which specified that the crew was at the complete disposal of the client for work extending from the Mississippi River to the Mexican border. In over 5 months the prospectors and their overloaded truck covered some 8000 miles of sandy or muddy country roads and trails (at that time paved roads ended some 20 miles beyond the towns) to carry out 53 jobs on 45 wells scattered over 22 fields. The drilling sites were often surrounded by a quagmire that required incredible effort to cross. Again, the wells were so widely spaced that correlations were practically impossible. Furthermore, the dearth of adequate production data made it difficult to establish empirical criteria, and the petroleum diagnosis was uncertain. Also, the many breakdowns and the frequent difficulty in lowering the sonde to the bottom of the hole would not have created enthusiasm in any client. With the depression as an additional consideration, Hugush decided not to renew the contract.

One crew remained in the area for a few more weeks to carry out scattered operations in small exploration drillholes. The United States offered a market of immense potential, but with surface prospecting at a standstill, and after the disappointments of the first logging tests, Schlumberger could hardly support a money-losing crew with its inadequate equipment. Temporarily, at least, it seemed wiser to concentrate on technical improvements, while taking advantage of the contract with the U.S.S.R., where electrical coring was in full swing. Business in the United States was postponed to better times.

However, even the failures were not altogether fruitless. At least a perception remained of the intrinsic merits of electrical coring. After the first shock waves of the depression had subsided, Schlumberger returned, offering greatly expanded and improved logging services (better cables, simultaneous recording of S.P. and resistivity, power-driven winches, etc.) in California (July 1932) and on the Gulf Coast (early 1933). This time it took only a few months of demonstrated know-how and intensive activity for the prospectors to establish logging firmly in the United States petroleum industry. Contributing to this result was an exhaustive communication presented to the February 1932 meeting of the American

Institute of Mining Engineers,[28] as well as the outstanding results obtained in the U.S.S.R., Venezuela, and other countries.

As had been the case in Venezuela and the United States, electrical coring at Grozny was made difficult at first by equipment ill adapted to depths exceeding 4000 feet. Moreover, the drillers, concerned over possible cave-ins or uncontrolled blowouts, hesitated to add clear water to the mud in order to facilitate the lowering of the sonde. Adequate cables and winches had been shipped from Paris, but slow customs clearance and rail transport delayed their arrival. Sauvage, the leader of the first Schlumberger crew, skirted the problem, as Bayle had done in Venezuela, by lowering the cable inside the drill pipe.

If a slow and not very efficient administration in the U.S.S.R. could create difficulties, it had, on the other hand, a great advantage over countries where operations are conducted by various companies, in that here there was a single owner and the diagrams could be compared freely across any geological structure. This circumstance, coupled with propitious sedimentary features, resulted in the production of many accurate correlations by electrical coring within a period of several months. Also, the world depression had left the U.S.S.R. untouched, and the First Five-Year Plan provided for a rapid growth of petroleum production; to that end, anything was welcomed that could increase the efficiency of drilling and production. In no time the Grozny geologists made a basic tool of electrical coring. While relying greatly on the guidance of their French colleagues, they took interpretation into their own hands and made the most of it. The results were especially good in the New Fields, where large producing horizons showed peaks of up to several hundred ohm-meters, while the surrounding shales did not exceed 4 or 5 ohm-meters. Such sharpness of contrast was explained by the fact that in these areas the formations had a very high oil saturation, whereas the remaining interstitial water was of low salinity. It was recognized early by the geologists that for some reservoirs the resistivity value reflected the order of magnitude of the production rate; it could be anticipated from the height of a peak whether the horizon would be gushing, and even whether the daily production would be, say, 500 tons of oil without water, 10 tons

[28] C. and M. Schlumberger and E. G. Léonardon, *Electrical Coring, a Method of Determining Bottom-Hole Data by Electrical Measurements,* A.I.M.E., New York meeting, February 1932.

of oil with water, or water only. Thereafter it became easier to convince the drillers that electrical coring was no waste of time, not only because it avoided the complications of mechanical coring, but also because it added safety and speed to accuracy in exploration and production. By October 1, 1930, 1 year after its arrival in Grozny, the Schlumberger crew had a record of 240 operations in 101 wells, of which 6 were wildcats, with measurements covering 105,000 feet of drillhole.

For the first operation on behalf of the Azneft Trust (Baku, October 1930), the same primitive equipment was at the site: hand-driven winch and nonrecording potentiometer. One wonders whether the Azneft people, though well aware of the value of electrical coring, wanted to test it under conditions that looked somewhat like a trap. The well designated for the operation belonged to the Surakhany field, one of Baku's best producers; but no one had told the prospectors that it was located on the flank of the structure, where the strata were known to yield nothing but salt water. Being accustomed to the Grozny diagrams with their huge peaks, the prospectors were rather upset when, foot by foot, they read very low resistivities varying between 0.5 and 1.5 ohms. This seemed to be a bad omen for the future of electrical coring in Baku, but for the Azneft people, on the contrary, it was definite proof that Schlumberger meant business. The well selected for the next test was located near the top of the structure, and, as expected, peaks appeared at the depth of the oil-bearing strata.

In fact, Baku provided conditions even more suitable for electrical coring than Grozny. The more than 500 wells drilled annually were less deep, and lowering the sonde proved less laborious. Nearly the entire production of some 20 sands interbedded with shales came from four main fields: Surakhany, Lenin district, Bibi-Eibat, and Kara-Chukur. These pay zones had variable thicknesses and were disturbed by multiple faults. Clear geological markers were few (all the beds from top to bottom were more or less the same, and there were essentially no fossils). These circumstances explain why, especially after the introduction of the S.P. curve, electrical coring could very rapidly make an invaluable contribution. In the spring of 1931 four crews were at work with a daily average of three to five operations, sometimes more; the strain of overwork was all the greater in that, for lack of recorders, the plotting was still done point by point, and only a few power-driven winches were available. In 1932 there were six crews with three recorders. The number of jobs exceeded 1200 in 1931 and 1800 in 1932.

During the same years the Surakhany and Bibi-Eibat fields were comprehensively resurveyed using electrical coring. Accurate and detailed structural maps defined the locations and throws of the faults and delineated the oil-bearing horizons, including some that until then had remained unknown. A close relationship was observed between the resistivity and the contents of a horizon: as the formation waters were highly mineralized, a fraction of an ohm-meter indicated an altogether aquiferous sand, whereas a few ohm-meters were a sure sign of an oil sand. Bolder than their Grozny colleagues, the Baku geologists reached the point where they could forecast from the diagram not only the order of magnitude, but also exactly what the initial production of a reservoir would be; they called this "translating ohms into tons." Their enthusiasm was such that, in spite of the warnings of the French engineers, they abandoned mechanical coring altogether and conducted development and even exploration wells under the sole guidance of electrical coring.[29]

Yet there were zones where the translation of ohms into tons could lead to errors, and where even a qualitative diagnosis was uncertain. It happened that in a field where electrical coring had just been introduced resistivity peaks marked all the sandstones, whether oil- or water-bearing. It took some time to realize that in such cases the mud had been prepared with fresh water, and the invaded zone amounted to a resistant mantle around the hole.[30] After this experience and a few other experiments had shown that diagrams alone would not solve all their problems, the Soviet geologists gradually returned to more rational operating rules wherein electrical and mechanical coring complemented each other.

On the other hand, electrical coring allowed for the use, in Baku, of the "bottom to top" operating method. Standard procedure in most oil fields was to exploit each horizon by a series of wells, systematically

[29] In the summer of 1933 Maurice Martin, then assigned to Baku, was told that in lieu of cuttings, cores, or geological cross sections the Azneft people showed long paper strips on which undulating lines were supposed to represent electrical parameters. He was greatly amused to find that the source of this "confidential" information was none other than Messrs. Herold and Uren, the reputed American oil specialists who had been invited by the Soviet industry for consultation.

[30] The same difficulty was later encountered in many other regions of the world. The solution was the use of a sonde with a sufficient investigation diameter.

spaced 50 to 100 feet apart. The separate exploitation[31] of each of the multiple superimposed horizons required the drilling of a huge number of wells (hence the dense forests of derricks on old photographs). To obviate such a proliferation, the Soviet engineers thought of drilling down to the deepest reservoir, lowering and cementing a string of casing to the bottom of the hole, putting the reservoir into production, plugging it once it was depleted, then putting the next upper reservoir into production, and so on to the top reservoir. Such a method required the accurate location and delineation of each reservoir, and for this electrical coring was eminently suitable. As holes had to be made through the casing and the cement at the depth of each productive horizon, the Azneft engineers designed a rather strange tool, consisting of a drill, which was lowered to the bottom and driven from the surface by cables and pulleys. It took several hours to pierce a single hole at depths never exceeding 2000 to 2500 feet. In most cases the pressure and permeability were such that a substantial flow of oil followed the first perforation; the tool was then abandoned and the wellhead shut in fairly quickly. This primitive method served in Baku for several years without any serious accident, until it was replaced by gun perforators of the Schlumberger type.

In 1931 and 1932 electrical coring, conducted by the Russians with or without French advisers, was expanded to the Donetz coal basin, the Maikop (North Caucasus) region, Georgia, the Emba Basin, and the Ferghana Valley in central Asia.

A contract signed on November 8, 1928, with Shell provided for the assignment of a crew to the Dutch Indies, but operations in Sumatra did not start until early 1930. The main producing sands were very friable. All cores recovered by the drillers were so fragmented and washed out by mud that it was difficult to detect the hydrocarbons with the methods then available.

This, indeed, was the main concern of the local geologists, but electrical coring was unable to give a dependable answer, although the results were excellent in regard to correlations and the delineation of sands (even without S.P.). The mission chief, who attributed this failure to the effect of deep mud invasion, saw a remedy in extending the lateral

[31] It is important to prevent reservoirs from flowing into one another, to control water intrusions, and to keep a precise production record for each pay zone.

sonde to lengths of 30 to 50 feet. It was not known at the time that, with a lateral sonde of such length, insulation faults could entail substantial errors in measurements. The curtailment of drilling operations resulted in cancellation of the contract (end of 1930) before the question had been clarified. Not until 5 years later was work resumed in the area.

A 4 month contract with Steaua Romana, a company for which Schlumberger had already conducted several years of surface work, introduced electrical coring into Rumania (March 1931). The team of Hubert Guyod and Roger Henquet had a power winch driven by the wheel of a truck. The formations (a sequence of shales and sands) were not unlike those at Baku. Experienced drillers, using advanced rotary rigs, drilled through these soft layers to a depth of 6000 to 7500 feet with a single surface casing, an operation requiring especially heavy, viscous mud (density above 1.5). The 60 pound weight at the end of the 1 ton cable was often too light to reach the bottom of the hole even after endless attempts. If additional weights propelled the sonde to bottom, pulling it out was likely to cause unfortunate cable breaks with ensuing unpleasant and costly fishing jobs. The prospectors fumed, demanding by letter and telegram a stronger cable on the order of 2 or even 4 tons. Paris recommended more care in splicing and elaborate lubrication to reduce friction, and even designed a detonator tool that could scuttle the weights if the cable became stuck. Finally, everything was straightened out. The prospectors received their 2 ton cable and recorders, and as they became more familiar with the borehole conditions operational efficiency increased. The company appreciated the results of the electrical measurements to the point of considering them indispensable, and so became more tolerant of mishaps; it even accepted the fact that the wells needed to be thoroughly prepared for logging operations by prolonged circulation of mud of lesser density.

By the end of 1931 Schlumberger services were being provided to all Rumanian and foreign companies: operations were in progress in the fields of Gura Ocnitei, Aricesti, Boldesti, Moreni, and so on. Cores were no longer taken in development wells; as in the U.S.S.R., diagrams were used in support of the "bottom to top" production method. The technical balance sheet was highly positive. Yet in Rumania, too, the depression curtailed drilling, and by the end of 1932 the number of operations did not exceed 200.

Phase Two: 1933–1940

An overall view

In January 1933, 8 not excessively busy crews were operating outside the U.S.S.R.: 1 in Pechelbron, 1 in Morocco, 1 in Rumania, 2 in the United States, 2 in Venezuela, and 1 in Trinidad.[1] By the middle of 1934 there were 20 crews, and at the end of 1935, 40—many of which were overworked. When World War II broke out, the number had reached 140, and more than half of these were in the United States. The others were spread over some 20 countries, with a special concentration in Venezuela and the East Dutch Indies. In comparison with the slow and unsteady progress of the 1920's, the Company's growth had become stable and continuous.

Paris was still the only research and development center, as well as the main source of equipment. By 1936 one fully equipped truck was produced by the workshop each month. Twelve-hour shifts and night work were often required for such an output. The Schlumberger Well Surveying Corporation (S.W.S.C.), a United States corporation established in Houston, Texas, in late 1934, was contributing to the procurement of the cables; it was also purchasing and assembling trucks for the crews in the United States.

The offices, workshops, laboratories, and testing stations, all at Nos. 40 and 42 rue Saint-Dominique, underwent a substantial and necessary expansion and reorganization after 1936. The testing station, located in the basement, included, among other things, wells designed by Marcel,

[1] According to the workload, a crew included one or two engineers and their helpers.

where instruments could be exposed to pressures and temperatures similar to those encountered at the bottom of boreholes. Three of them, one 45 feet deep, had been built in 1933 for tests up to 1500 PSI. Three others, for 9000 PSI tests at 400 degrees F,[2] were installed in 1936. Only the barrels and breeches of heavy guns responded to such specifications, and the manufacturing facilities of the French army and navy were called upon. Highly elaborated thermal treatments were required: it took 6 months to temper the steel in wood ashes.

Paul Charrin played a leading role in the management of operations: outside of his sojourns in the U.S.S.R., he divided his time between Paris and the United States, where Eugène Léonardon was resident manager. Jules Miller was in charge of accounting; René Clairin headed the engineering office. René Seydoux had joined the management of "La Pros" in 1931. New contributions were made in 1938 by Jean de Ménil, an experienced financier, and the young scientist Eric Boissonnas.

The number of engineers on missions, or in Paris ready to depart, had grown from 28 in early 1933 (the U.S.S.R. included) to 70 in July 1935 and was to exceed 100 in 1938. The engineering and manufacturing staffs had increased commensurately. In 1934 Schlumberger recruited some of the engineers who were to have long and distinguished careers with the Company: William J. Gillingham, Louis Magne, and Alain Morazzani, all of whom reached the top echelons, and Maurice Tixier, an outstanding logging expert. Also in 1934, S.W.S.C. began to recruit an increasing number of American engineers for its United States operations. The first of them was Harold B. Markam.

Measurement techniques and interpretation methods

From 1933 on, S.P.E. ceased its surface work activities except for incidental technical interventions. Electrical logging became its principal activity, complemented by other initially modest but rapidly developing processes: sidewell sampling and perforation (see p. 182 ff), teleclinometry, temperature measurements, and dipmeter surveys. Additional services technically related to logging involved water intrusions, well depth de-

[2] Such figures correspond to 15,000 foot boreholes, a depth that was far beyond the drilling record of the time.

terminations, water–oil contacts, and so on. Cables and winches also served occasionally for lowering and igniting explosive charges (torpedoes) for the purpose of fracturing very compact rocks and facilitating the drainage of oil or gas.

The practice of drilling oil wells with a degasified oil-base mud loaded with clays originated in the mid-1930's. Its purpose was to minimize the contamination of potentially oil-bearing horizons during drilling. Whereas an oil-base mud invades a reservoir by only a fraction of an inch, the depth of invasion may reach several feet or more with water-base muds, and affect the reservoir to the point where the production rate drops substantially.[3] Also, when drilling with oil, the center of core remains practically unaltered; this is not always the case with water.

These reasons explain why oil-base muds became popular, but their use remained limited by high cost and the generally disagreeable conditions encountered in their handling. As oil is an insulator,[4] there was no electrical connection between the standard lead electrodes and the formations. An induction logging system was the answer, but was not realizable with the electronic techniques then known. Therefore Schlumberger devised "scratcher" electrodes that rubbed against the borehole wall. One model, a kind of bronze wire cylinder brush, was tested with some success in 1937 in Venezuela and California. With sufficiently soft formations the resistivity diagrams resembled those obtained with standard electrodes in mud- or water-filled holes, and even some acceptable S.P. curves could be recorded. However, such logs lent themselves to correlations only. In hard formations, the contact of the electrodes was unpredictable and the readings were meaningless. Beset by other operational problems such as difficulties in lowering the scratchers into the borehole, the method found few applications.

Great emphasis was placed during those years on making available to the engineers accurate and systematic documentation on techniques and instruments. *Proselec* (see note 10, p. 110), an invaluable source of infor-

[3] The water of the mud displaces a large part of the oil or gas. As the adhesion of the water to the grains of the rock is usually greater than that of hydrocarbons (water is said to wet the rock better), the invaded zone acts like a seal, inhibiting the motion of the hydrocarbons toward the well once it is in production. Furthermore, the interstitial shale may swell under the effect of the filtrate, thus reducing the permeability of the invaded zone.

[4] There is always a small fraction of water suspended in the oil, but in the shape of separate droplets, and the mixture has no electric conductivity.

mation, was published regularly until World War II, along with a number of pamphlets and memoranda (*Water Intrusions,* October 1934; *Sondes and Weights,* December 1934; *Dipmeter,* September 1935; *Photographic Recorder,* 1937; *Gun Perforator,* 1939; etc.). With the hiring of American personnel, some of this literature had to be issued in English. The first English document, *Principles of Electrical Logging Operations* (January 1936), was followed by an issue of *Proselec* (November 1936) devoted solely to S.P. Both were written in Paris in the usual Schlumberger format and style. The writing of technical memoranda began in Houston with the introduction of photographic recorders. In this respect it is interesting to compare the two memoranda on the three-galvanometer recorder as written in French in Paris and then in English in Houston, by the same French engineer. The first, although concerned with practical directions, reserves a large section for the functional description of its various components; the second consists mainly of push-button-type, complete, and detailed instructions for use.

The electrode spacing of the normal sonde[5] was set in each region after a few tests and varied between 10 and 20 inches. Soft sand–shale series contain many rather thick, porous horizons, with little invasion, so that the measurements were close to the true resistivities. When the lithology and the water salinity of these horizons were fairly uniform,[6] a comparison with production results in a few wells led to a sort of calibration of the resistivities, and the basis for interpretation obtained in this empirical way could be extrapolated to the whole field. Yet in other regions, such as northern Texas and the Mid-Continent, where more consolidated producing zones occur, true resistivities were rarely read on the normal sonde diagram, mainly because of deep invasion, and it was difficult to differentiate between hydrocarbon- and water-bearing formations. For several years the long lateral[7] (12 to 24 feet) was used from time to time in these regions. There was nothing systematic in this approach, one reason being that the overworked staff had little time for the recording of a third curve. Around 1937 the number of crews was increased; they were better equipped and worked under more normal conditions,

[5] The Baku operators for years kept the same 7.5 to 12 foot *AMN* devices as prevailed when logging was introduced into the U.S.S.R.

[6] This is mainly the case with waters of very high salinity (over 100,000 PPM).

[7] In Trinidad a deep lateral sonde was frequently used to study very thick and deeply invaded sands.

and with competition becoming a factor, the long lateral sonde was offered as a standard service under the name "third curve with great investigation range." In spite of its limitations, it was favorably received and quickly became a standard tool.

Another improvement was introduced at the end of 1938 in the form of the second normal. Thereafter, and for years to come, standard equipment included the short normal (16 inch), the long normal (64 inch), the lateral, and the S.P. (Fig. 49). With the simultaneous introduction of the four-conductor cable, the operation took place in two stages: first the two normals and the S.P., then the lateral and again the S.P., so as to accurately adjust the depths between the two logs. Sometimes one run could be saved by making the lateral recording while lowering the cable into the well. This was a compromise made to gather as many data as possible in the shortest time. In many cases, however (e.g., consolidated formations or horizons made of alternating thin layers such as sands, shales, and hard beds), log interpretation remained uncertain. There was much room for improvement, but it would have to wait until after the war, when focused sondes would be put into operation.

Between 1930 and 1932 S. Stefanesco, already mentioned in the discussion of surface prospecting, had laid the theoretical basis needed to calculate the relationship between true and apparent resistivity.[8] The actual environment was simulated by a model in which the formation was assumed to be homogeneous, isotropic, and infinitely thick; the wall of the borehole and the limit of the invaded zone were represented by two circular coaxial cylinders, on the axis of which the electrodes were assimilated to points.[9] The curves thus obtained were plotted in logarithmic coordinates, with the sizes of the device (normal or lateral sonde) as abscissae and the apparent resistivities as ordinates. Since the curves gave the value of the apparent resistivity as a function of the investigation diameter, they were called "lateral sounding curves" by analogy with the vertical electrical sounding curves of surface prospecting. However useful in facilitating interpretation by imparting greater accuracy to the re-

[8] The same problem was dealt with at the same time in Leningrad by Professor V. A. Fok, but with a different mathematical approach.

[9] There is no strict analytical solution when the media involved in the measurement are separated by both cylinders and planes, as is the case in a formation having a finite thickness.

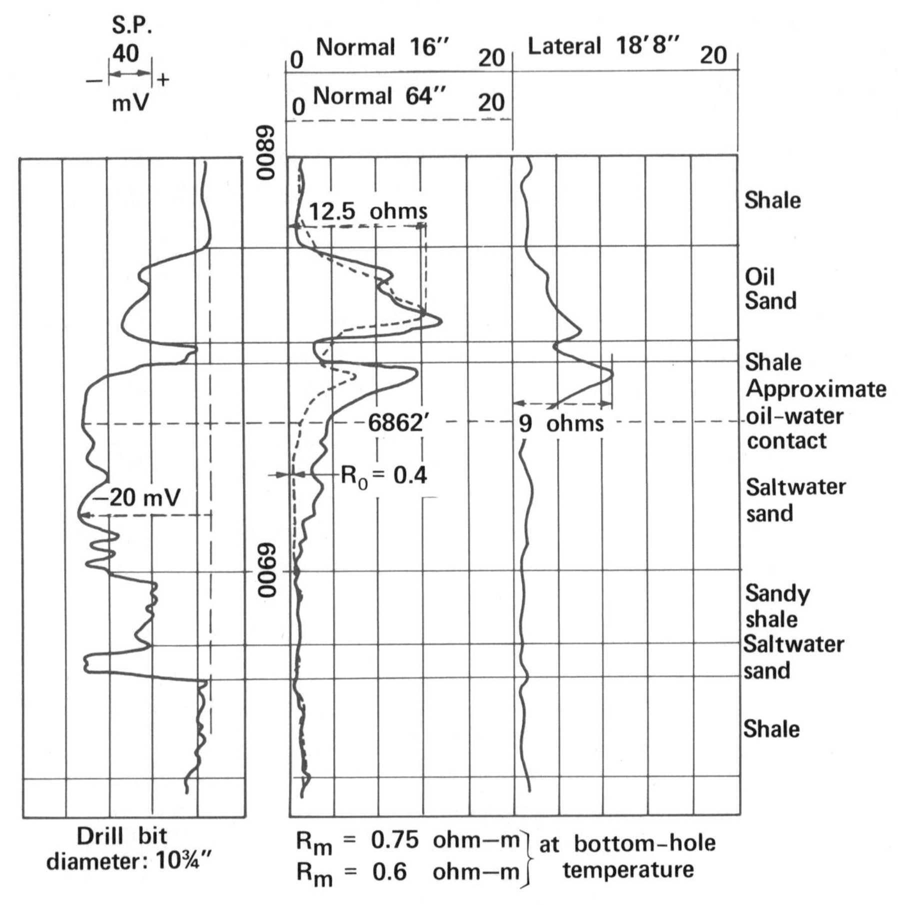

Figure 49. Log recorded with standard equipment: short normal, long normal, lateral, and S.P.

sponses of the sondes, they were of little help in actual cases in obtaining the precise values of the true resistivities. To do this would have required, among other things, that the formations be sufficiently thick and homogeneous, a situation almost never encountered.

What needed to be defined more precisely to improve empirical interpretation methods was the relation of true resistivity to saturation, which, in turn, required many measurements on samples in a specially equipped laboratory. Absorbed by technical developments and the demands of an industrial activity in full expansion, Schlumberger was unable to cope with such a task. The first studies on the subject were done in 1934–1935 at the Petroleum Research Institute of Azerbaijan.[10] They covered a great number of moderately consolidated samples of sands originating from several horizons of the same field, where a network of production wells was exploiting every pay zone.

The method was, in each case, to determine the porosity and then to proceed with a series of resistivity measurements by first impregnating the rock with water, followed by the injection into it of oil.[11] The results were recorded as a family of curves in which each corresponded to an average porosity value. Once the resistivity of the water had been measured for each horizon, the authors used these curves to deduct the oil saturations from the resistivities read on the logs. Combined with the porosity values, the thickness of the formations, and the lateral extension of each oil-bearing zone, these saturation values made it possible to compute the volume of oil (reserves) in each horizon. As a verification the same computation was resumed on one of the horizons on the basis of logs taken, on the one hand, between the end of 1932 and the beginning of 1933 and, on the other hand, between late 1934 and early 1935. Subtraction of the two evaluations of the reserves showed a drop of 1,230,000 tons against 1,154,000 tons for the same period according to production statistics—a difference of only 8 percent. Without minimizing the part played by chance—after all, the statistics were not dated with great precision—the accuracy of the method seemed reasonably confirmed. The authors had postulated that, in a practical sense, the logs gave the true resistivity of the horizons, and the result proved *a posteriori* that the assumption was reasonably valid. It may be added that they were unaware of the effect on resistivity of the interstitial clays; the horizons studied probably contained so little of them that this effect was negligible. This research marked an epoch in log utilization.

Observations made shortly thereafter in American laboratories on

[10] See I. Kogan, *Ispolsovanie Karottajnykh Dannykh dla Opredelenia Nieftianykh Zapassof (Utilization of Electrical Coring Data for the Determination of Petroleum Reserves),* Azerb-Neft. Khoziatsvo, Baku, October 1935.

[11] This operating method endeavored to repeat the natural process whereby, according to prevailing theories, petroleum penetrates the reservoirs.

rocks of various origins proved consistent with the Baku measurements.[12] Finally, in an article published in 1938,[13] Schlumberger presented a full discussion of the relations between reserves and true and apparent resistivities. The article was a forerunner of what would later be called "quantitative interpretation." Yet it still failed to give any numerical relation of a general nature between saturation and resistivity; this meant that application of the method required long and costly work for each formation. For this reason empirical interpretation processes continued to prevail for a long time.

In pursuance of his studies on S.P., initiated in the spring of 1932, Conrad devised a special apparatus for measurements to confirm the law on electrofiltration potentials. The main direction of his research, however, was toward electrochemical potentials. Actual ground phenomena were simulated by means of three glasses (Fig. 50), the first and third of which were filled with the same sodium chloride solution,[14] representing muds of R_{mf} resistivity, whereas the second glass contained a more concentrated solution (not exceeding, however, a few tens of grams per liter), representing formation water of R_w resistivity. The first and the second glasses were connected by a sand-filled syphon, the image of the permeable layer, in which the contact of the two solutions generated an electromotive force as in the concentration cells. The shoulders of the formation were simulated by a sausage-shaped curve of clay connecting the second and third glasses, between which another electromotive force, not fully explained at the time, occurred. The potential difference between the first and the third glass corresponded to the electromotive force which, in actual cases, generates the spontaneous currents. These spread through the ground and squeeze again inside the borehole; therefore, when the permeable formation is of such thickness that its resistance to the current is negligible, the ohmic drop in the mud column measures practically this whole electromotive force.

Many measurements made with varied concentrations and types of

[12] R. D. Wyckoff and H. G. Botset, *Physics,* September 1936; J. J. Jakosky and R. H. Hopper, *Geophysics,* Vol. 2, No. 1 (January 1937); M. C. Leverett, *Transactions of the AIME,* 1938, p. 1.

[13] M. Martin, G. H. Murray, and W. J. Gillingham, *Geophysics,* Vol. 3, No. 3 (July 1938).

[14] Sodium chloride is by far the most widespread salt in drilling muds and formation water; *mf* and *w* are current symbols for "mud filtrate" and "water," respectively.

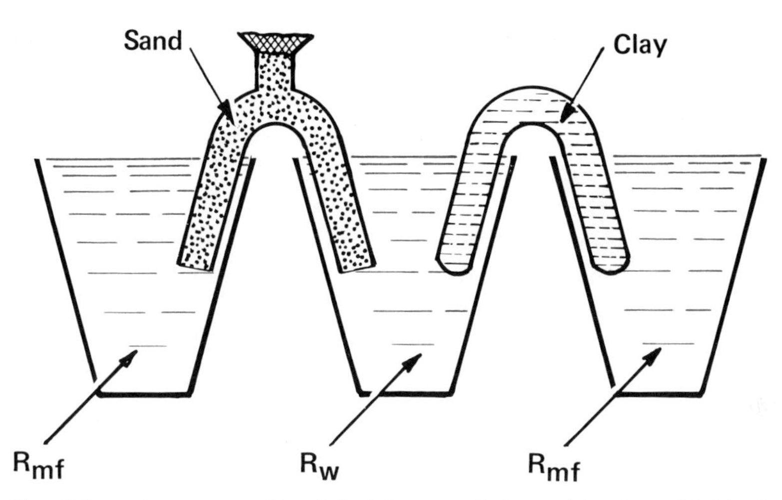

Figure 50. Schematic representation of the laboratory device used by Conrad Schlumberger to study the electrochemical component of S.P. (1932).

clay demonstrated that the potential difference between glasses 1 and 3 is expressed in millivolts by the formula

$$E = K \log_{10} \frac{R_{mf}}{R_w}.$$

The various values found for coefficient K averaged 17, whereas more recent studies have shown that its precise value is 71 at 25 degrees C. The reason Conrad found a much lower value was that in his experiments the clays were under less stress than those in the subsoil, where they sustain the pressure of the surrounding media. Although these results were known to lack accuracy, the conclusion was drawn that the electrochemical component of the S.P. was relatively small.[15] This idea prevailed for a decade, during which time nothing was done to better define the respective parts of the two main components of S.P. All that was known was that, in areas where mud and formation water resistivities were close and rather high, the electrofiltration phenomenon might be preponderant, as shown by tests in Burma (1934), Pennsylvania (1936–1937), and Califor-

[15] Yet Tanida, *Kyoto,* Vol. 4, p. 1333 (quoted in *Proselec,* November 1936), had given a K value of 58 for potassium chloride solutions (similar to sodium chloride) separated by a collodion diaphragm, whose electrochemical behavior is comparable to that of clay.

nia (1934–1939). On the other hand it seemed that, when the mud was relatively conductive, although less so than the formation water—as in Baku or southern Texas—the electrochemical phenomenon perhaps prevailed. In boreholes where the mud was more conductive than the formation water the S.P. anomalies became positive, a discovery that could be explained only by the preponderance of electrochemical phenomena.

At the same time, it was realized that the amplitude of the S.P. anomalies, as well as of the apparent resistivity, might be a function of the geometry and the resistivities of the media in which the current flows (borehole, invaded zone, undisturbed zone, formation shoulders), and that, in particular, it decreases opposite thin formations. It gradually became apparent that the spontaneous potentials, continuous and low as they are (usually below 100 millivolts), are exposed to perturbations of various origins: stray currents (often caused by operations in a nearby well), the vertical component of telluric currents, instability of the surface electrode, voltages induced by the terrestrial field in the slack cable between winch and well, uneven polarization of the metal of the sonde (bimetallic effect), heterogeneous mud column, and so on. The protection of S.P. measurements against such disturbances has been the subject of careful and continued studies.

It will be recalled that during his early work Conrad had observed the polarization induced in the ground by the flow of a direct current (I.P.) and had thought of utilizing it in the search for conductive minerals (see p. 31). From the introduction of electrical coring, he had advanced the hypothesis that rocks might behave like a system of resistors and capacitors, in which the water would be the armature and the mineral grains the dielectrics. Such a system would generate I.P. potentials reflecting the porosity and oil content of the rock. He hoped that these potentials might be substituted for S.P. whenever the latter lacked sharpness, and complement the resistivities in the identification of oil-bearing horizons. A few tests in the U.S.S.R. between 1930 and 1933 confirmed that I.P. generated measurable potentials, but no clear correlation could be established between those and other properties characterizing the formations.[16] Schlumberger was then so engaged in urgent developments that

[16] It had not occurred to Conrad that the clays might have a part in the mechanism of I.P. It emerged much later that their role was a preponderant one; hence the values he had measured had a meaning different from his interpretation.

the continuation of a study holding little promise was postponed indefinitely.

As logging spread over the world, new problems arose daily. It was noticed that the presence within permeable rocks of very fine particles, mainly clays, may substantially reduce the resistivity of oil-bearing strata. In certain regions like southwest Texas, the reduction could be so significant as to prevent petroleum diagnosis. Thus there was, by 1936, a first awareness that interpretation could be complicated by the presence of clay in the reservoirs—a prelude of the many studies made after World War II on the "shaly sands" problem.

Other problems arose in the study of limestone fields. It was found that S.P. diagrams gave only broad, approximate limits of limestone series, and showed only vague inflections through permeable or shaly intervals. In these limestone series it was also observed that, contrary to what occurs in sand–shale series, the pay zones were generally more conductive than the impermeable adjacent formations. Furthermore, since limestones usually have very high resistivities, most of the current was channeled along the borehole, which made the diagrams confusing; it was difficult to read the depth of the contacts between formation boundaries, and there was only a remote relationship between apparent and true resistivities. In most cases nobody could tell whether the conductive intervals corresponded to permeable zones or to marls, let alone differentiate water from hydrocarbons. Moreover, it frequently happens that limestone fields are topped by thick layers of rock salt, which saturate the drilling mud and result in still more featureless resistivity curves and the reduction of the S.P. curve to an almost flat line. Not very successful attempts were made to revive the sharpness of the resistivity logs by fitting insulating sleeves on the sonde between electrodes. In certain cases it was possible through thermometric measurements to locate roughly the potential gas-bearing zones, but such information remained of only local and contingent use.

The utility of logging in limestone fields thus continued to be very limited. In the 1950's this situation was remedied by the introduction of improved measuring techniques. It is quite conceivable, however, that if early logging tests had been conducted only in limestones the failure might have been final. Fortunately, Schlumberger's reputation had been so firmly established by the Company's success in logging sand–shale basins that, when the limestone problem had to be tackled, the partial failure of the method was ascribed by professionals to the particular circumstances rather than to any shortcomings of the technique.

Cables, winches, and miscellaneous tools

From 1928 (see p. 109) to about 1938 the types of cables hardly changed. The tensile strengths were 2, 4, and finally 8 tons. As these values meant more steel, increased tensile strength involved higher density, and the running in of the sonde became easier. In 1938 a fourth insulated conductor was added to the cable (see p. 163).

The protective braid was still made of a textile fiber: first hemp, flax, or jute, and somewhat later ramie for its higher resistance to abrasion and decay. Yet in spite of the utmost care in its preservation—drying in the sun, spraying for disinfection—the braid would be practically worn out while the insulated conductors were still intact. The action of oil-base muds was especially harmful. Friction tape was used to patch up threadbare or torn-away sections to the extent that the braid was gradually replaced over its full length. In many cases, especially when replacements became scarce during World War II, shreds of braid would dangle between winch and well like old clothes on a line, and the cables were called "rag lines." However unsightly, the situation in no way lessened the strength and insulation of the conductors and did not affect the accuracy of the logs. What the drillers did not appreciate were the bits of friction tape carried by the mud through hoses, pumps, and valves.

Certain formations yielded gas, or oil charged with gas, which under high pressure and temperature became embedded in the natural rubber. On emergence from the hole, the expansion and discharge of the gas produced open blisters in the rubber. This happened mainly when perforations brought an immediate flow of hydrocarbons, rather than in the course of logging operations. Shortly before the war the introduction of synthetic rubbers (Duprene and neoprene) as insulating materials began to obviate this shortcoming.

For years sondes were made of a rubber-coated section of tricable, and electrodes of lead wire coils. The weights were of cast lead or brass. To ensure that the measurements reached as closely as possible to the bottom of the hole, weights were placed above the *AM* electrodes of the normal tricable sonde; the lower end of the sonde carrying the electrodes was housed in a kind of cage, whose rubber-coated bars had adequate rigidity to maintain the electrode spacing when touching the bottom. About 1936 weights and sonde were consolidated by placing the electrodes on a long, insulated steel sleeve, and the first connectors (heads) were built; instead of lengthy and inconvenient splicings and unsplicings, the cable and sonde could now be readily connected or disconnected. For depth measurements the braid was marked at regular intervals while the cable was under tension when being pulled out of the hole.

Three models of winches were operational by 1934: a rarely used lightweight unit, a medium-duty type for depths around 4500 feet, and a heavy-duty model for deeper wells. The first two were mounted on the platform of the truck. The lightweight unit was hand driven. The medium-duty type had an electric motor drive with gearshift, but could also be driven by a rear wheel of the truck. The heavy-duty winch, as in current equipment, was an integral part of the chassis. It was driven by both rear wheels of the truck, firmly propped up by two hinged jacks. This winch was so sturdy and powerful that it could rupture a 4 ton cable stuck in a borehole. The first trucks delivered had a berth behind the driver's seat; the space gained by its elimination became the measurement cab with its two recorders, batteries, and other components: rheostats, switches, and so forth. There was not much working space, but the engineers were sheltered.

Later, in another model, the recording cab was separated from that of the driver (Fig. 51). Because of the inelegant shape of the cab, not unlike the outhouses behind many farm dwellings (Fig. 52), the truck was given a rather crude nickname by the field crews. Although satisfactory in hot climates, the arrangement was completely unsuited for the harsh winters of the central United States. Whereas the engineer was reasonably protected in his cab, the winch operator's only shelter was a collection of tarpaulins flapping in the wind and wide open in the back. In Kansas, Colorado, and Wyoming, heating was provided by a blowtorch under the seat. On the other hand, the roadability of the truck had been substantially improved by moving the winch drum (i.e., the heaviest part of the equipment) to the fore of the rear axle.

In 1937 the California Division acquired a heavy-duty winch with a 12,000 foot 8 ton cable capacity. It could be driven either by the wheels of the truck or by a power takeoff. The latter device, both more flexible

Figure 51. 1938-type truck: inside of the cab, displaying the two hand recorders, the cylindrical housing of the pulsator, and the slanted stem transmitting the cable motion to the paper drums.

and more convenient, became standard on all Schlumberger trucks, first in the United States and then all over the world.

In California, which was asking for installations polished to the brass tacks, efforts were being made to correct the industrial untidiness characteristic of many oil fields and production centers. The Schlumberger people had apparently felt the same urge, for it was there in 1936 that the first truck appeared with its winch under a shiny paneled body. The resulting enclosure was used for both winch and recorder operation. In 1935 blue, the color of the French automobile racing team, was chosen as the standard Schlumberger color. The trucks, newly painted, soon became the "blue fleet"—to be seen and recognized in every corner of the world where Schlumberger operated.

Among other accessories, the early wooden Roman balance carrying the sheave was replaced by a more compact, all-metal A-frame with a dynamometer indicating the tension of the cable. The A-frame, like the Roman balance, was attached to, and positioned on, the rotary table. When, after completing

Figure 52. The truck fully equipped. The winch has been located in front of the rear axle for better load distribution.

measurements, the cable was pulled out at full speed, it sometimes happened that the winch operator forgot to slow down in time and the weight or sonde was pulled into the sheave; this usually meant that the cable broke and the sonde and weight dropped to the bottom of the well. To enable the operator to sight the emerging sonde immediately and stop it in time, an improvement was devised in 1934; the cable was passed over a sheave suspended by the traveling block some 30 feet above the rotary table. In the same year a first "logging stuffing box" mounted on the wellhead made it possible to lower the cable in wells drilled under pressure.

The Advent of photographic recorders

In 1933, when logging operations were resumed in the United States, Venezuela, and other countries, the crews were equipped with hand-operated recorders. Before the two instruments (one for the resistivity, the other for the S. P. curve) found shelter in the cab of the truck, they were, as in the early days of surface prospecting, placed side by side on their tripods next to the vehicle.

Hand recording was simple enough: as the cable was pulled out of the hole, the engineers watched the curves being drawn and took note of all the perturbing incidents (equipment deficiencies, stray currents, and other parasitic effects). However, when there were several thousand feet of open hole to be surveyed and the recording took several hours,[17] the operators were placed under heavy strain; on top of this they later had to make clean ink copies of the diagrams. On the other hand, certain companies required that all logging data remain confidential, especially for "tight" exploration wells.[18] Even though secrecy was a tenet of the Schlumberger staff's professional ethics, these companies were reluctant to see the paper strips, with their wealth of data which could often, so to speak, be read like an open book, remain in the hands of the prospectors. Hence there were several reasons warranting automatic photographic recording: aside from its operational advantages, such a system would guarantee secrecy since the client could receive the recorded log on unprocessed film. Moreover, a number of automatic measuring devices for all sorts of parameters were already on the United States (especially the California) market, and clients who considered neatness important tended to look down on the somewhat makeshift array of two hand recorders with their plates and cranks.

Studies on a photographic recorder began in Paris in late 1933. The Russians were pursuing the same project. The principle was the same on both sides: to replace the operator by a servomechanism capable of rotating the knobs of the potentiometer. The French solution was based on a photoelectric cell. Triggered by the deviations of the needle around zero, it actuated a servomotor through a system of amplifiers and vacuum tube rectifiers. After laboratory tests, however, this arrangement was found to be too slow and was abandoned even before a real field prototype had been built. The Russians, for their part, persevered. About 1936 they came up with a huge engine containing a servomechanism that was much less agile than the operator's arm, so that the system's inertia limited the recording speed to less than 1500 feet per hour.

[17] The highest recording speed was then 1500 feet per hour.
[18] "Tight" in the sense of proof against any leak of information.

Doll, meanwhile, had begun designing an altogether different version, no longer aiming at the automatic simulation of manual work, or at utilizing the potentiometer and the null method—all inherited from surface prospecting. The resistance of the *M* and *N* logging electrodes is negligible in relation to the total resistance of the circuit, which remains practically constant over the whole section to be logged. Since, on the other hand, this resistance is very high in relation to that of the ground between the two equipotential surfaces passing through *M* and *N*, it was possible to use a galvanometer serving as a voltmeter, or rather a millivoltmeter. A first variation, begun in October 1934, included a high-speed mirror oscillograph-like galvanometer (made by Siemens) and a special pulsator. It produced a film on which the light spot shifted alternately from the S.P. to the resistivity curve (Fig. 53). A prototype was tested in California in the fall of 1935, but was used only a few months and then discarded because the oscillographic loops were not sufficiently sensitive.

In another variation the circuits were the same as in hand recorders, but the potentiometers were replaced by galvanometers (Hartmann and Braun, A.O.I.P.).[19] The 1936 two-galvanometer model, designed by Maurice Lebourg, worked with an alternating-current generator and allowed for the successive recording of the normal and the S.P., followed by the lateral (Fig. 54). As the recording of the lateral required as much as 0.5 ampere, the alternating current generated parasitic electromotive forces (mutual induction and capacitance coupling between conductors) in the circuit, whose compensation required a rather delicate adjustment. The instrument was brought to California, where, in spite of its shortcomings, it quickly replaced the hand recorders. In 1937 there was a return to pulsated current, with dead segments timed in such a way as to have the measuring circuit closed after the inductive and capacitive effects had practically disappeared (see p. 53). This was the year when the quadricable and the long normal were introduced; to record the latter simultaneously with the short normal, a third galvanometer was added, and a three-collector pulsator was designed (one for the current and one for each normal). The three galvanometers constituted the elements of as many cells; the cells were identical and could be interchanged by an

[19] Association des Ouvriers en Instruments de Précision (Association of Precision Instrument Makers).

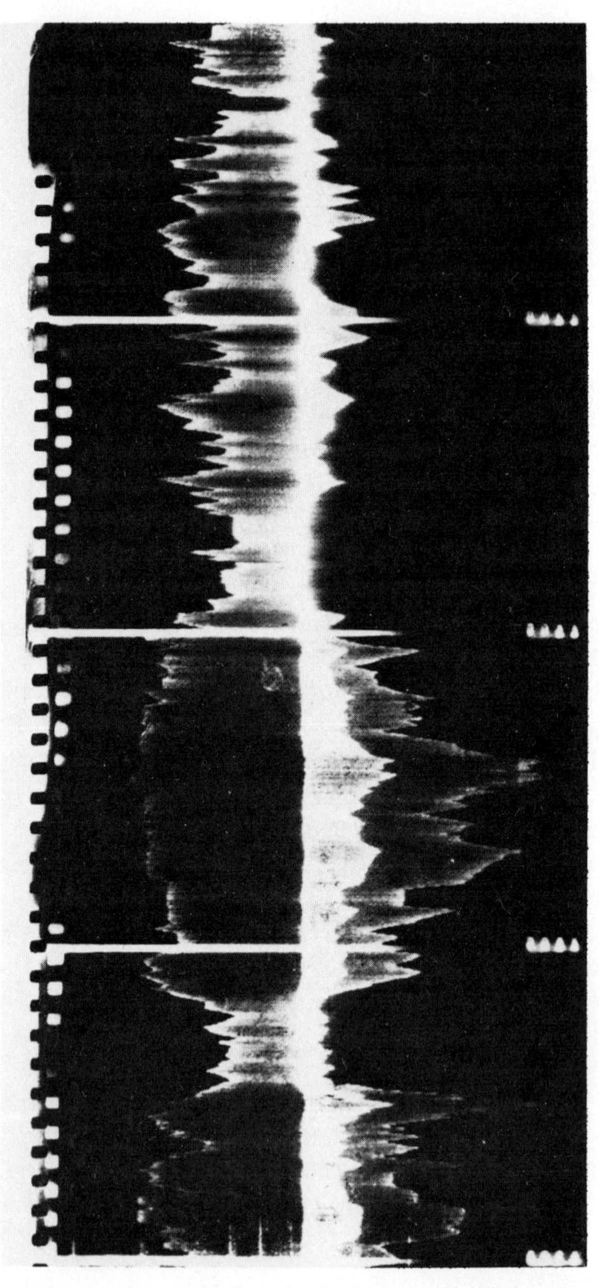

Figure 53. Facsimile of a log recorded with the high-speed mirror galvanometer. On the right-hand side, the two-scale resistivity; on the left-hand side, the S.P. (1935).

Figure 54. Two-galvanometer photographic recorder, as introduced in California in 1936–the first of its kind ever put into service.

arrangement of resistors allowing for the desired sensitivities and dampings. Each cell further included a filter (resistor, coil, capacitor) that absorbed most of the pulsated components. The three parameters were recorded side by side with the same depth scale for both halves of the film: the S.P. to the left, the two resistivities to the right. At the usual depth scales (1/500 or 1/600), the maximum recording speed, as determined by the inertia of the galvanometric cells, the characteristics of the optical circuitry, and the film emulsion, was three or even four times that of hand recording.

The requirements of the optical circuitry illustrate the difficulties encountered in the design of these recorders. With a single light source (an incandescent lamp) and through a system of prisms, screens, and lenses, the recorder had to perform the following optical functions: the light spots tracing the curves were duplicated for each track in a manner such that, when the first spot left the film, the second spot entered on the opposite side and ensured continuous recording; bold, lengthwise (vertical) lines separated the tracks, while thin, equidistant lines facilitated the reading of the values measured; crosswise, equally spaced lines marked the depths, whose 100 unit values appeared in figures; on a frosted glass viewer the operator could monitor the recording through six different symbols reproducing the motions of the galvanometers; finally, the long normal curve was recorded as a dotted line so as to be distinct from the short normal.

The truck engine made the galvanometers vibrate, producing on the films a bold, smeary line that promoted neither neatness nor accuracy. Putting the recorder on a football bladder proved to be an inadequate remedy, so a better solution was devised: the tripod carrying the table of the recorder was set on the ground under the truck, with three openings for retrievable legs cut into the floor of the cab. With this arrangement the recorder was physically isolated from the truck's vibrations, but other incidents occurred. A visitor in the cab might lean on the table and reconnect it mechanically with the truck; this resulted in distorted and vibrated logs at the corresponding depth. Sometimes, too, the tripod would be overlooked when the truck, after completion of the operation, began to move, and the consequences need no description. Another omission, more frequent and equally disastrous, was failure to lock the galvanometer for transport; the suspension wire would break, and the instrument would have to be sent to a center equipped for repair and calibration. In 1938 the study of a vibration- and shockproof galvanometer was undertaken. Initiated in Paris by Maxime Picard and continued at the cost of long effort during the war years, it was to be counted among the most significant and original achievements of Schlumberger (see p. 196).

A word must be said about the studies undertaken the same year in Paris, at Doll's initiative and with the major participation of Joseph Bricaud, for the purpose of achieving a system that could measure two resistivities and the S.P. with a single-conductor cable. The main purpose was to lighten equipment, but the problems were many. The best that might be obtained by using the electronic components then available, both at the surface and in the sonde, were resistivity variations instead of absolute values, and curves with baselines drifting with depth. Since such a degradation of logging standards was unacceptable, other solutions were sought, involving a sonde with a current source of double frequency, the signals being separated above ground by filters or resonance galvanometers. It took great ingenuity to produce components whose geometrical, mechanical, and electrical specifications were far above standard. Several prototypes were tested in Pechelbronn. After more than a year of effort, it appeared that no satisfactory result was in sight, and the project was shelved.

Temperature, well drift, and formation dip

Only minor improvements had been made on the well thermometer, which was used mainly to control cementing. The purposes of cementing are, first, to prevent oil in production from mixing with oil or water flowing from behind the casing, and, second, to seal off any communications in the annular space lest a high-pressure aquifer gradually invade a nearby oil-bearing horizon reserved for subsequent exploitation. It is therefore important to locate the top of the cement behind the casing; and since the setting of the cement releases a substantial amount of heat, this top can be determined by recording the temperature of the mud in the well a few hours after cementing.

The teleclinometer was intensively used in the U.S.S.R. (in the Baku area in 1933, 794 operations involving 11,653 measuring stations in 426 wells) and gradually introduced wherever there were logging crews. From 1940 on, however, the teleclinometer was almost completely replaced by the photoclinometer, a simpler instrument perfected by Marcel during the 2 preceding years. Centering devices brought the axis of the photoclinometer into practical coincidence with that of the borehole. It contained a steel ball rolling freely on a graduated spherical cup, and a magnet kept horizontal by a universal joint suspension. The whole was photographed at each station in a single exposure, which indicated the drift angle of the well and its azimuth with respect to the magnetic north. A series of consecutive pictures could be taken at intervals of 15 to 50 feet on a 35 millimeter film electrically controlled from the surface. The profile of the well could be easily computed from the data of each frame.

A first anisotropic—also called electromagnetic—dipmeter was sent to the U.S.S.R. in early 1934. The instrument was introduced in Venezuela at the end of the same year, and somewhat later in Rumania and the East Dutch Indies. Perfecting it for field service was a long and difficult process, requiring months of persistent effort and the know-how of the field engineers, especially Lucien Beaufort and Louis Bordat. Field tests were conducted by Doll in California (summer of 1935). A memorable episode occurred when the two dipmeters available for the tests became stuck in the wells, one after the other. This was a rather rough situation for Schlumberger's technical manager, who had come in person to test an instrument he had invented and built. As he himself stated, a staff prospector, for the same mishap, would have been blasted with a tirade of unprintable French and sustained a serious loss of face. It was Doll's good fortune that one of the dipmeters could be fished out without too much damage and repaired on the spot, and tests resumed in another well without further difficulty. The dipmeter was put into operation in California but could not compete with the equivalent and less expensive Sperry–Sun process.

Aside from the lengthy and delicate operation required, the dipmeter had other serious limitations: it could be used only where the formations were sufficiently homogeneous and anisotropic, which practically restricted it to shales; the accuracy of the measurements could be affected by the frequent occurrence of caves;[20] to compute the true dip, it was necessary to know the apparent dip from cores; finally, when measurements were made too close to compact formations, the latter's influence on the shape of the equipotentials could lead to erroneous dip values.

Another type of dipmeter was built at Doll's suggestion. This one had three small electrodes, placed in a plane perpendicular to the axis of the sonde and carried by three rubber arms at 120 degree angles, which maintained electrode contact or near contact with the wall of the borehole (Fig. 55). Three S.P. curves were recorded in sections straddling formation boundaries. In a section exhibiting dip the deflection of each S.P. curve at the level of a boundary occurred at a different depth. The amount of this S.P. shift was a function of formation dip and of geometrical elements (electrode orientation, well drift angle, and azimuth) as indicated by a photoclinometer mounted above the electrode assembly. Without any need for cores, the computation of these values gave the azimuth and

[20] A dip measurement at the depth of a certain formation consisted in computing the average of a series of readings made at intervals of a foot or so; the results of these usually did not fully agree.

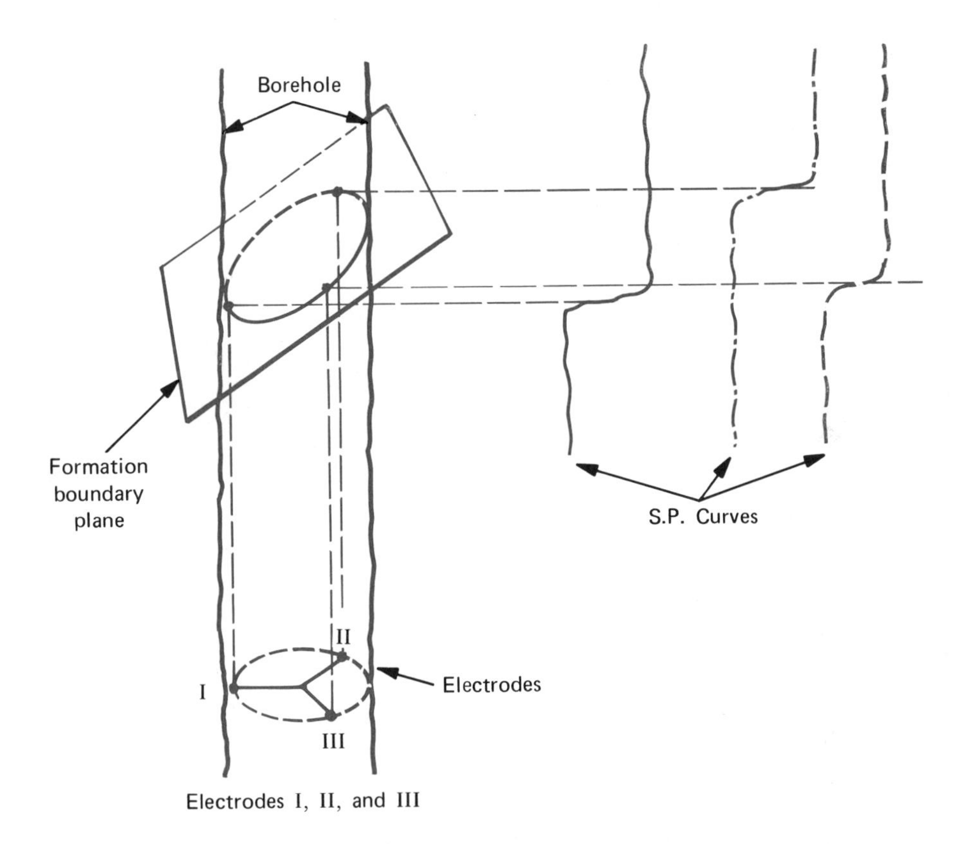

Figure 55. Schematic view of the S.P. dipmeter.

angle of the dip at each level where the boundary of two formations was shown by a sufficiently sharp S.P. deflection, that is, mainly at the top and bottom of the permeable layers.

A preliminary test, conducted in early 1936 in the Long Beach, California, field with a locally made electrode assembly, had demonstrated that the depth differences between the three S.P. curves could be evaluated with adequate accuracy. The tool was further perfected in France in 1938–1939. A. Claudet was instrumental in promoting its use in Louisiana in 1941; thereafter it was marketed throughout the United States.

The new dipmeter was not only faster, safer, and easier to operate than the electromagnetic dipmeter, but also gave more complete data. It would be followed by a series of instruments of ever-improved performance, called correlation dipmeters, based on the principle of comparison between three or four resistivity curves recorded along the wall of the hole.

Sidewall sampling and casing perforation

Logging had offered the petroleum industry subsurface cross sections of any length desired by the operators, in which three types of formations could be distinguished according to their electrical characteristics: shales, compact rocks, permeable and porous formations, and among the latter, in favorable cases, those containing hydrocarbons. However, the logs did not reveal any other rock properties such as mineralogical composition, shape and size of grains, porosity and permeability values, and types and number of fossils. Although logging had led to a reduction of mechanical coring, the latter remained, in many regions, indispensable in exploration and even production. Yet it happens that mechanical coring takes place during drilling and before logging: the decision on where to take cores is based, therefore, only on correlations with nearby wells, oil shows in the mud, or even remote geological extrapolations. None of these is a very accurate guide, and the result is that coring does not always occur exactly where it should. Hence there was a need for a tool which would take samples after logging, that is, in the walls of the boreholes. Some instruments had already been designed for this purpose,[21] but credit for the first workable tools goes to Marcel.[22]

On the other hand, well completion techniques were rapidly evolving toward running casing down to the bottom of the hole, then cementing, and perforating at the level of the productive zones. By 1932 this technique was already standard in Baku and in Rumania, and was spreading to the western hemisphere. The Russian twist-drill perforator was slow and undependable. In the United States the Lane Wells Company used for the first time in December 1932 a more convenient and efficient process: perforation by the firing of bullets through the casing. A similar device,

[21] Among others, the U.S. patent of S. A. Williston, filed in 1925, granted in 1931, and assigned to Sperry–Sun.

[22] Sidewall sampling was originally considered as having the primary purpose of refining and checking log interpretation.

though with a lower performance, was already in use in Rumania. The perforating market was, in fact, still a nascent one, and it was not too late for Schlumberger to enter the race.

Industry could only welcome good sidewall samplers and perforators as auxiliary services to logging, especially when their operating cost was reduced by the use of the same cables, winches, and trucks. Indeed, these new activities complemented the logging from which they had emanated, but their techniques and operations were quite distinct. For Marcel the special mechanical and ballistic problems entailed in designing the new tools opened a novel area of research suited to his frame of mind and afforded his creative ability full expression. A group of engineers—first among whom was Boris Schneersohn—and designers lent him valuable support during the many years when better and better models were being devised.

The first sidewall sampler (1931) was a long mandrel inside which a mechanism could swing a normally almost vertical lever around an axis. The actual coring tool was a hollow cylinder of alloy steel perpendicular to the lower end of the lever. The motion driving the coring tool into the formation and then bringing it back with a sample was not unlike the stroke of a pick. To swing the lever, various solutions were tried: an explosive charge, oil pressure, motor power. No prototype went beyond the experimental stage. In a later model (1933) the driving force was the hydraulic pressure of the mud column. A prototype successfully tested in Baku (1935) remained in the U.S.S.R.

The lever sampler was a heavy and complicated tool and could take only one core at a time. Marcel also had the idea of a sidewall sampler firing bullets. There has been little basic change in this tool since it was first completed in 1934. The core bullet is a hollow tube, projected into the wall by the deflagration of an electrically ignited powder charge. The bullet is connected with the body of the tool by two steel wires, coiled when the bullet is inside the barrel, and having a 16 inch range when uncoiled by the shot. These wires retrieve the bullet and its contents when the sampler is pulled out (Fig. 56).

The sampler was tested at Pechelbronn in April 1935, with a temporary arrangement of three barrels a few inches apart. Each barrel was connected to a conductor of the tricable and carried a 1 inch inside diameter bullet with a core length capacity of 1⅝ inches. Satisfactory

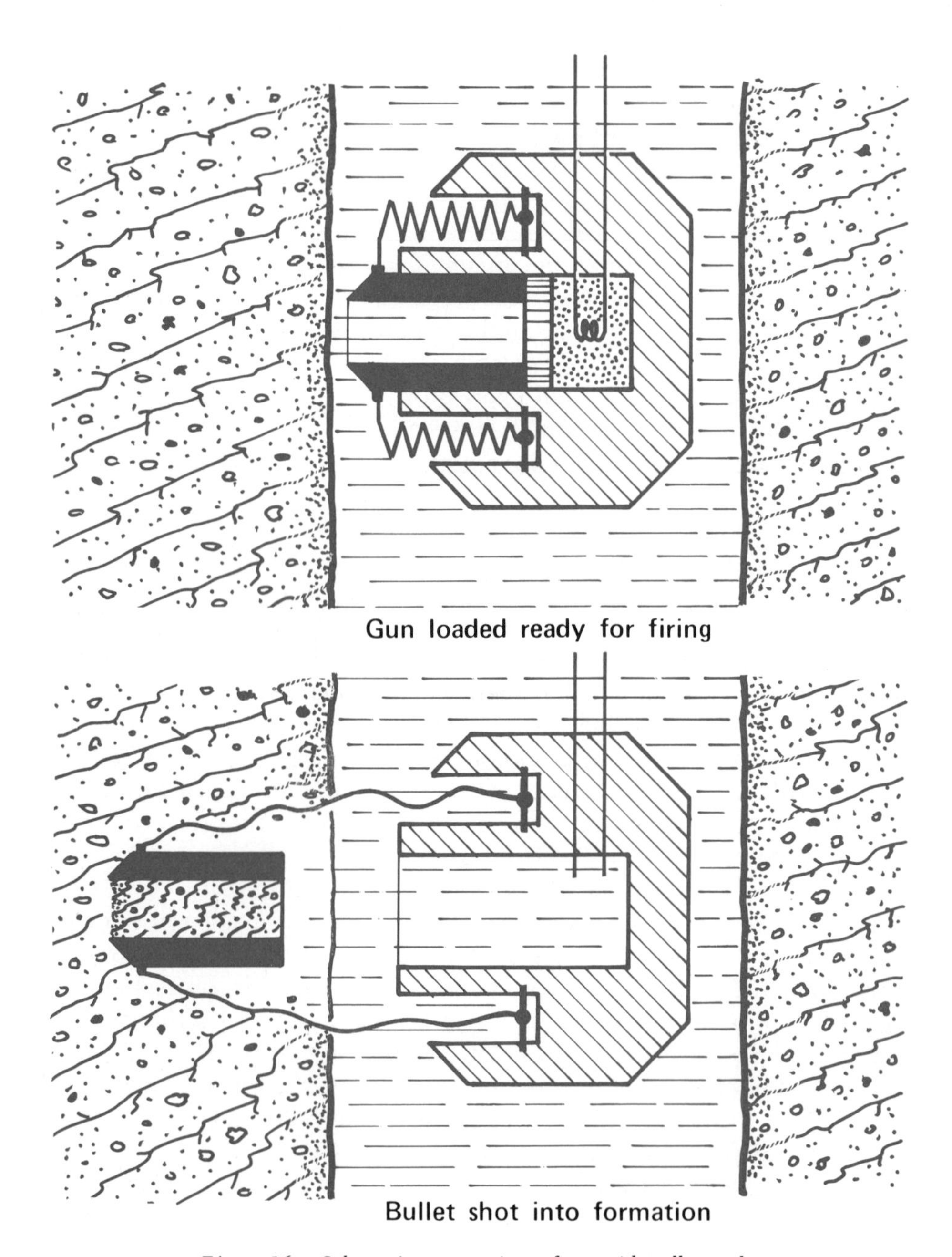

Figure 56. Schematic cross sections of gun sidewall sampler.

results led to the manufacture of a commercial tool: it was a strong mandrel, about 12 inches long, mounted at the bottom of a rigid sonde (Fig. 57) and fitted with three adjoining barrels. One conductor of the tricable crossed three series-connected igniters and ended on the body of the sampler, which served as electrode *A*. The igniters were designed in such a way that the measuring current was too weak to cause deflagration. After the electrical log—simultaneous S. P. and resistivity—was recorded, and while the sonde was still in the hole, the sampling depth was selected. The partial repetition of the S.P. log ensured that the sample would be positioned exactly where desired. To fire the three bullets separately, the igniters were calibrated so as to reach combustion temperature with increasing current intensities.

Dispatched to the United States and Trinidad in early 1936, the sidewall sampler was received by the drillers with mixed feelings; hollow tubes thrust into the wall aroused misgivings lest they become stuck and

Figure 57. The first sidewall sampler, used as a current emission electrode.

result in cable breaks entailing long and costly fishing jobs. As the first operations proceeded smoothly, these fears were promptly allayed. Operating conditions tougher and more varied than Pechelbronn's, however, led to a gradual change in certain design features. For instance, bullets were plugged by a small brass front plate which would snip off under the impact, so that the mud could not fill them and oppose their penetration into the formation. A whole set of front plates was required, and the choice—based on the powder charge, the well depth and diameter, and the nature of the formation—was a rather delicate one. Most of the time the plates snipped off too soon or not at all and were abandoned. Instead a hole was drilled at the bottom of each bullet, so that the mud could escape when the bullet penetrated the formation. Another improvement was to insert a disposable aluminum end plug in the bottom of the bullet.

Three bullets in one run were too few, especially when, on the average, one out of three was either lost or arrived at the surface empty because of faulty ignition, a very hard formation, the sample being flushed by the mud while coming out of the hole, or some other circumstance. Based on the same design, a six-shot unit followed. Finally, the 1939 model comprised three six-shot units fitted in the same body and operated by the quadricable (Fig. 58). A conductor went to the igniters of each of the three units, the circuit being closed by grounding. There were now two separate operations, each requiring a trip into the well: on the first trip the log was recorded; on the second the sidewall samples were taken at the depths indicated. The S.P. curve, recorded with one electrode connected to the fourth conductor, positioned the sampler.

All these improvements, together with the engineers' growing operating skill, made the sidewall sampler an efficient tool in the soft but cohesive formations constituting mainly the sand–shale series. The operational yield (i.e., the ratio of the number of samples retrieved to bullets loaded) reached an average of 70 percent. The data on formation characteristics gathered from such samples are not always as accurate as those from conventional cores because the former are much smaller; hence their analysis, whenever feasible, is less dependable. Such samples usually come from formations where oil and gas, if any, have been displaced by mud filtrate; or where the violent impact may alter the structure of the mineral grains and consequently the porosity and permeability of the formations. Nonetheless, because of its convenient operation and low

Figure 58. Sidewall samplers and perforators.

cost, the sidewall sampler found a wide field of application wherever sand–shale formations prevailed: California, the Texas–Louisiana Gulf Coast, Venezuela, Argentina, Rumania. The operation was often justified by the mere fact that hydrocarbon traces showed in the samples retrieved.

Since the results obtained were less favorable in friable formations, a sampler was designed that would prevent the loss of rock fragments when pulling out. A valve at the front end of the core barrel opened upon impact and closed almost immediately thereafter. Unfortunately, the tests of this so-called sand bullet proved disappointing. Most of the time it emerged empty, and the fragments retrieved were too few to characterize the formation. The yield was likewise poor in hard formations: in spite of all efforts to adapt the shape of the bullets, either they broke on impact, or, for the most part, the percussion cracked the rock and only a few worthless chips were retrieved.

In formations where retrieval is good, particularly in shales, sidewall cores can often be indicative of stratification. This, it was thought, could help in the computation of the dip, on the assumption that the actual positions of the cores *in situ,* especially their azimuths, could be determined. The idea was to complement the sidewall sampler by adding a device like the photoclinometer, but the difficulties, especially the vibrations produced by the firing, were such that the project was dropped.

In connection with the subject of sidewall sampling, Conrad's attempt in 1935 to build the so-called "sucker," a device to draw the liquids out of formations, is worthy of mention. At that time, however, what seemed to be a promising concept got no further than the drawing board. It was successfully taken up again 20 years later.

At the depth set for perforation, a well may be cased by two or three coaxial strings. The bullets were therefore expected to pierce one, two, or three steel casings—each more than 3/8 inch thick—and two or three sections of cement, while maintaining enough momentum to penetrate the formations and open drains. Furthermore, the bullet had maximum efficiency when fired perpendicularly to the casing wall, which left a bare 4 inches for the barrel and the powder chamber. To avoid excessive charges with the risk that their deflagration might deform or even rupture the powder chamber, Marcel's first idea was to have a longer barrel, parallel to the axis of the well, with a muzzle bent at a right angle. A preliminary test is said to have been made with a hot-bent shotgun barrel. At any rate a prototype was built in which the bend was lined with needle bearings to facilitate the turn of the bullet. After a few tests, this "gun to shoot around corners" (everyone, from workshop to field, made fun of it) was also abandoned. Moreover, since perforation with bullets was claimed in a patent granted to Lane Wells in the United States, it seemed preferable to look for a different concept. A hydraulically controlled punching tool, designed and worked out by Doll, was tested in Pechelbronn, the U.S.S.R., and Rumania (1935). From punch ruptures and other unlucky incidents it soon became apparent that its development would require a long period of work, and so the design studies veered toward a system with an electrically driven twist drill. Unfortunately the tool resulting from studies initiated in 1936 proved too complex for current field service.

Meanwhile, Marcel, backed by considerable experience already acquired with the sidewall sampling gun, had reverted to the idea of bullets. He designed a powder chamber surrounding a transversal barrel; in this way not only could the latter be given all available length, but also deflagration gases built up a very high pressure on the bottom of the bullet by the time it began to move (Fig. 59). The perforator based on this principle was made of three adjoining cylindrical steel blocks (Fig. 58). The barrels were screwed into chambers (four to eight) machined into each block at regular intervals (8 to 12 inches) along helicoidal lines. Each chamber received powder in bulk; the ignition system was much the same as for the sidewall sampler. Each block could fire a burst if desired: all that was necessary was to substitute an element with a discharge port between contiguous chambers in place of the usual igniters; the deflagration would then automatically spread to all the chambers after the first one had been electrically fired.

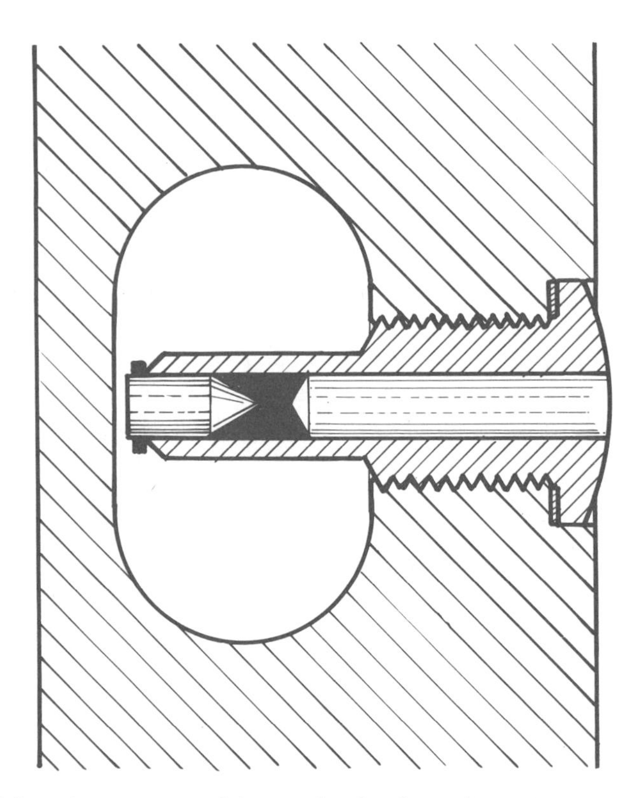

Figure 59. Schematic cross section of the annular chamber perforator barrel. The dark object in front of the bullet is a rubber plug ensuring watertightness before firing.

A first prototype was completed in a few months, and in 1936 two series, 100 and 85 millimeters, were ready. Strict safety measures had to be instituted, especially for handling on the well sites. The electrical ignition system was foolproof. A special tool used to screw the barrels into and out of the loaded chambers included a man-sized steel tube set on the muzzle of the barrel to protect the operator from burns in case of an accidental deflagration.

Thereafter, except in the United States, the Schlumberger perforators became operational wherever its logging crews were at work. Strong marketing incentives were the Company's already existing local facilities and the reputation which it had earned. Perforating services were offered in the United States (1938) after settlement of a patent suit and were successfully developed in spite of active competition.

The gun perforator underwent various improvements over the years. One of the most remarkable was the "lined bullet" whose base was inserted in a light metal sleeve, released immediately past the muzzle; the muzzle velocity and penetration power were substantially increased.

As perforation became the completion method for reservoirs of increasing thickness, the oil companies, as, for example, on Lake Maracaibo, requested the shooting of hundreds and even thousands of holes in a single oil well. In response to such a demand, it was not always sufficient to bring a battery of preloaded guns to the well; most often they had to be reloaded on the site, thus entailing many hours of continuous work. To facilitate and accelerate large perforation programs, Marcel undertook in 1937 the design of yet another model made of units assembled end to end, each comprising three radial barrels and three small-diameter chambers parallel to the axis of the instrument. With the chambers set at 120 degree angles, the barrels had the greatest length compatible with the size of the instrument (Fig. 60). The powder was contained in copper-lined cartridges; in bursting them, the gases built up the initial pressure required. Cleaning and reloading the chambers was both more convenient and speedier than for the other perforators. The assembly of the required number of units increased the number of bullets fired in a single run without any loss of penetrating power. The weak point, as demonstrated in rather extensive use of this perforator in Sumatra (1939), was the electrical connection between units. No correction could be made before 1940, and further development of the unit perforator had to be suspended during World War II.

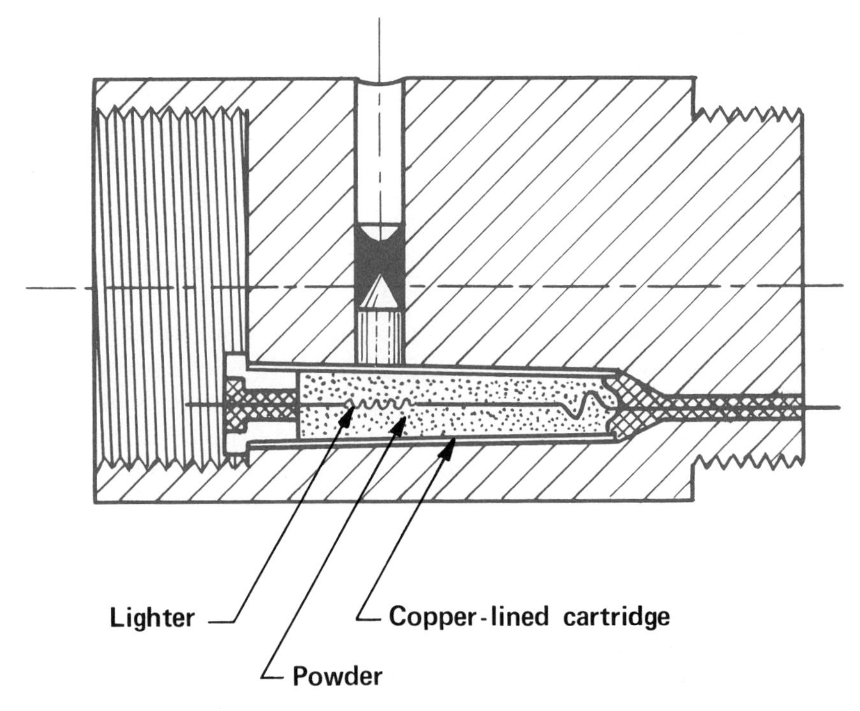

Figure 60. Cross section of an element of the so-called "unit" perforator.

Operations

Under a new contract with Shell Oil, one crew (Gilbert Deschâtre and Jean Legrand) was assigned to California in July 1932, and another (Jean Mathieu) to the Gulf Coast in early 1933. Shell Oil supplied the vehicles and drivers, and the crews were free to work for other clients. The service was paid for at a fixed rate per operation.

Initial operations got under way with a minimum of physical facilities. The prospectors' only office was at their place of residence, and space for the workshops and stores was rented from garages. In spite of such modest means, the potential market to be served included many clients scattered over an immense territory. Once the crews were able to survey several wells in the same field, they quickly demonstrated, particularly by the discovery of numerous faults, that correlations furnished by logs gave an accurate representation of the subsurface; in fact, the correlations often changed the prevailing concepts of the geologic structure. Enhanced by the successful location of several pay zones, these results made such effective publicity that after 1 year a commercial breakthrough took place, and the road was open. For greater operational flexibility Schlumberger canceled the Shell vehicle supply arrangement. From this time forward the Company became the free and sole master of its working tools, offering its services on a performance basis according to rates fixed on the actual depth reached by the sonde, the interval logged, the mileage to site, the stand-by time, and similar factors.

With four crews on the Gulf Coast, three in California, and one in Oklahoma, the volume of business had grown by 1934 to a point where an organization under American law was in order, and the Schlumberger Well Surveying Corporation (S.W.S.C.) was established under a contract of technical support with the Société de Prospection Électrique. Business was expanding rapidly (3225 operations in the United States in 1935), and seven districts were created: one in California, two in Texas, three in Louisiana, and one for the Mid-Continent, each led by a chief engineer. Legal and financial staff began to join the Houston headquarters.

There was a language problem with the American engineers, and a technical department was set up in 1934 for better liaison between the United States and France. An equipment department followed in 1936, whose main functions were to assemble the items coming from France on American truck chassis, to centralize procurement, and to store instruments and machinery. The offices and sheds of yesteryear were outmoded: land was purchased and construction begun, first in Houston and then all over the world.

From 1933 on, operations, while continuing in Trinidad, were resumed in Venezuela and included the Cumarebo (Falcon) and Quiriquire (Monagas) fields. The discovery and exploitation of new fields in Pedernales (1935), Temblador (1936), Oficina (1937), and Jusepin (1938), all in eastern Venezuela, afforded a new expansion of Schlumberger services in the country. With a crew in Ecuador (1935), and soon thereafter one in Colombia, 23 engineers were at work in this part of the world. There was also a new start in Rumania, where, under the leadership of Poirault, four or five prospectors had been assigned by the time World War II broke out. In Germany the first logging took place for Elwerath (February 1933).

Although local circumstances varied, the introduction of Schlumberger services into numerous other countries had one common aspect: in spite of agreements concluded at "the highest level," the crews were usually received by uninformed and reserved people. Two examples out of many may be cited. Under a contract with the Burmah Oil Company of London, Sauvage conducted, at the end of 1934, some 10 logging operations at Yenanguang, Burma. Yet the local geologists, confined in their routine, showed such little interest in his cross sections and structural map that it took the intervention of a young engineer just arrived from England, who fully understood the meaning of the data, to talk his colleagues into using the logs as working tools. Similarly, Gabriel Guichardot and Elie Paulin went to Argentina under contract to Shell Oil, but 2 years of effort was required to enlist the cooperation of the heavy-handed Y.P.F.,[23] in spite of satisfactory results in the Patagonian oil basin.

Nevertheless, logging was spreading from year to year: Assam (1934); Morocco (1935); Austria, Colombia, and Japan (1936); Sumatra, which in 1938 became a rather large operation with 10 engineers; Poland and Hungary (1938); Iraq and Kuwait (1939). Such expansion on an international scale could not avoid competition, and in its footsteps followed patent suits and infringement actions. Even though, in 1942, Schlumberger lost the benefit of its basic patents, such conflicts were salutary inasmuch as they demanded increased efforts to compensate for a measure of relaxation into which, at least in the United States, the Company had been lulled by a period of exceptional security and prosperity.

Thus, before World War II, Schlumberger teams were already all over the world. Each engineer, whether alone or serving as the chief of a mission (the word "mission" had not yet become a clear-cut concept), had physical, technical, legal, and commercial responsibility. The crew, the basic unit described in a brochure captioned *Operations sur le Terrain* (*Field Operations*), was in fact a kind of spontaneous outgrowth, a logical result of similar experiences in different parts of the world. Consisting of one engineer and two driver–helpers,[24] the crew was on duty around the

[23] Yacimentos Petroliferos Fiscales. The first operation in Argentina (and in the southern hemisphere) took place on November 30, 1934.

[24] According to the brochure, they were "conscientious, sober, serious-minded, punctual, dedicated, healthy, and honest workers, who could be depended upon at any time." For

clock, 7 days a week, ready to answer the client's call and drive the truck with its required equipment to the drilling site. Whatever the duration of the operation (with hardly a chance for a meal or a rest), it could well be one in a continuous sequence of equally demanding tasks. In many regions permanent readiness was expected for the servicing of far-away wells requiring expeditions of several days and employing various conveyances. Most of the time the Schlumberger engineer would thoroughly examine the diagram with a geologist or engineer present on the site, and his professional pride was great when the log pointed to a probable productive zone otherwise undetected during drilling. Back at his base he had to draw fair copies of the logs, clean and, if required, repair the equipment, and only then think of sleep. Sometimes an engineer–trainee was assigned to the crew, and the helpers were only too happy to relegate to him the most tiring and dirtiest part of the work. It was his initiation. Only much later would he deal with more refined tasks, such as cables, film, and control panels.

It was a tough but exciting job, completely unpredictable as to work schedule, yet, outside of work, unhampered by any constraint. Once in a while, of course, fatigue, isolation, and harsh climate produced clashes, but as a general rule teamwork and comradeship prevailed in the small group without impairing discipline and mutual respect. In countries where language was originally a problem, contact with assistants and drillers soon made the prospector fluent in profanity—which he would then use elsewhere in total innocence. From Venezuela to Indonesia, from Assam to Oklahoma, these hardworking crews were proud of a life so different from that of others. They could depend only on themselves, but they went along, solving their problems, and on the whole giving their best at the cost of great effort.

---

this reason, their salaries were slightly above average. "With all that is expected of servants," asked Figaro, "does your Excellency know many masters worthy of being one?"

Phase Three: 1940–1945

France

Depleted by mobilization at the beginning of World War II, the Paris center endeavored mainly to supply the missions, all of which were continuing their operations because most of the field engineers had earlier been classified as on "special duty." In June 1940 the headquarter's staff fell back on Clairac in southwestern France, but returned to Paris during the summer, leaving behind a small nucleus at Saint-Gaudens with the staff of the Compagnie Générale de Géophysique, which was undertaking electrical prospecting in southern France. As this was in the unoccupied zone, this echelon managed to maintain a few contacts with overseas centers, especially in the United States, and to assist demobilized prospectors, or those who had escaped from prisoner of war camps, in rejoining the missions.

The Paris technical staff was reduced to some 10 engineers and 30 technicians, draftsmen, and workers (compared to about 80 before the war), all disheartened by events and none too eager to go back to work. It took all the imagination and energy of Marcel to get things going again by setting up research objectives for the staff and to maintain morale as well as he could. Food and transportation shortages, air raids, police raids, poor heating, gas and electricity failures, and, later, bombing raids—none of this was conducive to creative thinking, particularly since stimulating contacts with the field had essentially vanished. Yet a few projects were advanced, especially the vibrationproof galvanometer, which, with many improvements, remains a part of today's recorders.

The principle of this instrument was to fill its housing with a liquid of such density that the Archimedes forces were equal to the weight of the coil. Since the latter then behaves like an integral part of the liquid, it does not tend to move with respect to the liquid or to the housing when the system sustains an acceleration.[1] It is, in particular, unaffected by truck vibrations; moreover, the torsion wire (or rather ribbon) is no longer under tensile stress (thus cannot break), and it becomes unnecessary to lock the coil. Since the centers of gravity of the coil and the liquid coincide, the angle of the coil with respect to the vertical is immaterial, a valuable asset whenever it is impossible to station the truck on a strictly horizontal platform.[2] Additional desirable features distinguish the instrument. It is small enough for nine units to be housed in the recorder. Its sensitivity can be accurately adjusted by a magnetic shunt as demanded by circumstances. It can sustain 100 times its design voltage and still return perfectly to zero.[3] Even informed observers marvel at the sturdiness of this galvanometer under the roughest handling.

The inventor, Maxime Picard, had to overcome tough theoretical and technical problems before completing such an outstanding instrument. While meeting the specifications set for volume, accuracy, sensitivity, and damping, he had to find a liquid that had the desired density, viscosity, and refraction index and was, furthermore, neutral with respect to the metal of the housing.[4] The equilibrium of the coil inside the liquid necessitated highly delicate adjustments. The reduction to zero of optical dispersion in a system working with white light required study of the placement of lenses between the mirror of the coil and the window of the housing, which led to endless and inconclusive calculations. The problem was solved by closing the window with a spherical lens centered on the axis of the coil. Much work was still required to bring the galvanometer to final perfection. This was not achieved before the war ended. Felix Barreteau, a young engineer recruited in Paris, made a decisive contribution.

Another long-term project was the improvement of the remote transmission system. The monocable solution, studied before the war, was

[1] Only a rotation around the axis modifies the relative coil-housing position.

[2] This advantage is even more valuable in marine operations aboard rolling ships.

[3] This is equivalent to an ordinary 110 volt light bulb being unaffected by a voltage of 11,000 volts.

[4] What proved impossible was to prevent the etching of the metal by the liquid in the long run and the ensuing formation of microscopic gas bubbles sticking to the connection of the wire and modifying the torque and the response of the instrument.

again put on the drawing board with the objective no longer merely lighter equipment, but also shorter operations by simultaneous recording of the lateral and other parameters. Marcel then reformulated the problem in broader and more flexible but ambitious terms, namely, to record four parameters with three conductors, five parameters with four conductors, and so on. Two years were spent on this effort, made more difficult because all testing had to be done in the laboratory since Pechelbronn was occupied by the Germans. The work proceeded by trial and error with great imagination and skill, but doubtful success, until the idea for a new device, the sequential (later called the chronological) system, originated in the mind of Bricaud. The idea of multiple frequencies was abandoned; the measurements were to be made according to the standard technique, but, at each depth, one after another and in a given order. The result was obtained by appropriate timing of the pulsator, allowing signals to be switched into the proper circuits. With the Houston technical services participating,[5] the project was completed in 1946. An important question to decide was the number of conductors required in the armored cable: on the basis of tests conducted in France, Bricaud again proved conclusively that a correct simultaneous recording of S.P., two normals, and one lateral required six conductors. This was a far cry from the lighter equipment advocated earlier by Marcel, but the system was operational and served for many years.

Other studies sought to improve minor technical features of perforators, winches, the photoclinometer, and the S.P. dipmeter. Another problem tackled during the first postwar months was the rational layout of the various control instruments, such as rheostats, switches, and ammeters, used in all operations (logging, perforating, sidewall sampling, etc.). Except in the United States all these instruments had, until now, been lodged in the cab according to the operators' preferences and without any fixed rule. *Le combiné* (a multipurpose wiring system) was designed to obviate such diversity: it was a kind of big trunk housing all surface electrical controls, but if the look was neat and orderly, there was a drawback to its multiplicity; the slightest malfunction entailed a total, time-consuming and laborious disassembly to locate the trouble. "Le combiné" was unable to prevail against the system soon to be adopted in

[5] Even though the state of electronic technology in the United States at that time might have supported a multiple-frequency solution as the natural one, the fact remains that modern electronics makes extensive use of chronological systems.

the United States, whereby each operation was controlled by a separate and interchangeable panel.

A project initiated before the war and subjected to rather thorough tests involved a device operating at the lower end of the cable for the purpose of injecting cement rings behind the casing and preventing selectively the circulation of fluids between formations. There were to be guns similar to those of the perforator, containing a cement slurry or any other suitable sealing liquid; once the holes were punched, the liquid would be projected by a piston driven by a gas enclosed in a high-pressure chamber. The testing station saw the end of this project.

Migaux, who, until late 1942, was the manager for administration and operations, devoted his forced leisure to a thorough study of the mechanism of electrofiltration. Although he had no laboratory at his disposal and had to rely solely on a large body of documentation, theoretical reasoning on the basis of his high competence in physics led him to a remarkable analysis of the phenomenon. It is regrettable that this work was never published.

Finally, it will be mentioned for the record that at the beginning of the winter of 1940 Marcel, to ensure a modicum of comfort to the staff, designed a heating vest with electrical resistors.[6]

The United States

After the Munich Agreement of September 1938, S.W.S.C. began planning for the worst by expanding its local supply sources and manufacturing facilities. At the same time, Paris shipped to Houston the drawings and specifications of all the equipment items being built or planned. Notwithstanding these precautions, after the 1940 armistice, with Paris completely cut off, Houston had to face the huge task of providing for the needs of 60 American crews, not to mention those in other parts of the world.[7] Difficulties increased after Pearl Harbor, when the whole of United States industry was mobilized for the war effort and submitted to

[6] A patent was even granted, captioned "Improvements to Individual Heating Systems."
[7] These had become solely dependent on Houston. The war made it necessary to consolidate them into legally independent companies, which still exist: Schlumberger Overseas for the Far East, and Surenco for Latin America except Argentina, where the Compagnia de Investigaciones Geophysicas Schlumberger was established.

Draconian regulations. The War Production Board ordered rationing and priorities for strategic raw materials: steel, nonferrous metals, and rubber, as well as for a substantial number of manufactured items such as automobiles, trucks, and tires. Although the massive slowdown in drilling, for lack of tubular goods, entailed a drop in Schlumberger's activities for a few months, petroleum production was soon given high priority, and the Administration granted the necessary steel allocations. An increased demand for logging followed; but, deprived of part of its staff and hampered by shortages, the Corporation had great difficulty in responding in spite of its reserves in cables, tires, truck chassis, and other items. None of these difficulties, however, prevented Houston—with special thanks to Robert Leger and Maurice Lebourg—from answering the call of duty. Certain key items like pulsators, photoclinometers, and optical recorders happened to be available in numbers about sufficient to supply the crews, but galvanometers had to be copied. The servicing of all this specialized equipment was done by a few technicians trained before the war; they were the nucleus of a group that, when the war ended, was able to work in close liaison with Paris.

The problem of cables was especially critical. Their composition (alloy steel and copper wire, high-grade rubber) required hard-to-get priority allocations. Repairs with friction tape became standard practice. This was no longer a question of patching a few bruises, but rather of substituting thousands of feet of textile braid. An electrical wrapping machine using wide friction tape was built to speed up and improve the repairs.

From 1943 on, Schlumberger services were granted top priority, and more liberal allocations became available. But with over 100 worn-out trucks in the United States alone, it remained extremely difficult to supply the crews until the war ended, and the amount of maintenance and repair work reached unprecedented dimensions, both in the workshops and in the field.

When Doll arrived in Houston at the end of 1940, he initiated a research and development program. His memorandum of March 21, 1941, records his study on S.P. in boreholes. Based on reasoning showing his acute understanding of the current's tridimensional distribution, the study leads to an appropriate analysis of the effect brought to bear on the shape and amplitude of the anomalies by the geometry and the resistivities of the media where S.P. currents flow. This became the nucleus of a comprehensive and rigorous treatise which Doll was to present in a long 1948 communication (see p. 229), a basic reference document for the interpretation of S.P. logs.

Another research project on S.P. proposed by Doll was pursued for a few years by André Blanchard. The principle was to submit the mud to quick and continuous impulses, thus producing periodic variations of the pressure and hence of the electrofiltration potentials. The hope was to record by this process, at the level of permeable formations, a "vibrated" S.P. characterizing them distinctly, whereas ordinary S.P. gives only vague indications, as in limestone fields. The project never got out of the laboratory.

Finally, in 1943, a device to tag formations by radioactive bullets represented the first application of electronics by Schlumberger to either surface or downhole equipment. It did not reach the field until 1946.

From 1942 on, Doll and several members of the Engineering Department devoted most of their time to work for the War Department. A nonprofiit organization, Electro-Mechanical Research (E.M.R.), was established at the initiative of André Istel to support the war effort; its stock was shared equally by Schlumberger, Doll, and two other French partners. The principal achievement of the new organization was a mine detector carried in front of a jeep. Up to May 1940, Doll had already done considerable work on this project on behalf of the French Ministry of Armaments after being released from the army on special duty for this purpose. At the end of 1942, he assembled a team[8] to resume the project, and within a few months an operational prototype had been completed. The detector proper, located far enough forward not to be affected by the metal parts of the vehicle, rested on wheels with special soft rubber-lined rims, so that the ground could be scanned smoothly. To prevent the whole thing from being blown up when driving over a mine, the wheel pressure was reduced by a lever and countersprings. The electronic circuit was designed in such a way that the presence of a mine ahead would automatically and instantaneously set the brakes of the vehicle; the same happened in the case of circuit failure, an essential safeguard for equip-

[8] The team was composed of M. Lebourg for the mechanical part and Ch. Aiken and G. K. Miller for electronics.

ment whose very vocation was to court danger. This system also compensated for the slow drifts of the signal caused by the distortion of the winding under high temperatures. Another electronic component, based on the phase discrimination between emitter and receiver, eliminated the stray signals produced by mechanical distortions of the detector or by certain features of the terrain, like magnetic soil (e.g., a granite-paved or slag-surfaced road) or seawater on a beach. About 200 detectors of this type, some tank mounted, were made by the U.S. Army.

Toward the end of the war, Doll improved his detector by the introduction of an original feedback circuit. An essential feature was a variometer tube, which separates the in-phase component of a signal almost instantaneously and completely from the out-of-phase component —up to 100 times larger—and conversely. This phase selection system became standard in the manufacture of hand detectors for antipersonnel mines. In recognition for its financial contribution to the project, Schlumberger retained full commercial rights to patents granted relevant to logging, particularly to those applying to the above detector system, which was later to become a part of induction logging.

Finally, research work was conducted by E.M.R. on the automatic guidance of missles.[9]

In January 1942 G. E. Archie of the Houston office of Shell Oil Company published an article[10] that became the basis of quantitative log interpretation, which Schlumberger had announced in 1938 (see p. 166). The article first presented the results of many laboratory measurements on sandstone samples of very diverse porosity and permeability from the Gulf Coast. From his observations the author had deduced a simple relation:

$$R_0 = FR_w$$

where R_0 is the resistivity of the sample fully impregnated with mineralized water of resistivity R_w, and F is the "formation factor." Archie had further observed that F is a function of the type and characteristics of the rock, particularly its porosity, and that this function is expressed by the formula

[9] See *New Weapons for Air Warfare,* Little, Brown & Company, Boston, Chapter XXVI.
[10] "The Electrical Resistivity Log as an Aid in Determining Some Reservoir Characteristics," *Petroleum Technology* (T.P. 1422), 1942.

$$F = \frac{1}{\phi^m}$$

in which the exponent m usually varies between 1.8 and 2.2.

In addition, the author had analyzed the results of measurements published by other researchers (see note 12, p. 166) in which the oil present in the pores intervened, and had worked out the average formula:

$$R_t = \frac{R_0}{S_w^n}$$

where R_t stands for the resistivity of the sample, S_w for its water saturation, and n for another exponent which, in most sands and sandstones without interstitial shales ("clean sandstones"), seems to be close to 2. The saturation was then given by the equation

$$S_w = \sqrt{\frac{R_0}{R_t}}$$

or, again,

$$S_w = \sqrt{\frac{FR_w}{R_t}}$$

Archie underlined the approximate nature of these formulae and cautioned against using them in cases more complex than those upon which his study was based. He also called attention to the fact that measured resistivity values frequently required corrections (effects of the borehole, the invaded zone, etc.). The article contained examples of applications to oil-bearing formations, where these corrections had been relatively unimportant.

Supported by a vast experience and the prestige of Shell, the Archie study was above suspicion of any commercial bias and aroused much greater interest than would have been accorded a similar Schlumberger publication. The proposed formulae offered a convenient way of evaluating a reservoir; aside from confirming the value of logging and its large commercial potential, they had a definite technical interest for Schlumberger. However approximate, the Archie formulae helped in the understanding of various instances of success and failure; by clarifying the problems, they suggested the developments necessary to obtain good results under conditions heretofore considered adverse.

One of the most remarkable studies made in the wake of Archie's publication was that of Tixier, then the head of Schlumberger's Rocky Mountain Division. To compute the saturation, F and R_w had to be measured on samples of rock and formation water, respectively. Not only were these measurements sometimes of doubtful significance, but also they usually required rather long and costly, and sometimes unfeasible, operations. Tixier's main idea was to substitute logs alone for these measurements, by using the resistivity R_i of the invaded zone and the deflection of the S.P. curve.

The application to the invaded zone of the Archie formula gave

$$R_i = \frac{FR_m}{S_i^2}$$

where R_m stands for mud resistivity, and S_i for water saturation (there always remains a substantial amount of oil or gas that the mud filtrate cannot displace).

When this equation was combined with that for the undisturbed zone:

$$R_t = \frac{FR_w}{S_w^2}$$

the result was

$$\frac{R_i}{R_t} = \frac{R_m}{R_w} \cdot \frac{S_w^2}{S_i^2}$$

where F has been eliminated.

As the horizons studied by Tixier in the Rockies were usually very thick, had average porosity, and hence were rather deeply invaded, he assumed that after certain simplified corrections the readings of the short normal and of the long normal (or of the lateral) gave the values of R_i and R_t, respectively. On the other hand, the many comparisons he made between the amplitude of the S.P. deflection and the values of R_m and R_w (as measured on samples) led him to conclude that the electrofiltration component was practically negligible, and that the electrochemical component of the S.P. could be expressed by

$$E = -110 \log \frac{R_m}{R_w}$$

Finally, Tixier established empirically between S_i and S_w an average relation which led to the formula for the calculation of S_w:

$$\frac{R_i}{R_t} = 1.18 \frac{R_m}{R_w} S_w$$

This was far from rigorous: in particular, as the water content of the pores diminishes gradually from the borehole to the undisturbed zone, parameters R_i and S_i are rather vaguely defined averages. However, these approximations did not prevent the Tixier method from being used in many cases to take the greatest possible advantage of the logs, and this far beyond the Rocky Mountains.

Tixier also deserves credit for having shown experimentally that a reasonably approximate value of R_w can be obtained from the S.P. An article by Humble's W. D. Mounce and W. M. Rust[11] brought a contribution to this problem based on laboratory measurements concerning the electrofiltration and electrochemical components. In the postwar years the interest engendered by all these works motivated extensive research that confirmed the predominance of the electrochemical component and gave more accurate values for the coefficient in the formula defining it.

[11] "Natural Potentials in Well Logging," *Petroleum Technology* (T.P. 1626), September 1943.

PART
THREE

After 1945

Phase One: 1945–1957

The situation and the prospects at the end of World War II

Only in the western hemisphere, and primarily in the United States, was Schlumberger able to maintain a continuous presence during the war years, in spite of serious difficulties.

The services offered were few, and electrical logging (resistivity logging and S.P.) was the primary activity. Aside from the Shell Oil Company, few companies used the quantitative interpretation methods advocated by G. E. Archie in 1942. It must be said, however, that most of the problems relating to sand or sandstone reservoirs having little shale content and drilled with comparatively freshwater mud could be solved by standard four-curve logging: one S.P., one short normal, one long normal, and one deep lateral. Only the difficulty in procuring equipment slowed development in the Gulf Coast, northern and eastern Texas, northern Louisiana, California, part of Oklahoma, the Rockies, and Illinois. On the other hand, the results in limestone and dolomites, or in wells drilled with salt-saturated mud, were practically limited to well correlations.

In 1937 the settlement of a lawsuit on logging patents resulted in Schlumberger receiving a license from Lane Wells to perforate in the United States, while Lane Wells received a license to perform electrical logging. Each company was confident that it could make serious inroads into the business of the other. However, Schlumberger had a decided advantage in that it had been perforating for several years outside the United States and had already developed the equipment to do the job. Lane Wells, on the other hand, had to start from scratch developing its own logging equipment. Consequently, Schlumberger began commercial perforating operations in 1938, 2 years before Lane Wells could start its logging services.

Taking advantage of its lead in logging, Schlumberger claimed that, although depth measurements with a cable were not always perfect in absolute value, they were still relatively consistent; and, therefore, it is preferable to lower the perforator with the same cable which was used for logging. The argument was a somewhat weak one because the conditions in a cased well, filled for perforation purposes with a generally low-density mud, were quite different from those at the time of logging (i.e., open hole and a heavier mud). However, it had considerable appeal where the productive sands were thin, 5 to 10 feet, and it led to the development of substantial activity in the Louisiana and Texas Gulf Coast. With greater well depths, these questions of depth correlation became increasingly thorny and required a new approach.

On the other hand, Lane Wells had good equipment, their armored cable was a particular advantage over Schlumberger's rag line, and many companies preferred to split the business by giving Schlumberger the logging and Lane Wells the perforating; hence perforating never developed into as large a business as had been expected.

Schlumberger's auxiliary operations included sidewall sampling, which was very successful in alternating sands and shales; temperature recording, specifically used to locate the cement top and the rarely occurring actual gas intrusions in the drilling mud—a technique improperly applied for locating oil zones in Kansas; and, finally, dip measurements, much appreciated by geologists, especially in areas where the generally quiet oil field tectonics had been disturbed by the upsurge of salt domes. Although the anisotropy dipmeter was still being used in certain parts of South America, only the correlation dipmeter was employed in the United States, for it gave not only the azimuth of the dip (as did the anisotropy dipmeter) but also its magnitude.

The truck had been somewhat modernized since its initial standardization in 1936. It was an International or Mack weighing about 12 tons gross. The winch accommodated 12,000 to 15,000 feet of four-conductor, rubber-insulated, and textile-braided cable.

The electrical equipment, designed for resistivity logging and sidewall sampling, was luckily, because of the nature of the measurements, well suited for temperature logging; but the dipmeter operation involved a few tricks and additional apparatus in the cab. Perforating required only the cable and the winch, with the firing devices connected directly to the winch collector.

In addition to the standard or heavy truck, there was a model for fields where the depth of the sedimentary rocks did not exceed 4500 feet, as in Illinois, Kansas, and eastern Oklahoma. Generally similar to the Mack or International type, though lighter and less powerful, the equipment was mounted on a Ford chassis.[1] Several of these trucks endured some of the roughest winter operations in Illinois. The terrain difficulties encountered in the muddy Louden and Salem fields were attested to by the fact that, for certain months, the trucks remained in the area and were pulled by caterpillar from well to well; as a result, the gas consumption of these vehicles with little recorded mileage was 1 gallon for each mile charged to the client. Perforating, which included on-site transportation of heavy, loaded guns, utilized a pickup truck to assist the logging truck.

Needless to say, by the time the war ended, all of the equipment was in rather sad shape. Nothing had come from France since 1939. In spite of the shortages, considerable equipment had been made in the United States because of the crucial need for logging in top-priority petroleum production. Basic items like photographic recorders, galvanometers, pulsators, and photoclinometers had to be drawn from prewar stores. The American-made truck chassis, engines, and transmissions were also giving out, because replacements were allocated only in cases of dire need.

In South America, where communication with Paris had been cut off and contact with the United States was difficult, things were even worse. Some local procurement was attempted, particularly in Argentina, and great ingenuity was shown by engineers and assistants in stretching the life of the equipment, but it was reaching the end.

While the apparatus required a thorough renewal, it was also time to take a fresh look at a number of pending problems and to sort out the wartime technological developments that could be applied to logging. Both the need for renewal and the opening of new avenues warranted fast and proper action: fast, because the clients who had shown forbearance and understanding when Schlumberger failures stemmed from wartime

[1] The chassis was too short for the winch to be installed in front of the rear axle.

shortages in men and material would no longer tolerate them; and proper, because the new course to be charted would determine the direction of the future for years to come.

At the end of the war the economic climate in the United States was highly favorable. Notwithstanding temporary or local vicissitudes, the petroleum industry was clearly off to a new start and new record performances, first in the United States and then all over the world. The time of shortages was over; all available energy resources had to be expanded to produce rapidly the means to operate the cars, refrigerators, and toasters that everybody wanted. For Schlumberger the market was open and the challenge tremendous.

New available technology and techniques invited bold approaches. In 1939 electronics was limited to radio; now it was all-embracing, from aeronautics to nuclear industry. Whereas it would have made little sense to try to accommodate the equivalent of a 1938-model superheterodyne in a battered sonde, it had become quite conceivable to incorporate in the sonde circuitry elements like those of a proximity fuse strong enough to withstand the muzzle acceleration of a shell. In 1939, when Doll conceived the induction sonde that was to revolutionize resistivity measurements, it probably was technically unfeasible; but in 1946 he was able to construct it with the components of the mine detector he had developed during the war.

Psychologically, too, the climate was stimulating. Although unprepared initially, the United States was convinced that the war had been won by its industrial might, its organization, and its techniques, but, above all, by the moral and material superiority of the American way of life. Never had there been more self-confidence, and it was already taken for granted that landing on the moon would be a mere question of money and time. At every level the same confidence permeated companies that wanted to forge ahead and, for that purpose, kept abreast of new technical and other developments.

For Schlumberger this was nothing new: technical development was its very life. But close attention was now given to the latest trends in management and to the favorable effect of the general climate on client relations. To these new directions the company owed not only its survival, but also the impetus that was to increase its activity tenfold and deeply affect petroleum exploration, drilling, and production.

The administrative reorganization

One of Schlumberger's remarkable achievements of that time was to simultaneously and successfully conduct an administrative and a technical reorganization. Since the former affects the latter, it is worth brief mention. Here again it was in the United States that the move was launched before it was carried out in the rest of the world.

The administrative reorganization of 1945 began at the top. In 1943 Pierre Schlumberger, Marcel's son, became controller of the Schlumberger Well Surveying Corporation; in 1944, executive vice-president; and, in 1946, president. One year earlier, Roger Henquet, whose attractive personality made a deep imprint on this period, had been appointed vice-president and general manager. In 1945 he returned to the United States after having received commando training in England and been parachuted into France, where he had worked before D-day with the French underground. Influenced by his prior life in California, he maintained a true Hollywood style, which, coupled with his love for snap decisions, made him a flamboyant leader of the rather heterogeneous team, which was ill at ease in peacetime reconversion.

In the field the reorganization consisted of a regrouping and standardization of assignments. Aside from the geographic divisions already mentioned, some centers or districts had developed that, although lacking divisional status, would nevertheless report directly to Houston. All of these independent units were promptly incorporated into the existing operational system. For the first time the responsibility and authority of division heads and district chiefs were clearly defined and standardized. In 1948 increased growth mandated a more efficient management structure, and a new reorganization took place that consolidated the existing divisions into regional units called areas.

In 1945 what had previously been the Houston exploitation unit was enlarged into the Field Operations Department with more specific functions. Its first leader, Henquet, was replaced in 1949 by a vice-president who, supported by a technical staff, was fully responsible for operations. The area heads reported to him, and only he issued instructions to the field crews. This streamlining was much appreciated by everyone because it greatly facilitated relations with the Houston headquarters.

The Engineering and Research Department was diverisfied in 1947, with each function operating separately. The Research Center in Ridgefield, Connecticut, was established in the same year under Doll, while engineering headquarters remained in Houston. Ridgefield was now in charge of long-range research on instrumentation as well as interpretation; its laboratories became operational in 1948. The staff engaged in engineering and research grew between 1944 and 1949 from 22 to 83; it was to reach 224 in 1955. During the same period the Equipment Department experienced similar growth in personnel: 82, 260, and 354.

With a growth rate of such magnitude, a Personnel Department was required, and the new unit was created in 1948. Under the leadership of Ame Vennema, it performed outstandingly. One of its major contributions was the publishing of a management manual in which everybody's terms of reference, responsibility, and authority (from the president to the sweeper) were properly defined. The department maintained an employee benefits policy that had existed since the inception of the Company under the direction of Conrad and Marcel. Employees' profit-sharing had begun in 1937, and in 1945 a profit-sharing trust was created in which the shares of every employee were invested. Its insurance, holiday, vacation, and pension system, as well as its profit-sharing trust, have put the Company far ahead of many American firms in terms of total employee benefits. In 1948 the Personnel Department organized a training program for engineers and, later, a series of refresher courses.

To round out the picture, mention should be made of the newly created Sales Department, a rather odd name since, in addition to its responsibilities concerning high-level client relations, price lists, and sales promotion, the department was also in charge of the technique of log interpretation. In close cooperation with the Ridgefield laboratory, it disseminated all available information on the subject as well as on the many and diverse new tools that came on the market.

Thus, between 1945 and 1950, the Schlumberger Well Surveying Corporation established in the United States a structure well suited to its volume of activities and fully equipped to handle new developments. Testimonies to its importance are the facilities in Ridgefield (1948, Fig. 61) and Houston (1953, Fig. 62), which have served as architectural and functional models for offices, laboratories, and manufacturing plants. Yet the real strength of S.W.S.C. was less obvious: it rested on the many small centers scattered over the United States (54 in 1948, 119 by 1956), which

Figure 61. Ridgefield, Connecticut, Research Center.

conducted 5000 monthly operations in 1948, 6000 in 1950, 8000 in 1951, and 10,000 in 1953.

The reorganization had multiple effects. Originally, engineers were assigned to various departments according to need rather than background and experience. As the years passed, however, the demand for greater specialization resulted in an engineer or a physicist being recruited on the basis of his suitability to a particular post. The romantic character of the Schlumberger engineer was fast becoming obsolete; no longer was he a jack-of-all-trades who talked to the client, interpreted the measurements, kept the payroll, salved the customs inspectors, and did the cooking. Efficiency gained, and eventually this "Americanization" spread, first to South America and then to the rest of the world. "Prospectors" of an earlier day were to be found only in Paris, at 42 rue Saint-Dominique; lend an ear, and these veterans of "La Pros" were quick to tell you how things were "in the good old days."

The new organization also had psychological advantages. As the field engineers became more aware of their responsibilities and prerogatives

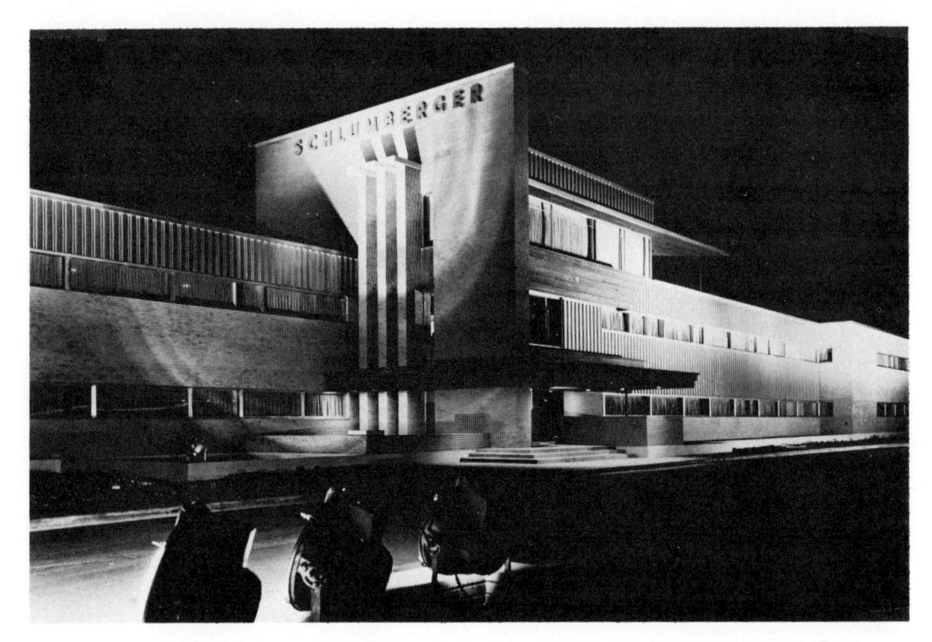

Figure 62. Headquarters of Schlumberger Wells Surveying Corporation in Houston.

and were informed monthly of the financial results of their divisions or districts, they became increasingly aggressive promoters, though never "peddling their wares" at the cost of their technical performance. In fact, their frequent contact with clients gave them greater insight into the latters' problems and needs, and this, in turn, enabled them to clarify these viewpoints to their colleagues in research or engineering. Additionally, the engineers were given new opportunities to voice their opinions, either privately or at frequently scheduled meetings. As always, the observations coming from the field were given the greatest attention and consideration by Doll and Marcel.

The reorganization of Schlumberger Companies gradually expanded to South America and then throughout the world. Under the limitations imposed by geographical separation, this task was undertaken by Jean de Menil, a man of unbending willpower and tenacity concealed beneath an exterior of tact, human respect, and personal charm. In the process of effecting necessary changes, he had to break with tradition and put an end to privileges, even if it sometimes meant resorting to sheer strength.

Thanks to his endeavors, by 1952 South America—scattered as the various units were, it nevertheless boasted a strong esprit de corps—had been reorganized in a manner similar to that of the United States.

The geopolitical realities of the eastern hemisphere hardly lent themselves to a divisional regrouping. The road from Nigeria to Gabon went through London and Paris; and although some early consolidations could be achieved in the Middle East, reorganization often had to wait for better communications or a concentration of drilling operations. Only after 1956–1957 did Guy Baboin succeed in establishing an administrative structure modeled after the one in the western hemisphere.

Another equally important development, also initiated by de Menil, gave a worldwide uniform status to field engineers: direct compensation, fringe benefits, and standardized living conditions wherever possible. There was now greater flexibility in worldwide staff transfer, which promoted a stimulating mixture of individuals and a fruitful exchange of technical skills.

The Series 500 and 700 Trucks

In this discussion of technical developments, special mention must be made of a new truck[2] that first appeared in the field in 1946. Not only was it a superb working tool fitted with many new instruments, but also the future of the Company was almost irrevocably committed by the originality of its design features.

It had become clear that standard logging could lead to many other activities. The "auxiliary" operations (perforation, sidewall sampling, temperature and dipmeter surveys) were selling well, and the divisions and districts could see from bookkeeping records that these operations were financially very attractive since they used existing equipment, personnel, and organization. Electrical logging was still considered the mainstay of the Company, but there was a growing awareness that petroleum exploration had opened new doors in measurement techniques. As early as 1945, it was apparent that the drilling of every oil well in the United States and soon elsewhere would involve conventional logging,

[2] "Truck" refers not only to the vehicle, but also all the operational equipment it carried: cable, winch, recorder, power supply, and so on.

but whatever the ensuing expansion of the market, overall drilling activity would remain the determining factor. Therefore the only prospect of expansion lay in offering a series of measurements in each well that would provide a coherent data system for all the variables of the problem. To think in terms of electrical logging only, and not consider these additional measuring systems, would deter progress. This was the basis on which the new truck was designed: it had to be suitable for a whole new range of operations other than those of the past. Although the concept focused mainly on electrical outfitting, it affected the entire project.

Thus, before the Series 500 truck had even reached the drawing board, it had a diverse group of progenitors: all the field engineers received a questionnaire seeking their opinions on a broad range of choices.[3] The final choice was the Mack 18 ton chassis, with a forward cab and tandem dual-wheel rear axles. It provided ample room for the recording cab, the winch, and its accessories. The eight driving wheels distributed the load over an area wide enough to minimize the risk of getting stuck in the mud, a constant nightmare of oil people. The profile of the body was reminiscent of the experimental California trucks in its emphasis on elegance, although its real beauty lay in the way it was perfectly designed for the work expected (Fig. 63). "Strictly functional" would be the expression used today to describe it.

The winch was an integral part of the truck. It was operated from the recording cab through advanced control systems (air brake, air throttle, air clutch), and a dashboard that, in addition to giving the usual information on the engine performance, also indicated the velocity and tension of the cable.

Film processing had earlier been limited to the equipment used by the photographers of yesteryear: a box and black cotton sleeves. Now it took place in a closed nook with a black curtain, a red lamp, a timer, and a stainless steel sink. Instead of operating blindly, there was a film of commercial standards, meeting the growing demand for a field record of high quality. The photographic recorder was new. Designed in Houston, it was the mechanical and optical heir to its predecessors. It had retained the same film drive and the same optical circuits for the production of

[3] To the question "Should the recording cab have a seat for the client?" one field engineer, probably harassed by trifling queries of his own client, replied, tongue in cheek, "Yes, ejectable."

Figure 63. Series 500 truck.

light spots, but there were now nine galvanometers instead of three, and two films instead of one; also, the film width had been increased.

Why two films? The explanation lies in the basic functions of the log: well correlation and formation analysis, for which different depth scales are desirable. For correlations a smaller scale, usually 1 : 1000 or 1 : 500 (in English-speaking countries 1 or 2 inches per 100 feet, i.e., 1 : 1200 or 1 : 600), showing a rather long section in a single look is preferable. Detailed formation study requires a larger scale, 1 : 200 (or 5 inches per 100 feet, i.e., 1:240).[4] Therefore the recording of two logs, one over the entire length of the open hole for correlations, and the other at a larger scale, limited to sections deemed to be of interest, was in order. With the hand recorder two simultaneous logs could be recorded: only two pencils, actuated by the potentiometer's crank, that traced the curves on two rolls of grid paper revolving at different speeds were needed. To obtain two

[4] A larger scale, 1 : 50, is sometimes used for certain operations (dipmeter, Microlog, Microlaterolog), as well as for certain detailed correlation studies, provided that the vertical resolution of the device allows it.

logs at different scales with the single-film optical recorder, two runs were required; this entailed loss of time and greater operational hazards, since the sonde had to be lowered again to the bottom of the hole. The difficult problem of impressing two films, unwinding at different speeds, was solved by splitting the light beam reflected by the mirror of the galvanometer, in order to obtain a spot on each film. The time saved justified this complication when the operating costs of drilling rigs rose to the point of making standby time prohibitively expensive.

Another substantial improvement was the vibrationproof Picard galvanometer (see p. 195–196). This galvanometer was compact enough for nine of them—more than necessary for the techniques of the time—to be housed side by side in the new R9G ("9 Galvanometer Recorder"). As early as 1947 this equipment gave Schlumberger a system which proved so satisfactory that, with a few modifications over the years, it has remained basically unchanged to this day. This is a remarkable achievement, considering that it was impossible at that time to foresee the range of measurements that would eventually supplement electrical logging.

The final form of the technique was determined by development work in Paris and the United States. Although the simultaneous recording of two depth scales was time saving, production of the three resistivity logs in a single run would have been a still greater asset. Two choices were open: multiple frequencies, or the sequential or chronological system, on which Bricaud was putting the final touches (see p. 197). The advantage of the former alternative was the possible use of a single-conductor cable, but it required much electronics in the sonde, a nearly damning condition. In light of today's sondes, say a dipmeter and its transistorized cartridge, such reluctance toward bottom-hole electronics may seem strange. In 1945 it probably would have been considered only as a last resort, and certainly not as a substitute for a mandrel, ringed in places by electrodes directly connected to the cable conductors. Therefore, despite the problems of a six-conductor cable, the chronological system prevailed because it rested on proven techniques. The various sequential connections were made by a new revolving switch, more complicated and sophisticated than the old pulsator, yet of the same basic, rotating segment-collector type. Designed and built in Paris in record time, its near-capacity operation was marginal, and hence it was the least dependable unit in the chronological system. It was replaced in 1957 by a cam pulsator with opening and closing circuit breakers (points), which gave a much better performance.[5]

To those who used it, the memory of the chronological system is almost as nostalgic as that of the great sailing ships or steam locomotives. Indeed, this was the final and most efficient form of electrical resistivity logging, until it disappeared[6] and was replaced by "focused" measurements—a great step forward for the theoretical and practical interpretation of results. Its decline began in 1955 and gained momentum in 1960, and by 1965 the chronological system had been abandoned, except for a few specific correlation problems mainly in shallow, geological reconnaissance coreholes. The merits of conventional resistivity logging went far beyond the success of the Schlumberger system: it opened the way to the whole body of physical parameter measurements in boreholes, that is, to *all* kinds of logging. Partially based on such measurements, a new science was born: petrophysics, an essential tool for the efficient production and effective conservation of oil deposits. Although accurate statistics are not available, it is safe to say that conventional resistivity measurements have resulted in the discovery of more oil and gas than any other method of borehole geophysics.

With regard to the previous discussion of the Series 500 truck, it can be said that, once the main choices had been made, design and technology offered many solutions. One of them, a modular arrangement of the controls, had been proposed in the United States; in France, another one was based on their unitization. The differences may have been symptomatic of national idiosyncrasies; but French and American engineers had worked hand in hand on the American solution, and it was only postwar circumstances that had prevented American engineers from sharing in the French proposal. Technically, both were valid; however, a fact not fully realized at the time was that the improvements of the French solution followed a more traditional line. They presented a system conceived for conventional resistivity measurements and possibly adaptable to other applications. On the other hand, the American version led to a system broad enough to accommodate a whole family of measurements, conven-

[5] It was machined with a higher accuracy than the camshaft of a racing-car engine.
[6] At least in the western world. Conventional resistivity measurements with lateral sondes of various radii of investigation were still in use in the U.S.S.R. in 1970.

tional resistivity being only one of them. There is hardly any doubt that the American solution was the better and the more progressive, since it added the flexibility needed for auxiliary operations. The modular concept of instrumentation probably resulted from a closer contact between engineering and field, and it certainly gave Schlumberger the opportunity to offer many new services between 1947 and 1955. The idea was simple enough: for each parameter (resistivity, radioactivity, temperature, dip, etc.) a sonde and a control panel were designed to suit the case. Data gathering and transmission utilized the procedure best adapted to each kind of measurement, without the concern for standardization which, until then, had led to the use of pulsated current for all operations. Only the cable, the recorder, and the generating unit[7] were used in all operations. With such a system, enabling a crew to perform a new service without any other adaptation was merely a matter of providing it with the appropriate sonde and control panel.[8]

All this may seem obvious in retrospect, yet it was less so at the time when a decision was required as to which direction to follow. Marcel was a lover of fine mechanics, and many of his engineers had fond memories of equipment performing at capacity limits. For this reason they did not readily accept the idea of a truck offering a surplus capacity in every respect. But times for individual feats were gone, and modern trends were calling. The spokesman for the American system was Doll, who was able to lay aside his own preference for electronic solutions in favor of dependability. He retained the Paris proposals with their indispensable items like the chronological system, the vibrationproof galvanometer, and the pulsator, but rejected any outdated instrumentation biased toward resistivity measurements. The same discrimination prevailed for all the American proposals; with reflection, explanations, persuasions, and, when needed, technical and moral authority, he succeeded in synthetizing his choices into an outstanding solution.

As previously mentioned, the chronological system required six con-

[7] A small 2 kilowatt, 110 volt, 60 cycle alternating current unit providing power to actuate the various individual measuring devices (sensors) and to mechanically control the borehole apparatus.

[8] This is still being done, but interest in the method has lessened since tool combinations became necessary to perform in a single run an increasing number of borehole operations. Several control panels must then be connected in parallel, a requirement that entails technical and human problems.

ductors and hence an armored cable. Anyhow, the textile braid had outlived its usefulness. Granted, the cable had changed over the years, from three to four conductors, from rubber insulation to hydrocarbonproof neoprene, and from cotton to rayon braid; but its large diameter required huge winches, and its coating wore off quickly. With a 4 ton tensile strength, this cable could tolerate abuse without permanent damage. The steel wire conductors took all the stress and practically never broke, so that the only electrical faults came from insulation cuts or bruises. Such leaks were detected by an ohmmeter, located by the guard ring method, and then cold-repaired with adhesive rubber tape. As the diameter of the cable was not critical, bulging patches mattered little.

Utilized in various ways by competitors and for perforating by Schlumberger, there was a cable with a load-carrying armor and a central copper conductor, but it was only a monoconductor. There are complex problems in the engineering and manufacturing of a hexacable. The strands must be calculated in such a way that the various components of the cable—conductors and internal and external armors—have compatible elongations, to prevent possible disasters: depending on the case, the conductors may break, or, if they become permanently elongated, the release of the stress of the armor may produce a loop in the conductor, piercing the insulation. But the most spectacular effect comes from an elasticity difference between armors. Under certain conditions and after excessive stress, the internal armor reverts to its original length, whereas the external one remains permanently elongated. All the armor wires then take the same shape and remain about equidistant, so that the whole thing bulges symmetrically into what in the field is called a "bird cage." This is a major accident: the wires of the external armor must be cut and tied lest they untwist, until a final workshop repair puts the cable back into shape, because, as discovered promptly by the operators in the field, a bulge interferes with the perfect winding on the winch, an absolute requirement for the proper handling of armored cable.

Much time and experience are required to fix conductor breaks and electrical leaks. Concerning leaks, the advantage of the armored over the textile-braided cable is that leaks in the former are so obvious and enormous that no measurement is possible. This is, then, a clear "all or nothing" situation, greatly preferred when flawless measurements are sought.

The cable is such a basic component that continuous engineering studies are needed to improve its specifications, quality, and performance. In accordance with a suggestion of the Paris group (F. Barreteau) in 1960, a seventh conductor was added at the center. Simple as the idea seemed, for a long time it had been deemed impossible. The seventh conductor was straight; it became the center core around which the other six conductors were helically wound, and unless adequate mechanical precautions had been taken, would have broken when overstrained. Another substantial contribution at that time was the reduction of the cable diameter from 13.2 to 11.6 millimeters without affecting its tensile strength. This made it possible to wind a longer cable, designed for the general trend to deeper drilling, on the same winch with no weight increase. Additionally, it was easier to lower the cable through a blowout preventer into a well under pressure: since the force opposing the motion of the cable is proportional to the square of its diameter, the slightest decrease in the latter makes an appreciable difference.

The Series 500 truck had been designed and equipped to handle an armored cable. Large-diameter sheaves prevented sharp bends between winch and hole. The standardized location of these sheaves on the derrick made it possible to measure[9] the cable tension accurately. In addition, by knowing the tension, the elastic elongation of the cable and hence the exact depth of the sonde could be calculated.[10]

Another accessory, a "quick-mounting" head for the connection of the sonde with the cable, played a primary role in outfitting the Series 500 truck. As long as electrical logging and an occasional auxiliary operation were performed, a semipermanent connection was adequate. Using a

[9] A conventional strain gauge, requiring an electronic measuring device, was used.

[10] This important depth problem is a delicate one when the accuracy sought is 1:10,000. Metering the number of turns of a sheave, however well calibrated, is not an adequate solution. Neither was the tangential measuring wheel, conceived and made by Marcel to give the scale of the logs by unwinding a length of film proportionate to the travel of the sonde in the hole. Therefore measurement of the depth is made with a cable that has been calibrated with a surveyor's chain every 50 meters (or 100 feet) under a known and constant tension. The exact depth is obtained by counting the number of marks passed by from the origin (usually the rotary table), and making the correction for elastic elongation. Between two marks the measurements are interpolated by using the tangential wheel. On the textile-braided cable the marks were rings of friction tape. This was not possible with the steel cable, but it was discovered that the steel of the armor could be marked magnetically either by using a coil or by mere contact with a good horseshoe magnet; the mark is then detected when passing before or through a coil.

combination of cable connectors, rubber tubes, and string bindings, a trained team required hardly an hour to substitute a thermometer for the electrical logging sonde. A perforating operation had to be speedy, however, because three or four runs were needed if there were many bullets to shoot. For example, if there were 80 holes to be perforated, four guns with 24 bullets each had to be run in and pulled out in succession. This had led, as early as 1936, to the design of a rapid coupling for the assembly of the cable with the gun or any other instrument. Once the electrical connections were made, tightness was achieved by a rubber or neoprene gasket compressed by a double-threaded collar, the "differential sleeve." The six-conductor cable not only involved more electrical contacts, but also required insulation between the metal head and the armor of the cable. The new "cable-head," designed and manufactured in Houston,[11] was a key item for versatile equipment.

The problems just described may seem trifling and unglamorous. Yet, with a technique like electrical logging, founded on basic ideas and discoveries, the development of the most remarkable inventions could be delayed by inadequate attention to practical know-how. Without doubt the technical lead enjoyed by Schlumberger owed much to the patient craftsmanship which produced instruments that endured higher temperatures, withstood higher pressures, and insured better insulation. Nobody was more aware of this than Marcel. All one had to do was to see him in the field, carefully observing every detail, every strategem, every failure. Not only did he put his experience into practice, but also he brought his convictions home to his collaborators (Fig. 64).

The outstanding performance of the Series 500 truck, as designed and realized mainly by Maurice Lebourg, Roger Legeron, Pierre Dubost, and their team, has never been questioned. When commissioned in 1947, it gave engineers the field unit they needed to match their competitors, who, until then, had had the advantage of more modern-looking, though less dependable, equipment than Schlumberger's. The truck was sturdy, well designed, and easy to operate, with a well-laid-out cab (Fig. 65). From the start its chronological system gave technically flawless logs, whereas it took several years for the competition to achieve a reasonably dependable electronic system.

[11] The first model, the seven-pin head, had seven contacts and was designed in 1947 by Maurice Lebourg. A second model, designed by Charles Senouillet in 1953, had ten pins.

Figure 64. Marcel Schlumberger in a cab on a truck.

All of the mechanical solutions introduced with the Series 500 truck have remained in use with only slight modification. A lighter truck with a shorter cable, mounted on an International chassis, was commissioned in 1948 as a Series 700 and was produced in large numbers. Today most trucks are International models with a forward-tilt cab, weighing 20 tons and carrying 25,000 feet of cable. For very heavy duty service, there is a special 25 ton model with three driving axles, one front and two rear.[12]

The same concern for dependability had inspired the design of the "Offshore Skid Unit, Type C" (O.S.U.-C), used on offshore drilling plat-

[12] Instead of dual wheels on the rear axles, this truck has single wheels with tires of very large cross section (1600 × 20), which provide much better traction on sand.

Figure 65. Cab with photographic recorder.

forms or on *Glomar*-type ships. The winch, in accordance with maritime fire regulations, is diesel driven through a hydraulic transmission. Its flexibility and operational precision are excellent. The recording cab is roomy and well fitted. The unit is suited for offshore conditions, where the operating cost of the platform is so high that no breakdown, no delay, can be tolerated (Fig. 66).

When chronological logging, which required a six-conductor cable was abandoned, Schlumberger, with the help of modern electronics, could have turned to a more conventional, more easily made, and sturdier single-conductor cable. There were good reasons, however, for retaining the hexa- or heptacable.

First, the chronological system was phased out slowly and gradually,

Figure 66. O.S.U.-C unit for offshore wells.

rather than simultaneously in all Schlumberger field operations. Second, after a somewhat difficult start, the cable had become fairly dependable, and there was no pressing reason to change. And although the six conductors (even for measurements other than conventional resistivity) were not indispensable, they had the great advantage of requiring a minimum of downhole electronic equipment in the sondes or cartridges, the bulk of complex equipment remaining above ground. Today the argument may appear to be slightly antiquated, but the question had to be considered in the light of tube electronics, where even miniaturized tubes raised serious technical problems of space, temperature, energy dissipation, and shock- and vibrationproofing. Dependability was also essential, since trouble shooting on surface equipment was much easier and speedier than pulling the sonde out and lowering it again after repairs. Another argument against placing costly electronic devices in the hole was that, if failure of a gasket provided the slightest hydraulic leak, the inside of the cartridge would be exposed to very high pressures (thousands of pounds per square inch) and totally destroyed.

It is interesting to note that recent technical developments have reawakened an interest in a cable with six or seven conductors. The trend toward recording as many parameters as possible in a single run requires increasingly higher frequencies. Yet, beyond 20 kilocycles, any frequency that is transmitted through a cable, several kilometers long, armored with magnetic steel, is severely and abruptly limited by losses. However, it is possible to successfully exceed this limit by using the appropriate combination of conductors. Thus, in the frequency modulation technique of the high-resolution dipmeter, signals of 100 to 175 kilohertz can be transmitted by using hexacable or heptacable.

The postwar progress in interpretation

The impact of Schlumberger's technical development on modernizing over a few years the measuring methods in boreholes was a logical sequel to the progress made in interpreting the results.[13]

Giant strides had been made by using electrical measurements to assess the characteristics and contents of rocks since the first, strictly qualitative results were obtained on the Gulf Coast, and more systematic studies had been completed in the U.S.S.R. Mention was made earlier of G. E. Archie's remarkable research, well documented by his long experience and numerous laboratory measurements (see p. 201). His study of 1942 established a firm basis for quantitative interpretation. Such work was in line with the thinking that prevailed at Schlumberger; its merit lay not only in its clear presentation, wherein each parameter was assigned to its proper part, but also in its adherence to the standards required for setting such petrophysical "laws." Whatever the *a posteriori* attempts to justify them theoretically, they remain inherently statistical in their nature.

A closer analysis of the interpretation process covers the background of the various contributions in an effort to facilitate an understanding of how the concepts have evolved. Two main phases must be considered.

[13] Only much later (1958), pressed by clients and seismologists, did Schlumberger develop a technique that appeared to contribute nothing to interpretation: the "sonic" logs, which measured the velocity and attenuation of acoustic waves in geological formations. In time, however, these measurements, too, became part of the interpretation methods and played an important, though initially unforeseen, role.

First, the exact values of the physical parameters must be determined, for example, the true resistivity of a formation, based on its apparent resistivity (see p. 127), and corrected wherever necessary. The second phase, based on the corrected physical parameters, aims at determining the characteristics of potentially productive formations, with emphasis on economic factors such as porosity, oil percentage, and oil mobility.

Since the first phase relies on the behavior of the tools utilized, it would normally be handled by the service company taking the measurements. The second phase, however, could be dealt with by the producer (client) on the basis of his own experience. This did not mean that the data gathered by Schlumberger's clients were ignored in the Company's research and engineering programs, but Schlumberger believed that its competence and responsibility should be limited to the correct determination of the physical parameters. At least at the top level, this concept prevailed for many years. Only the features of each instrument and the corrections to be made on each log were published. The separation between the two phases made things difficult for the field engineers, who worked closely with the oil people and could hardly stop halfway through the interpretation. A quick review of some important technical publications shows how thinking in this regard changed during this period.

In September 1945 H. Guyod, a former Schlumberger engineer, wrote a series of articles on electrical log interpretation in *Oil Weekly* (which later became *World Oil*).[14] His was the first in-depth coverage of this subject, wherein he described the various resistivity sondes: monoelectrode, normal, and lateral, giving the particulars of each instrument, and indicating when and how closely the true resistivity was represented by the readings.

These articles were widely read in the petroleum industry and came as a surprise to Schlumberger's field engineers. They firmly believed that this information, which was nothing new to them, should be available to the industry. Their increasingly close collaboration with their clients created an obligation to provide explanations so as not to seem evasive. It was no longer deemed fair to supply logs without commenting on their peculiarities, anomalies, or asymmetries. Originally, there may have been

[14] *Oil Weekly,* December 3, 10, 17, and 24, 1945.

some reason for keeping a good part of the technology and even the technique of resistivity measurements confidential; but since the competition was now fully cognizant of those alleged secrets, there was no longer any point to such reticence. What really stung the field engineers was that the disclosure had come from outside the Schlumberger group. Something had to be done, and in 1946 "departure curves"[15] were published to compute the true resistivity subject to certain conditions: the assumption that a formation was infinitely thick[16] and that the electrode spacings, the borehole diameter, the invaded-zone diameter and resistivity, and the mud resistivity were known.

Even before this publication, departure curves were widely known among the interpretation experts who were beginning to staff the oil companies, although their application, at least in the United States, was limited. They were difficult to use and often unreliable when major corrections were required. Therefore, instead of serving to compute true resistivity, the curves were used to indicate cases in which the apparent resistivity came close enough to be applied as a substitute. Obviously, it was necessary to develop systems and tools that either would measure the true resistivity directly or would provide smaller and more convenient corrections than did the departure curves.

In February 1948 Doll presented to the A.I.M.E. Congress a paper entitled "The S.P. Log: Theoretical Analysis and Principles of Interpretation."[17] This was a basic study, not so much about the nature of the S.P. phenomenon, as on its current and voltage distribution. Generated by the electromotive forces arising from the contacts between different media, these currents propagate in both the permeable formation and its shoulders, and close through the mud column. Since the S.P. curve is a continuous record of the potential of an electrode moving in the borehole, the deflection between sand and shale can measure, not all of the electromotive forces involved, but only those reflected by the ohmic drops in the mud. The flow of current is schematically represented by the

[15] Originally called "lateral sounding curves"; see p. 163.

[16] Departure curves for formations of finite thickness, based on approximate calculations, were published in 1949. This problem was more complex, since other parameters had to be considered: the resistivity of the shoulders (themselves assumed infinitely thick) and the thickness of the formation studied.

[17] See *Journal of Petroleum Technology*, September 1948.

lines drawn on Figure 67, which clearly indicates how the relative resistance value of each circuit component—permeable bed, shale, mud column—influences not only the S.P. value but the shape of the log as well. The S.P. value approximates the total electromotive forces when the resistance of the circuit section inside the mud is proportionally high compared to that in the bed and its shoulders, as is always the case in thick

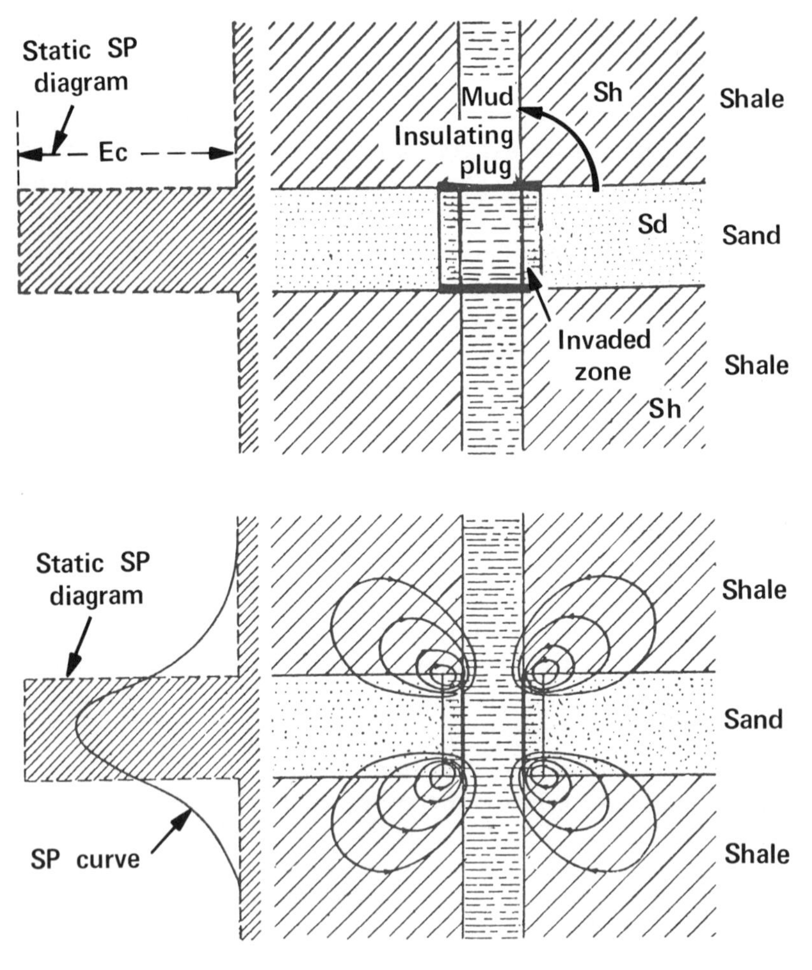

Figure 67. S.P. diagram current flow. (From H. G. Doll, "The S.P. Log: Theoretical Analysis and Principles of Interpretation," Journal of Petroleum Technology, *September 1948 (Courtesy SPE of AIME).*

formations.[18] As to shape, study of the tridimensional current flow shows that the sharpness of the deflections decreases as shoulder or formation resistivity increases.

To measure, under any circumstances, the total electromotive forces with electrodes confined to the mud column, an open-circuit operation would be required. As this is impossible, Doll defined a "static S.P." as the difference of potential that would prevail between two points, one of them either above or below the permeable bed, and the other facing this bed, if insulating plugs could be placed at its top and bottom (Fig. 67).

This "static S.P." remained an abstraction, but made it possible to conceive S.P. readings that were free of any interference from the current distribution inside the borehole and its immediate vicinity. As a result the readings could be standardized and mutually comparable. Going further into this distribution study, Doll presented a satisfactory explanation for the heretofore totally disconcerting behavior of the S.P. in high-resistivity limestone systems: it takes only one or two thin, permeable layers in a large and compact limestone series for the S.P. deflection to be stretched out and slowly peak over its whole thickness (Fig. 68). This analysis provided little hope that some accuracy could be achieved in the delineation of permeable intervals, which were revealed, at best, by slight curvature variations of the log. What the analysis did uncover was the valuable fact that, since the conventional S.P. could not be relied upon to locate these intervals, some other way had to be found. In the meantime excessive errors in interpretation would be avoided. As a matter of fact, there was a nearly overnight change in the perspective which engineers operating in predominantly limestone fields had of S.P.

Great attention had also been given to the study of shales, especially thin layers, laminations, and, at the limit, shaly sands. Although the behavior of the S.P. in such formations, long acknowledged as important potential oil sources, was less puzzling than in limestones, log interpretation was still difficult.

In July 1950 Doll published another paper, actually an offshoot of the earlier one, entitled "The S.P. Log in Shaly Sands,"[19] in which he discussed the relationship between shale content, resistivity, and the S.P.

[18] With a good approximation it can be stated that the ohmic drop in the mud represents the total electromotive forces in a sand 10 feet thick with a resistivity not exceeding 10 times that of the mud.

[19] In *Journal of Petroleum Technology,* Vol. 2, No. 7 (July 1950).

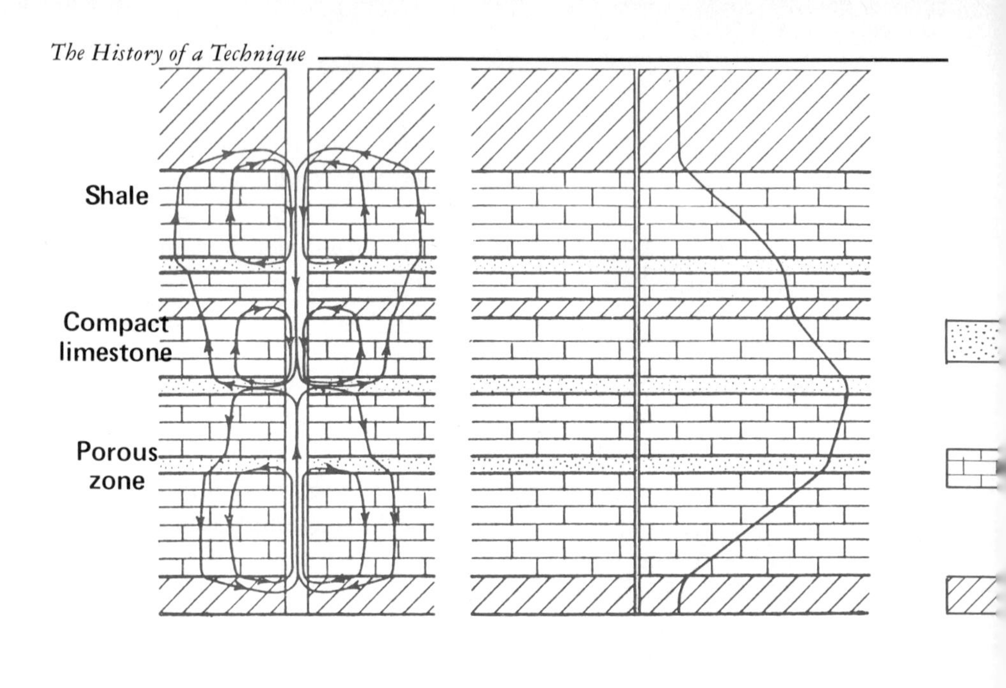

Figure 68. S.P. in limestones. (From H. G. Doll, "The S.P. Log: Theoretical Analysis and Principles of Interpretation," Journal of Petroleum Technology, *September 1948). (Courtesy SPE of AIME).*

reduction in shaly sands as compared to the "clean sand"[20] value. The importance of this study lay in its applicability to the shaly sands of many large productive basins (Texas, Louisiana, Venezuela, Nigeria). Both papers were outstanding in their presentation, clarity, and discussion of the subject matter.

At the time there was great interest in the importance of the static S.P. within the general framework of quantitative interpretation. Although Doll had precise ideas of his own on the matter, he preferred to keep them to himself, thus remaining faithful to the position that Schlumberger should concentrate its efforts on providing only the exact value of the physical parameter. As described earlier, Tixier demonstrated in 1944 that, at least in the Rockies, the electrochemical component of S.P. was dominant. From then on, it became possible to evaluate the resistivity

[20] The name given to sands without any clay content.

(R_w) of the formation water, a fundamental datum that in turn led to the calculation of R_0, the resistivity at 100 percent water saturation of the porous formation, provided its porosity is known. Any true resistivity value above R_0 indicated hydrocarbon saturation, in a proportion that could be deduced from the ratio between true resistivity and R_0, at least in a clean formation.

It was a research physicist for Gulf Oil, M. R. J. Wyllie, who gave a general formula for the electrochemical S.P. component in October 1948 in "A Quantitative Analysis of the Electrochemical Component of the S.P. Curve."[21] This study, based on theoretical considerations and experimental data, established, as a function of temperature, the relation between the static S.P. value and the logarithm of the ratio between the resistivity of the mud filtrate and the resistivity of the formation water (R_{mf}/R_w). This formula became even more valuable when experience showed that, as in the Rockies, the electrofiltration component was nearly always negligible. Here was a general method giving R_w and R_0.

Between 1949 and 1953 Doll made several major contributions to technical literature wherein he explained the principles of his newly designed instruments, which were already on the market. These studies included discussions on the quality of measurements, their expected degrees of approximation according to operating conditions, and the main corrections to be made. The tools themselves will be described further on; the purpose here is to give a chronology of the papers published to show how fast the logging industry was progressing. The titles of four of Doll's studies and the dates on which they appeared are as follows:[22]

June 1949	"Introduction to Induction Logging and Application to Logging of Wells Drilled with Oil-Base Muds"
February 1950	"The Microlog: A New Electrical Logging Method for Detailed Determination of Permeable Beds"
November 1951	"The Laterolog: A New Resistivity Logging Method with Electrodes Using an Automatic Focusing System"
January 1953	"The Microlaterolog"

[21] See *Journal of Petroleum Technology,* January 1949.
[22] The four studies listed were published in *Journal of Petroleum Technology.*

For Schlumberger field engineers, however, living close to the problems of production, it was no longer possible to restrict themselves to providing precise values in ohms or millivolts. At last, in 1949, Tixier was given the green light to publish two articles, after he had disclosed their substance months before:

"Evaluation of Permeability from Electric Log Resistivity Gradient"

and, more importantly,

"Electric Log Analysis in the Rocky Mountains"[23]

Finally, in 1954, the Ridgefield Research Center, a pillar of official thinking, published under the names of H. G. Doll and M. Martin an article entitled:

"How to Use Electric Log Data to Determine Maximum Producible Oil Index in a Formation"[24]

The title was self-explanatory; the subject matter was no longer physical parameters, but barrels of oil.

In the same year André Poupon in Ridgefield, Milton Loy in California, and Maurice Tixier in Houston jointly published an important study:

"A Contribution to Electrical Log Interpretation in Shaly Sands"[25]

Aside from the description of new instruments, Schlumberger contributions to specialized literature between 1950 and 1957 may appear modest, but the role the Company played in the dissemination of information was, on the contrary, far from unimportant. This dissemination developed in stages. Training courses for Schlumberger engineers were first organized in the United States. After 1948 confirmation to a post required prior technical tests with the emphasis on log interpretation. Although initially this indoctrination took place in the various centers,

[23] Both articles appeared in *Oil and Gas Journal,* June 16 and 23, 1949.
[24] *Oil and Gas Journal,* July 5, 1954.
[25] *Journal of Petroleum Technology*, June 1954.
Technology, June 1954.

where a senior engineer was assigned to train the beginner, the need for a more uniform approach soon became evident, and a training school was opened in Houston under the direction of André Allégret. By 1956 Surenco in South America and SPE-Overseas in Europe had their own schools. Meanwhile, in 1950, 2 week refresher courses were conducted in Houston; a third week, entirely devoted to interpretation, permitted each Schlumberger engineer to invite one of his clients. Success was such that the attendance soon outgrew the facilities, and separate courses had to be offered to clients only. Even this soon proved inadequate, and in-service training had to be organized in each division, where the Schlumberger staff, assisted by specialists from Houston, held full-fledged seminars. Such an educational approach had a wide influence.[26] It provided an opportunity for direct contact with clients, industry-wide, to discuss their specific problems, to assess the practical performance of new tools, and to evaluate the methods proposed by various authors.

It seems fitting to complete the record by mentioning the contributions of some of those who left Schlumberger, either to open their own offices as consulting engineers, to join oil companies as logging experts, or to become competitors. Whether they left on friendly terms or slammed the door behind them, they took with them all they had learned, but this did not affect their relations with their former colleagues.

Guyod was hired by a servicing company. Charrin and Castel created a company of their own. A. A. Perebinossoff became a logging expert for Socony. R. G. (Bob) Hamilton, a long-time Schlumberger leader in the Mid-Continent, became a logging consultant in Tulsa, Oklahoma. Louis Chombart, who, with Maurice Lebourg, had introduced an automatic recorder into the United States, opened an office in Wichita, Kansas, where he became an authority on carbonate rocks; among others opening their own offices were Bob Seale in Dallas, Texas, because he wanted to give more time to hunting and fishing, his true vocations, and Bob Kelso in Houston. John Walstrom, who among other activities had contributed to the writing of Doll's paper on S.P., joined Standard Oil of California as a logging expert and became a leader in the profession. While in Venezuela, Leendert de Witte, more of a physicist than a field engineer, left Schlumberger for Continental Oil Company, where he made significant

[26] It was estimated in 1958 that over 9000 participants had attended the seminars during the 3 preceding years.

contributions to the basic study of the interpretation of shaly sands. Frank Millard and Ray Braeutigam took their Schlumberger experience to Carter Oil and Sinclair, respectively. Hamilton Johnson, after 10 years with Schlumberger, first joined the competition and later turned to teaching.

There were many more who left Schlumberger for a variety of reasons. By establishing logging services within the client companies, organizing lectures and seminars, and publishing many articles, all of these men made major contributions to the interpretation and dissemination of logging interpretation techniques.

The new resistivity logs

The idea had prevailed between 1945 and 1950 that any problem of interpretation could be solved on the basis of the S.P. and resistivity measurements. Though this concept proved to be overoptimistic and even erroneous, it nevertheless made an essential contribution to the science of logging.

Before developing new tools and putting them into operation, it had been necessary to gain a clear understanding of the various rock types involved, to grasp the mechanism by which porous formations are invaded by the mud filtrate, and to speculate both on the best way to obtain the desired measurement with no or minimal corrections and on the benefit of measuring the resistivity of the invaded zone in addition to that of the virgin zone. In a short time this kind of thinking spread beyond the Schlumberger research group. The field engineers played an important role, along with the geologists and engineers of the client companies (some of whom were already specializing in log interpretation) and the universities and technical institutes. Although Doll no doubt deserves full credit for the basic theory of resistivity determination in a permeable formation, which took into account the parameters—borehole, depth of invasion, shoulders—it is also true that he sought the advice and experience of those dealing daily with interpretation. Adjusting theory to practice was basic to his design of the array of tools which was to give Schlumberger a substantial technical advantage and reaffirm its leadership in the field.

There are two kinds of difficulties in measuring the true resistivity of a porous formation, that is, the resistivity of the undisturbed zone beyond the invaded zone. In the case of fresh mud the invaded zone is much less conductive than the virgin zone and acts as a screen to conventional resistivity measurements; conversely, with salt muds the invaded zone is much more conductive than the virgin zone, and most of the current flows into the former while barely entering the latter, which thus escapes measurement. For a three-dimensional current flow these two cases represent the versions of a well-known electrical problem: how to evaluate a resistance (X) that is inseparable from a spurious resistance (Y); the solution is, when possible, to connect X and Y in parallel if X is small with respect to Y, and in series if X is large.

Induction and Laterolog, respectively, were the solutions. Realization, however, was difficult, especially for induction, with its requirement for high frequencies and electronic controls almost beyond the possibilities of that time.

Schlumberger engineers, as well as their clients, placed almost as much importance on the resistivity of the invaded zone as on that of the undisturbed zone. The reason was that the invaded zone, which has all the petrophysical characteristics of the formation, could easily be analyzed because it was saturated by water of known characteristics, namely, the mud filtrate, samples of which could be taken at the surface. Experience with the short normal, which under favorable conditions reads close to the invaded-zone resistivity, confirms the advantage of knowing this value in all circumstances.

It was realized that some oil always remains trapped, even in a zone well flushed by the filtrate. Moreover, the actual values of this residual saturation could be determined by laboratory analyses of conventional cores and sidewall samples.

Therefore analysis of the invaded zone seemed highly promising; and since the technical problems involved were much less serious than those in the virgin zone, this became the first objective of the postwar program. From this thinking and research came the microlog.

The Microlog Technique

In 1945 the R_i/R_t ratio method (R_i for the average resistivity of the invaded zone, R_t for the true resistivity) was already in current use. It had been introduced in the Rockies by Tixier, and with a few extrapolations its value was soon proved by the direct application of Archie's formulae.

In cases of a thick layer, a shallow invasion, and a rather low resistivity contrast between invaded and virgin zones, it was correctly assumed that a normal sonde of sufficiently long spacing was adequate for the measurement of R_t; as for R_i, what was needed was a sonde of spacing short enough to include only the invaded zone in the measurement. Yet a normal sonde cannot be shortened beyond a certain point without the R_i measurement being overwhelmed by the influence of the borehole mud. An elegant solution proposed by Doll was to apply the electrodes against the borehole wall by means of a rubber pad, which would also insulate them from the mud. The spacing between electrodes being substantially shorter, the measurement would extend only to a very small volume immediately against the wall and entirely within the invaded zone. As standard electrical logging equipment could be used, there was no serious electrical problem to overcome. This was not quite the case, however, for mechanical problems: many developed because of the behavior of the sonde and pad in the hole, but without critically interfering with the completion of field tests with an experimental model.

Since electrical logging equipment allowed for the simultaneous recording of several curves, three small round electrodes were inserted in the pad 1 inch apart. The objective was the recording of three curves with different depths of investigations: one normal AM_1 (1 inch), one normal AM_2 (2 inches), and one lateral AM_1M_2 (Fig. 69). Furthermore the experimental model could simultaneously record the borehole diameter through a makeshift electrical device actuated by the spring of the pad.

The first practical tests took place in May and June 1948 in Gulf Oil New Mexico wells. The results were analyzed in Houston by a group of experts under Doll, and all participating have retained a vivid memory of the event. Bricaud, who had conducted the field tests, was so elated that he called that day one of the brightest days of his life. Allowing for the favorable conditions under which the tests took place, the outstanding feature of the logs was the nearly perfect concordance among the various resistivity curves, with the exception of those zones indicated as permeable by the S.P., over which a large and constant separation between curves could be noted; the curve with the deepest investigation reading the highest. The separation and parallelism of the curves had a very characteristic aspect. This result, obvious as it may seem today, afforded some headaches at that time, all the more so because there had been an error in the presentation of the experimental logs: the measurement of

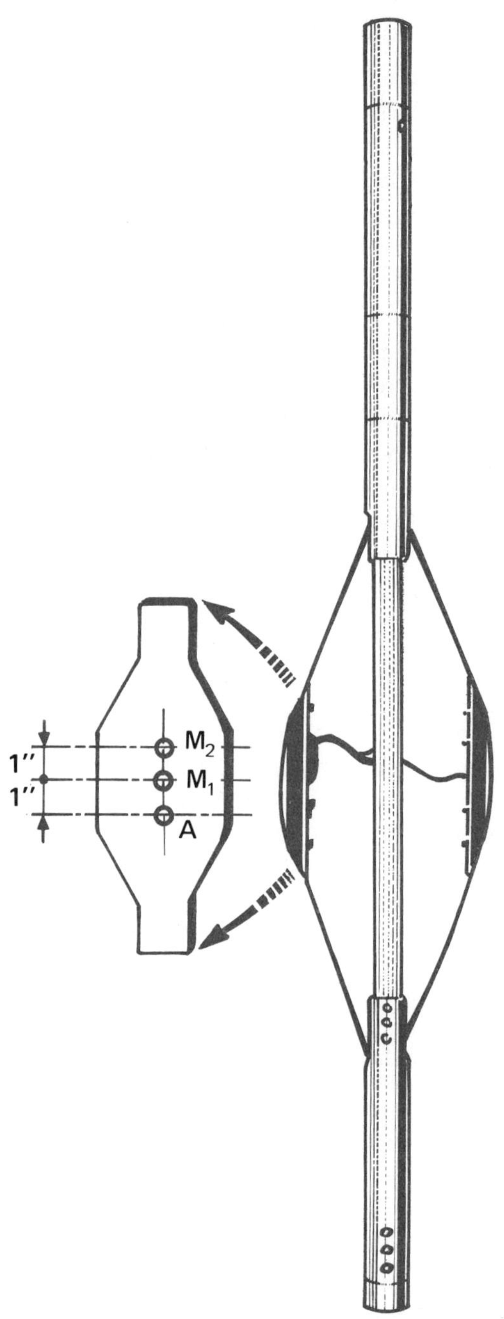

Figure 69. Microlog pad and sonde.

the diameter suggested a widening of the hole in the section with curve departures, whereas a correct reading would have indicated a narrowing. Bricaud had simply reversed the diameter scale. It remained for Doll to explain both the curve departures and the borehole diameter reduction in front of the permeable zones. The cause was the mud cake—the clogging layer built up against the wall of the permeable formation, by the solid particles suspended in the mud. The thickness of the mud cake that separated the pad from the borehole wall—0.1 to 1 inch—was by no means negligible with respect to the spacing of the electrodes, and a miniaturized two-media problem reappeared.

To be sure, the mud cake was nothing new to mud control specialists, nor should it have been to Schlumberger engineers, who daily collected some of it on sidewall samples. Yet there was surprise and perhaps even some discouragement. The two-media problem was no happy omen. Since experience with conventional logging and departure curves gave scant hope that it would ever be feasible to obtain the R_i value through adequate corrections, the question arose as to whether the idea of measuring the resistivity of the invaded zone directly should be abandoned. No one, at that time, could have anticipated what a remarkable future lay ahead for the method just devised.

Even though the desired resistivity was not always measurable, and there was little hope for a direct evaluation of the formation factor and hence the porosity, the fact remained that the new device located the permeable zones with an accuracy heretofore unknown. Indeed, the S.P. curve identifies permeable zones in many cases, and Doll's recent study had shown how to increase its accuracy. Yet the very same study pointed to a large area where the S.P. gave uncertain results. This was typically the case for consolidated rocks like sandstones and limestones, for which the new method proved fully applicable. The much too specific name "Formation Factor Logging" was changed to "Microlog," a vaguely scientific-sounding and hence commercially attractive designation.

For the detection of mud cakes and, therefore, permeable zones, the Microlog is a first-rate instrument. In consolidated formations, where the best that can be expected from the S.P. is to delineate permeable zones with an accuracy of a few feet and where its ill-advised use can lead to intolerable thickness overestimations, the Microlog locates the boundaries within an inch. Because such accuracy facilitates the understanding and interpretation of other resistivity measurements, the Microlog may be said to have played, in consolidated formations, the same role as S.P. in sand and shale sequences.

After rather unspectacular field tests on the Gulf Coast, where S.P. usually worked well, the Microlog was tried in Oklahoma with such success that Gulf Oil used it on a wildcat in Alberta Province, Canada. To everyone's surprise the new technique revealed a thick, permeable, oil-bearing section that had hitherto escaped the geologists in their examination of cores and cuttings. The incident created quite a commotion. Tixier, then operating in Wyoming and Colorado, requested and obtained the prototype; he proposed Microlog services to his clients with immediate success.

The Microlog was fully consecrated in Scurry County, western Texas, where an important discovery in 1949 was followed by feverish drilling activity. The reservoir was a rather shallow limestone reef where oil had migrated at random according to local variations of porosity and permeability. As the depth of the oil-water contact was well known, all that remained to do to reach the oil was to find permeable zones high enough in the structure. As often happened in shallow United States fields, most wells in Scurry County were drilled by independent producers, some of whom quickly ran out of funds and needed bank support for the production phase. As security for the loan, it became customary to submit the Microlog diagrams to the bankers. Even though the local banker knew little about geophysics, he understood enough to recognize the positive separation of the curves and correctly evaluate the thickness of the permeable beds. From that point on, in the way of bankers, he computed the size of the loan. The petroleum world was greatly impressed by such trust in the new method on the part of Texas bankers, who were not known to rush headlong into rash ventures.

At any rate, the story shows the ease with which the Microlog can be interpreted qualitatively—an ease that facilitated its success. and it is worth noting that, in association with other measurements, the Microlog continues to operate almost in its original form. Only two curves are now recorded, the deep (micronormal) and the shallow (microlateral), with the same spacing as initially: 1 inch between electrodes. The presentation has also remained the same over the years, as its users have opposed any change. One marvels at such lasting success: perhaps there were some lucky decisions, but what passes as luck in such cases is actually a reward of full technical background. What this success story mainly conveys is that by 1948 complete mastery had been achieved in conventional resistivity measurement systems.

Of course, this was only a first step, as no one was content to consider the Microlog merely as a mud cake detector, however valuable that function might be. The hope still was that, with the help of two measurements having different radii of investigation and appropriate departure curves, it would be possible to evaluate the resistivity, now called R_{xo} of the fully flushed zone. This approach, although applicable in some specific conditions, did not lead to a general solution for R_{xo}. The corrections were so large that the results could not be accurate and did not justify the hard work required. A technique that would give a good approximation of R_{xo} by direct reading was needed. The striking similarity with conventional resistivity logging suggested to Doll that, since the Microlog had proved the value of sidewall pad measurements, the principle of focusing, lately developed for the "Laterolog" (see p. 257), would also be applicable to a pad device. This was to become the "Microlaterolog." However, it required a downhole electronic technology that could not be realized before 1952. Meanwhile, the Microlog was improved over the years, the main effort bearing on the mechanical system of this forerunner of the increasingly numerous and important pad-mounted tools.

The year 1952 saw the first hydraulic pads, invented by André Blanchard, who, from his years with Michelin, had retained a fondness for rubber. To allow for better contact with the borehole wall, he had designed an oil-filled rubber pad. The system proved to be of interest inasmuch as the absolute measurements remained substantially unchanged, while the mechanical behavior was better, and the results were more consistent.

The next step was the improvement, in 1954, of the sonde itself. Until then it had been a rather crude tool made of a mandrel centralized by two opposite spring blades, one of which carried the pad (Fig. 69). The system was simple but had serious drawbacks. Because the pad rubbed against the wall on the way down when no recording was being taken, as well as on the way up, it wore out rapidly. In deep and difficult wells a damaged pad could mean a lost job. The first solution was completely makeshift; the springs were tied against the sonde body with a piece of rope, expected to be worn out by friction and to break by the time the tool reached the bottom. Then (it was hoped) the two springs would be released, the pad pressed against the wall, and the log recorded on the trip up the hole. Satisfactory as this remedy may have been to engineers who recalled the pioneering days of "La Pros," it was plain tinkering. The Engineering Department produced something better, the so-called "warhead": the sonde was locked in closed position when being lowered downhole, and the springs were released before logging by electrically firing a small explosive cartridge from the surface.

Even this was hardly satisfactory. It did not eliminate unnecessary pad wear when the sonde was pulled out after logging, or when it had to

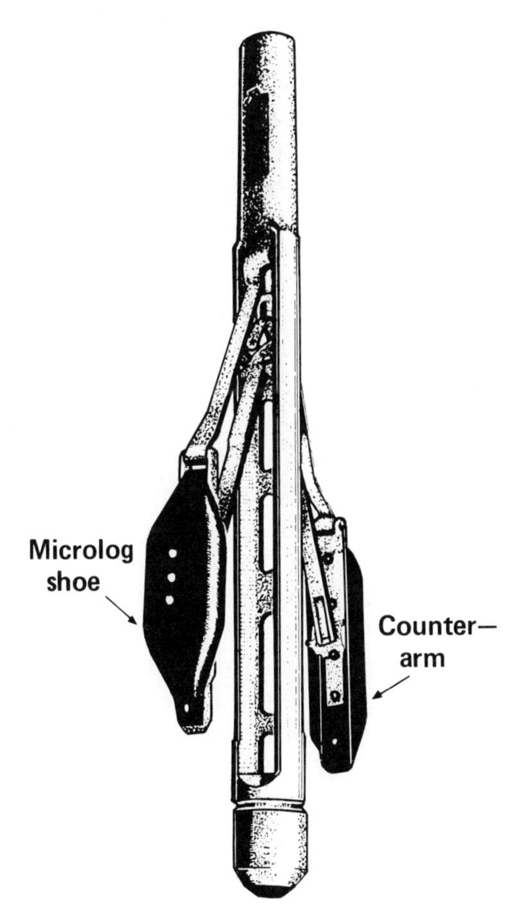

Figure 70. Motorized microlog sonde.

be lowered again to check or repeat a log. There was another very serious shortcoming. It took only a few cuttings caught between the sonde body and the springs to block their release or, alternatively, their closing, and on the way up, the system could become wedged in a keyseat or restriction of the borehole.

The true solution was to replace the springs by a remote-controlled, bidirectional, articulated arm. The downhole environment, however, caused complications. The problem was solved in 1954 by M. Lebourg and R. Q. Fields (Fig. 70) in Houston. Through a hydraulic system a small electric pump controlled the opening and closing of the tool, while an appropriate kinematic configuration ensured that the pad remained parallel to the borehole wall. This was the first in a series of "power sondes." Its cost was about eight times that of the conventional sonde, a feature that the financially responsible field engineers did not appreciate. But it did become the starting point of a technology that, unobtrusively, was to give Schlumberger an unquestionable superiority and a lead difficult to overtake. Over the years many and diverse tools made their appearance that called for tight contact between their specialized components and the borehole wall. All of them proved to be outstanding in performance and dependability.

Induction Logging

In a first stage, general considerations on resistivity measurements had led to the development of the Microlog. The same considerations showed also the need to know accurately the resistivity of the undisturbed zone. It might even have been logical to tackle the latter problem first; yet the existing system, however imperfect, gave substantial results, and improved measurement of true resistivities beyond the invaded zone posed far tougher practical problems than those encountered in developing the Microlog, which, in fact, was nothing but an extension of the resistivity log.

As long as the invasion remained moderate, the conventional measurements—short and long normal, lateral—gave satisfactory results, as on the Gulf Coast and in Venezuela, Trinidad, and California. The objectives there were highly porous sands drilled with fresh mud, where the discrimination of salt water and oil beyond the invaded zone was almost always possible with the long normal. Systematic difficulties occurred when drilling became deeper in these same regions, and also when a more thorough interpretation was attempted where the more consolidated sands turned into sandstone (northern Texas and Louisiana, southern Oklahoma, etc.). In both cases the depth of invasion increased as porosity decreased.[27] Longer spacings would have reduced the vertical resolution, and no great improvement could be expected from increasingly rigorous corrections: even in thick beds the laborious use of departure curves brought dubious results. The truth was that in such difficult cases, when it was impossible to reach a correct fluid analysis, Schlumberger was hardly better off than its competitors: without fluid analysis the only remaining use of logs was well-to-well correlation, for which competitor logs were equally good. Under such conditions, technical superiority was not a decisive commercial advantage.

The overriding problem was undoubtedly the fact that, according to the salinity of the mud, the invaded zone could be either more or less resistive than the undisturbed zone; hence any resistivity contrast could be encountered. Theory, confirmed by experience, shows that one and the same measuring system is not suitable in both cases, and the question may be asked why the system requiring what seems to the layman the most difficult technical solution was developed first. There were several reasons for this. Not least was the "challenge" of the oil-base muds (see p. 161), introduced in the 1930's and appearing for a while to have great potential for expansion. As electrodes insulated by oil cannot conduct current, the technique precluded conventional logging—an aspect that would have been a heavy blow for Schlumberger. As a matter of fact, only cost and operational difficulties[28] prevented these muds from spreading.

To make a resistivity log it was necessary to replace the electrical contact with the mud by induction, which, by virtue of the time-varying electromagnetic fields, would permit the circulation of currents in the

[27] This is only an apparent paradox, as filtration is controlled by the mud cake; whatever the formation, the amount per unit of time of the liquid filtering through is constant. Consequently, the lower the porosity, the greater is the volume required to absorb a given amount of filtrate.

[28] Anyone who has worked on a well drilled with oil-base mud appreciates the "tidiness" of operating with water-base mud. There is another drilling method that normally excludes conventional logging: percussion or cable-tool drilling of empty holes, but scant attention has been paid to this system, which was in declining use even in the 1930's, whereas rotary drilling with oil-base mud threatened to spread widely.

formations and their measurement. At first it was only a dream—a Jules Verne fantasy, according to Doll, because the technology required for such a system was simply nonexistent. This did not prevent the devotion of a great deal of thought to the matter. It soon became apparent that induction logging, mandatory for oil-base mud drilling, might also be the way to skirt the "electrical barrier" of a resistive invaded zone in water-base muds. This is one of the basic features of induction logging, as stressed by Doll in his 1949 paper (see p. 233). In conventional logging the current lines cross boundaries between regions exhibiting different resistivities; hence it is impossible to assess individually the influence of each portion in the volume measured because a change in resistivity affects the current lines not only where it occurs, but also over their whole path. On the contrary, the currents generated by induction being circular and coaxial with the sonde, whose axis, in turn, is that of the borehole, the cylindrical symmetry of the formation about the hole results in each current line remaining within a medium of constant resistivity, at least in horizontal or subhorizontal beds. Furthermore; as long as frequency and conductivity are not too high, these various elementary current loops have only a negligible mutual action, and the effect proper of each region may be considered separately, the total signal being merely the sum of the individual signals. Once the share of each portion of the volume involved in the measurement—borehole, invaded zone, transition zone, undisturbed zone, shoulders—is exactly known, correcting the readings to obtain the true resistivity becomes much easier than with the departure curves used in conventional logging.

Still better, this theoretical simplification made it possible to envisage more refined systems than the two coils. It became feasible to foresee and calculate the characteristics of a system of coils increasing or decreasing the proportion of the signal emitted from a given region: the system is called "focused," and it can be focused vertically or radially in space. To a certain extent, the lateral depth of investigation of the instrument may be increased while maintaining a good vertical resolution; in other words it became possible to read "in depth" inside a rather thin layer. At least in theory, these totally new possibilities gave induction logging a vast superiority over any alternative hitherto known, particularly in boreholes drilled with low-salinity mud. The physical properties mentioned above and the ensuing mathematical simplifications led to an early development of the theoretical basis of induction logging by Francis Perrin[29] and André Blanchard, followed in 1948 by André Poupon.

During World War II, as noted earlier, Doll applied the technique of induction to a jeep-mounted mine detector (see p. 200). Heartened by this success and well informed concerning the possibilities of modern electronics, he started to develop the induction sonde (Fig. 71). The transmitter coil, usually fed with a 20 kilohertz current, produces a variable magnetic field generating eddy currents, which follow circular coaxial paths in the formations surrounding the borehole. These currents create a secondary magnetic field, which, in turn, induces an alternating voltage in the receiver coil. The intensity of the eddy currents and of the signal induced in this coil increases when the formation resistivity decreases. Contrary to conventional logging, which records resistivity, induction logging measures its reciprocal, that is, conductivity. The theoretical unit is the siemens-m/m^2, the reciprocal of the ohm-m^2/m (called the mho by electrical engineers as a reminder of the link between resistance and conductance units). To avoid decimals in the majority of cases where formation resistivity exceeds 1 ohm, it was decided to use the one-thousandth submultiple, or millimho.

Although only a limited knowledge of Faraday's laws is needed to understand the mechanism of induction, it may be useful to point out some difficulties. The only voltage of interest is that induced by the eddy currents generated in the ground, yet the voltage resulting from the direct coupling between transmitter and receiver coils can be 1000 times stronger. The approach then is to generate a voltage equal to that of the direct coupling and subtract it from the total signal. However, this subtraction still leaves a residual error signal, caused in particular by the thermal drift. To sort out the useful signal one must use the fact that the voltage induced by eddy currents lags 90 degrees behind the one induced by direct coupling. Thus, since the system requires great stability under thermal and mechanical stresses, the simplest induction sonde is dependent on refined electronics.

Because of this complexity Doll had to win a psychological battle before designing his first version of an induction log. Except for his close

[29] To F. Perrin goes the credit for the concept of the "geometrical factor," permitting calculation of the response of the sondes and precise analysis of the induction logs.

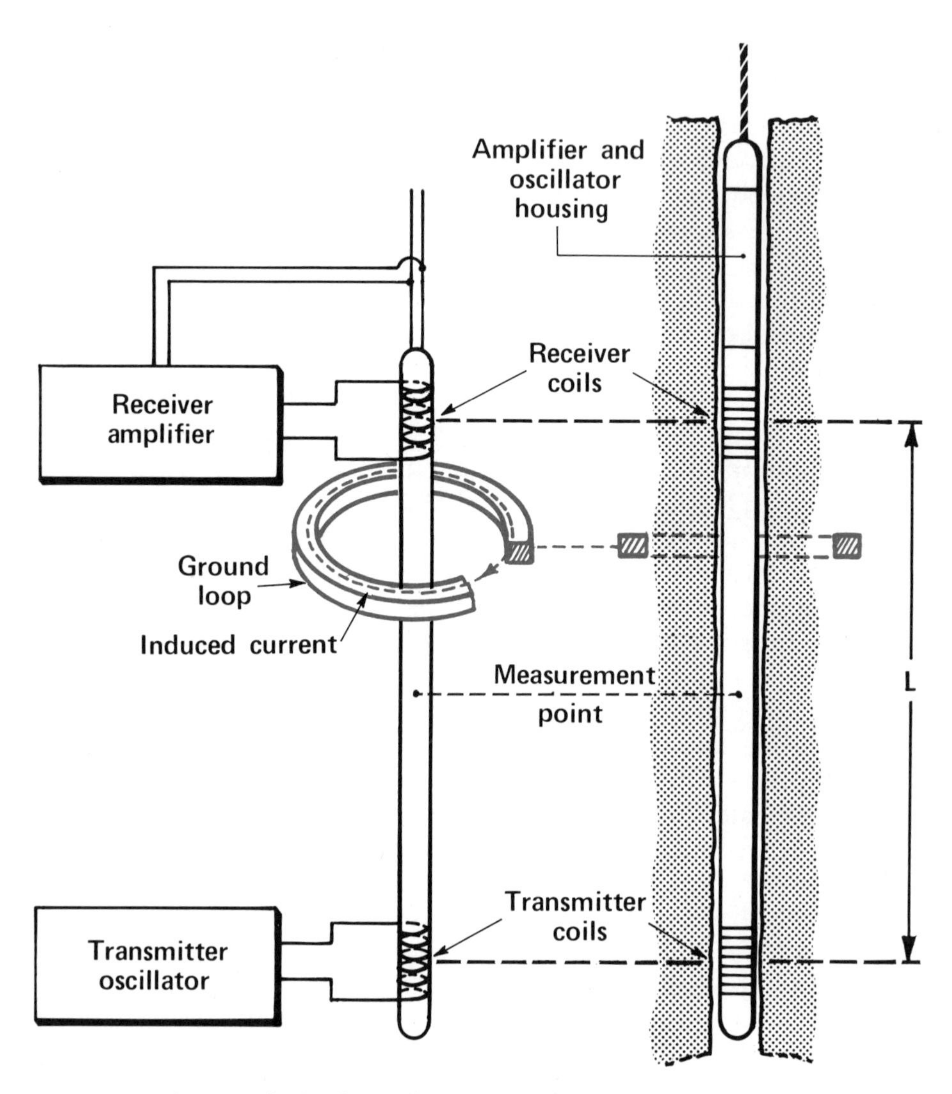

Figure 71. The principle of induction logging. (From H. G. Doll, "Introduction to Induction Logging and Application to Logging of Wells Drilled with Oil-Base Mud," Journal of Petroleum Technology, *June 1949). (Courtesy SPE of AIME).*

co-workers,[30] no one was on his side. On the minds of management, field, and research staffs alike was the considerable technical and financial effort just invested in the new truck, the new chronological system, and the new recorder; they wanted a breathing spell before embarking on other and more costly technical ventures. What must also be mentioned is that the advantages of induction were not quite as obvious as they became later, and the field engineers were ill prepared for its electronic complexity. Many thought it foolhardy to substitute a complicated, expensive, and without doubt unpredictable apparatus for the simplicity of three electrodes mounted on a mandrel. The saving grace was the wells drilled with oil-base mud, where, at least as an auxiliary, induction found its rationale. They constituted, not a large market, but one substantial enough to be taken seriously. What probably tipped the balance was the fact that Henquet, the new assistant general manager, had worked for a long time in California and was aware of the serious oil-base problem there. In 1945, objections and doubts notwithstanding, the prototype was put on the drawing board.

The first induction log was recorded on May 3, 1946, by O. H. Huston, assisted by R. T. Wade, C. Bailey, and R. Theis, in the Humble Faulk No. 7 well in the Hawkins field near Tyler, Texas. As was true of the logs of September 5, 1927, at Pechelbronn and March 6, 1929, at Cabimas, there was something almost touching in this unassuming little curve. Drilled with oil-base mud, the well did not allow for corroboration of the results by a conventional log; the only possible comparison could have been with a poor log taken with scratchers (see p. 161). The induction (Fig. 72) indicated the presence of an oil zone toward the bottom of the hole, yet even today one finds it difficult to believe that this log was the first in a series that was to revolutionize the industry. Considering that in normal muds conventional logging in the area gave results of high quality, one can only pay tribute to Doll's optimism and faith and his justified conviction that induction would win. Although marketed in some centers from 1948 on, the log was hardly utilized except in oil-base muds until 1952. In fact Doll's paper of June 1949 was specifically captioned "Introduction to Induction Logging and Application to Logging of Wells Drilled with Oil-Base Muds" (p. 233), but it again recorded and em-

[30] Especially O. H. Huston and G. K. Miller, who took a prominent part in the design and realization of the original equipment.

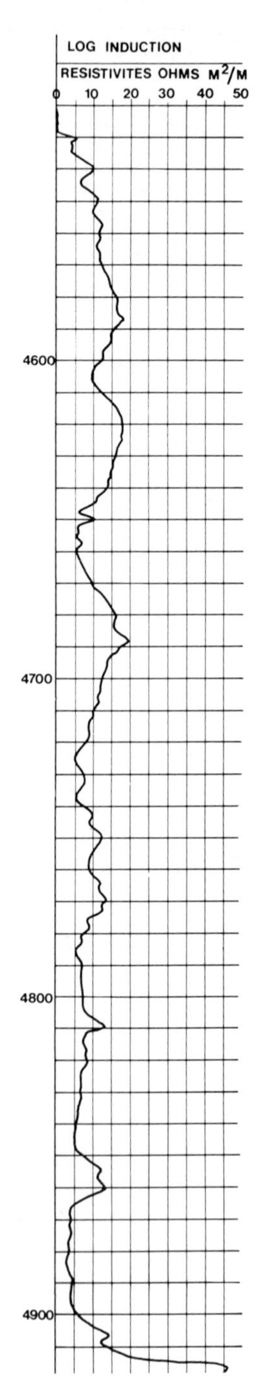

Figure 72. The first induction log.

phasized his conviction that the system would be equally valuable in water-base muds.

It is worth noting in the same paper the progress achieved since 1946. A sample log is compared with a good conventional log of the same well: considering the scale of the induction log, the correspondence is excellent. Conductivity, represented on a linear scale, increases from right to left, while resistivity is by convention always shown as increasing from left to right. The two curves thus move in the same direction, although in terms of resistivity the scale of the induction log is reciprocal and the variations of the smaller resistivities are emphasized.[31] This somewhat unusual presentation in no way affects the quantitative usefulness of the results. As long as induction was confined to wells drilled with oil-base mud, this question of scale hardly mattered; but when it came to water-base muds, direct comparison with conventional logs was required.

However, this came later. It took several years to produce more dependable and better focused sondes with low sensitivity to a borehole filled with conductive mud. Meanwhile, Doll had reconciled the field staff to his ideas, and the engineers began to appreciate the advantages of induction over the long normal under the operating conditions prevailing along the Texas–Louisiana Gulf Coast, where most of Schlumberger's activities were concentrated. Commercial success, however, was far from immediate. First, in fields with sands of high porosity and shallow invasion, satisfactory results were obtained with the long normal, supplemented by the lateral when needed—if the formations were thick enough. Indeed, the advantage of induction would have been clearer in formations 6 or 10 feet thick, but the use of the long normal had become so deeply rooted that the thickness effect could be assessed even without the assistance of departure curves, or allegedly so. On the other hand, the induction log required a special run, costly in time and money. It was

[31] The difference between 1 and 2 ohms is

$$1000 - \frac{1000}{2} = 500 \text{ millimhos.}$$

Between 100 and 200 ohms the difference is only

$$\frac{1000}{100} - \frac{1000}{200} = 5 \text{ millimhos}$$

and between 1000 and 2000 ohms it is 0.5 millimho.

therefore out of the question that the induction would be substituted for the conventional log, on which the S.P. immediately gave the lithology, and the comparison of the short and long normal provided a good qualitative appraisal of the production potential.

After being unreservedly accepted in oil-base mud drilling, induction began, after 1952, to spread to some conventional wells. This happened in northern Texas and Louisiana, where the reservoirs consist of sandstones of much lesser porosity and deeper invasion than in the southern parts of these states. The greatest benefit of this heartening initial success was the conversion to induction, first of Schlumberger unbelievers and then of the Company's clients. Only a good S.P. and a short normal complementing the induction were required to relegate the conventional log to a museum piece.

The solution was far from easy. One characteristic of the induction sonde is that its response is distorted by the slightest metallic influence, whether from electrode or conductor. Only in 1956 was a downhole tool completed that combined an S.P. electrode, a conventional 16 inch short normal, and a five-coil induction sonde with vertical and radial focusing. The presentation of the log was modified to conform with that of the conventional electrical log. On the right-hand track, the induction curve was recorded on a linear conductivity scale; it replaced the lateral. On the middle track, the reciprocal of the induction appeared as a linear resistivity curve. This curve was dotted and replaced the (dotted) long normal. It was obtained through a "reciprocator," an electronic system which converted the induction signal arithmetically.[32] As in the past, the short normal was recorded in the middle track and S.P. on the left (Fig. 73).

In spite of its higher cost, its more difficult operation, and the still quite satisfactory conventional systems, this new form of induction logging soon prevailed all over the Gulf Coast. Any remaining hesitation was overcome by superior results and the confirmation of theory by practice: once the major oil companies saw the light, all the others followed suit.

Such unanimity, however, had its drawbacks. All the former doubters became enthusiasts and demanded induction even in situations where the new system was obviously less efficient: whereas induction is highly suitable for measuring a conductive formation behind a resistive barrier, it is much less so in the case of a resistive formation behind a conductive

[32] $R = 1000/C$, with R in ohms and C in millimhos.

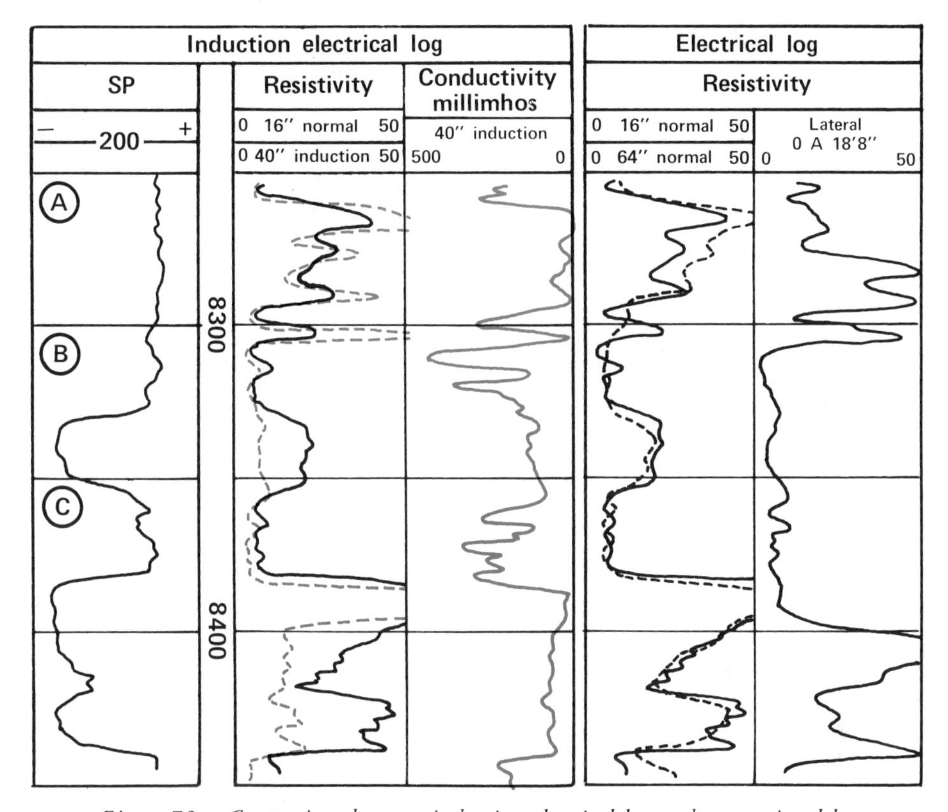

Figure 73. Comparison between induction–electrical log and conventional log.

invaded zone. It was not long before Schlumberger and the client interpretation specialists reverted to a more rational approach: that of utilizing the Laterolog, by this time commercial, in cases where induction was deficient.

Induction logging attained its definitive form in 1956, and since 1957 its use has become worldwide wherever conditions have been favorable. Such was its success that competitors had to develop their own induction systems and rediscover not only the technique but also the know-how required for making and operating the equipment. They succeeded and partially overcame their lag, but Schlumberger, as for conventional logging, had pioneered the service which, in a short 10 year span, had changed the course of the logging industry.

The Laterolog Technique

From the beginning of logging the technicians had been greatly preoccupied with bed resolution, particularly that of very thin beds. As early as 1927 Conrad had devised a system which, through the use of elongated, so-called guard electrodes, forced the current of a central electrode to penetrate laterally into the formation it faced. As shown earlier (see p. 124 and following), by opposing the spread of the current this device improved the delineation and detail of the formations. Conrad had described another system, where point electrodes would give the same effect, but unlike the former, it never went beyond the conceptual stage.

The failure of conventional logging to provide a sharp representation of the beds is found mainly in cases where there is a great contrast between the resistivity of the formation and that, much lower, of the mud. The boundaries between layers become blurred, and even the lithological interpretation of the log is dubious. In consolidated sandstone or limestone reservoirs, even with fresh mud, there is a large resistivity contrast. The situation becomes considerably worse when the shoulders are evaporites of practically infinite resistivity and the well is drilled with salt-mud. Western Kansas is a typical case where Paleozoic limestones lie under hundreds of feet of rock salt, too deep to justify the cost of an intermediate casing. Cave-ins produced by dissolution of the salt can be prevented only by allowing the drilling fluid to reach saturation. Its resistivity is then less than 0.1 ohm, which is a ratio of 1 to 1000 when the reservoir resistivity is 100 ohms. Similar conditions prevail in the important Permian Basin of western Texas, except that the salt occurs in several thinner beds. There, again, conventional resistivity logs are poor; and for lack of salinity contrast between mud filtrate and formation water, both salt saturated, the S.P. is useless.

Little activity took place in these regions until 1950, but, as in most of the Mid-Continent, an important potential market was there ready to be exploited. None were more convinced of this than the few Schlumberger engineers who were struggling to promote their logs in these problem areas that made the Gulf Coast look like an El Dorado by comparison. They were well aware of the weaknesses of the conventional system, and tended to think that the Engineering Department considered logging to be destined exclusively for the Gulf Coast. According to the Schlumberger tradition of uninhibited comment, they were rather outspoken in their views.

When Marcel visited the United States in 1945, he bore the brunt of their criticism. Since experience in the region demonstrated that the lateral curve showed at least some stratigraphic features, provided that its peculiarities could be sorted out, he conceived the idea of combining two lateral sondes head to tail. The assembly, being symmetrical, gave a symmetrical response on lithology and eliminated the most serious shortcoming of the lateral sonde. However, it was quickly evident that the apparent thicknesses of the resistive beds were reduced at their top and bottom boundaries by the basic sonde spacing. To preserve the lithological usefulness of the system, it became necessary to reduce the basic spacings AO_1 and AO_2 (O_1 and O_2 being the middle of M_1N_1 and M_2N_2, respectively), thus lessening the depth of investigation to the point of making the system sensitive to variations in the diameter of the borehole (Fig. 74). The elimination of the lateral's distortion had also removed its main quality, the depth of investigation. All this was confirmed by experience. Whereas this "limestone sonde" had been immediately commissioned in the Permian Basin, its only use was found in log correlation. Although some clients continued to request it even after 1950, it remained confined to western Texas, a fate that is not very exciting for a logging technique.

The solution lay elsewhere. The delineation of resistant beds in a well drilled with a salt-saturated mud was incompatible with the spreading of the current toward the more conductive shoulders; it was therefore necessary to direct or force the current into the formation of interest or, to use the term that was soon to characterize all the new measurement systems, to "focus" it. This is what the guarded monoelectrode sonde had tried to achieve. The two voltage measurements required were ill adapted to the continuous recordings of the 1950's, but this was not an insuperable problem. More serious was the fact that the effectiveness of the focusing depended on the selection of the shunt. At this point Doll came forward with a solution that was both original and technologically feasible. To prevent the current from escaping through the conductive borehole, he thought of plugging it—electrically, of course.

There was actually nothing new in the idea of electrically plugging the hole, as Doll himself had already introduced it in his analysis of the

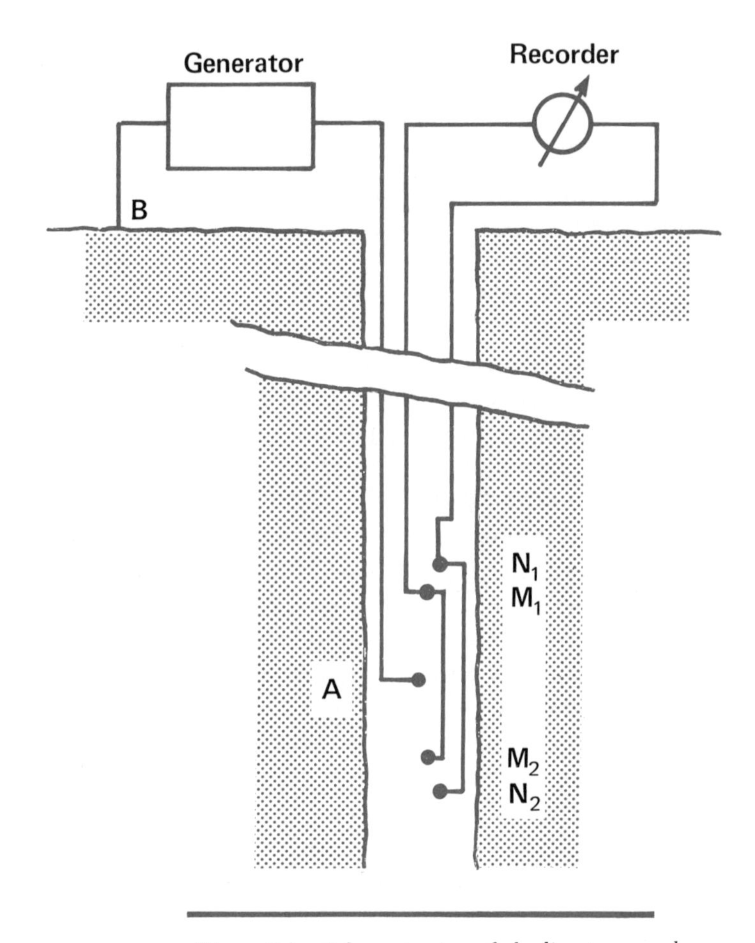

Figure 74. Schematic view of the limestone sonde.

tridimensional S.P. current. In his 1948 paper he had defined "static S.P." as the potential difference that would be created if the current were interrupted by two imaginary "electric plugs" at the top and bottom of the bed. In another study he had gone so far as to realize such plugs in the so-called "Selective S.P." system, directly derived from his reflections on S.P. in formations of very high resistivity, with the purpose of locating porous zones in such formations. Meanwhile, an easier solution to the problem had been provided by the Microlog, but the principle of the plug remained: to plug the current between the two electrodes M and N, auxiliary electromotive sources are used, with their outputs adjusted so as to balance the potential difference between M and N.[33] This was the system that would be used for the automatic focusing of measurements in salty muds.

The sonde array is made of seven electrodes (hence the name "Laterolog 7"): three current electrodes, A_0, A_1, and A_2, and two pairs of measure electrodes, $M_1M'_1$ and $M_2M'_2$ (Fig. 75). To prevent the current emitted by A_0 from spreading vertically and to force it into the resistive formation, the borehole must be electrically plugged, at the level of pairs $M_1M'_1$ and $M_2M'_2$, by emitting from A_1 and A_2 currents (called "bucking currents") adjusted to permanently balance the potential difference between the electrodes of each pair. In fact, the result is even better than merely plugging the hole, since the plug overflows into the formation, where it maintains the current emitted from A_0 in a well-delimited horizontal sheet. As for conventional logging, resistivity is measured on the basis of the potential common to $M_1M'_1$, $M_2M'_2$.[34] A little masterpiece of simplicity of ingenuity, Laterolog 7 was designed by Doll in almost finished form. It then became possible to study its characteristics, since the only problem was to superimpose the effects of several point electrodes and to optimize their geometry. The same held true for the calculations of the currents, which, as could be expected, had to be very strong at the focusing electrodes in the difficult case of a thin and resistant layer between conductive shoulders. Serious technical problems arose because for the monitoring of these heavy currents it was necessary to use the very small potential differences appearing between M and M',[35] which had to be reduced and maintained at a value near zero. In 1950 electronics was not sufficiently advanced to solve this "gain" problem by downhole amplification at frequencies intentionally equal to those of conventional electrical logging. The possibility did not even present itself that, for the measurement of resistivity, anything other than the conventional and proven pulsated current could be used. The decision was a fortunate one; although the pulsator was taxed to the limit of its capacity, it allowed time for the rapid and technically sound realization of the first Laterolog.

[33] The limiting case is MN infinitely short; the potential gradient in the vicinity of M is then said to be null.

[34] In practice, electrodes M_1 and M_2, M'_1 and M'_2 are short circuited, as well as A_1 and A_2.

[35] This is the potential difference existing between the short-circuited electrodes M_1M_2 and $M'_1M'_2$.

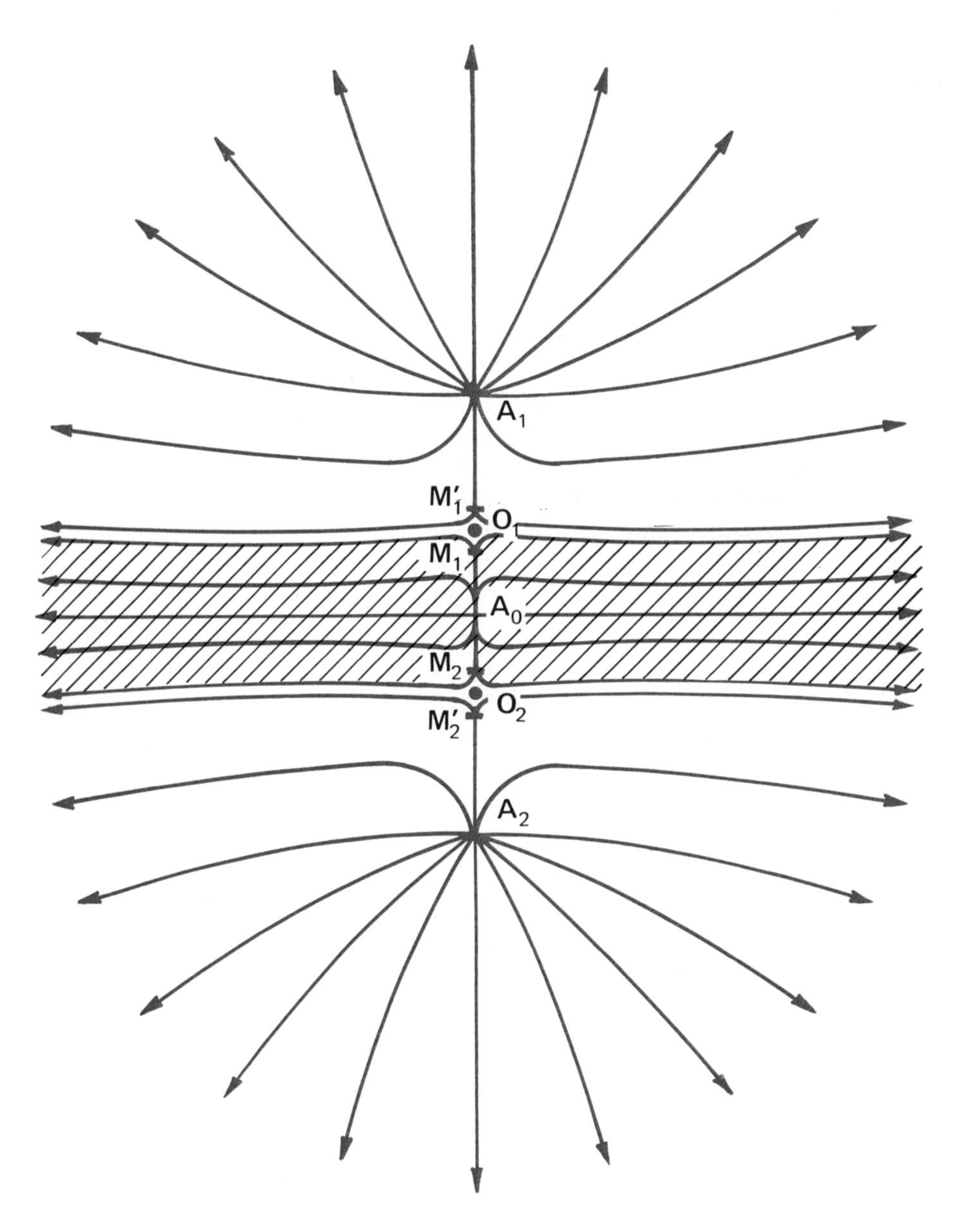

Figure 75. Schematic view of laterolog 7. (From H. G. Doll, "The Laterolog: A New Resistivity Logging Method with Electrodes Using an Automatic Focusing System," Journal of Petroleum Technology, *November 1951). (Courtesy SPE of AIME).*

Because of the difficulty of adjusting the bucking current downhole, all operations took place above ground, making it necessary to send the weak signal *MM′* to the surface and to inject the strong bucking current from the surface. From the perspective of the 1970's, the operation may seem somewhat primitive. However, it gives some insight into a rather dependable technology borrowed from the armaments industry as it had developed at the end of the war. The *MM′* signal was electronically amplified and drove a small motor in either direction. The motor actuated a rheostat (similar to a manual operation) which, in turn, controlled the bucking current. A mobile pointer on the control panel indicated the motion of the rheostat and informed the engineer on the damping of the system. If the pointer vibrated, the system was underdamped; it was overdamped when the pointer moved slowly. A log with sharp contacts required a slight underdamping, obtained by adjusting the "gain."

Since the bucking current was pulsed, it could influence the monitoring signal by induction. This is the pernicious effect of direct coupling, highly instable because of self-amplification. Only by a thorough study of the chronological system and of the inductive coupling between cable conductors was it possible to work out a solution and balance the circuits. Of course it would have been very desirable to amplify the *MM′* signal within the sonde; but, as already noted, a sufficiently stable electronic amplifier at such low frequencies and under prevailing conditions was beyond reach. A partial remedy was to place in the sonde a "square wave transformer"; the name brought shudders to electricians, but the device boosted the signal 10 times.

All this required the highest precision. To mention only the pulsator, it had, on the one hand, to be perfectly balanced so as to avoid any stray signal; on the other hand, it was expected to switch currents up to 2.0 amperes on and off without fireworks. This required absolute cleanliness. The answer to these constraints was the high-precision cam pulsator, developed in 1957. A further complexity introduced into the sonde is worth mentioning: in addition to the square wave transformer, it was fitted with a relay through which the electrode configuration could be changed from the surface, the purpose being to switch from Laterolog 7 to the limestone sonde—the very tool it was supposed to supersede![36]

[36] The switch may have been without much technical interest, but afterwards the relay

The first experimental log was recorded on April 11, 1949, north of Houston. Though the results were satisfactory, they did not arouse much enthusiasm because they were obtained in a fresh mud well and offered little advantage over a good conventional log. In August of the same year, a demonstration involving several instruments, including the Laterolog, was organized in an important Gulf Oil well near Ardmore, Oklahoma, in the presence of high-level experts: here again, nobody was convinced of the Laterolog's superiority over conventional logs.

So far, the objective had been only to test the tool operationally. The real testing started in western Texas in February 1950 and continued in Canada in May. Finally, the Laterolog was put into operation in Kansas in August 1950. Wherever drilling took place with salty mud, the success was overwhelming, and conventional logging was quickly eliminated. When Doll published his November 1951 article in the *Journal of Petroleum Technology,* the Laterolog was already an established technique.

Such success was like a whiplash to competitors, who in the same year produced their own version of the tool under the name "Guard Electrode Log." The system was not unlike that of Conrad's old guarded monoelectrode sonde. Focusing was achieved by nullifying the potential difference between the central electrode and the guard electrodes. The technology was all electronic; but either because the components then available lacked stability, or perhaps because it was more difficult to determine the potential difference between current electrodes than between measure electrodes, the competitors had tough going until they were able to demonstrate technical performance which matched that of Laterolog 7. In any case, when confronted later with the problems of very thin layers, Schlumberger was to experience the same difficulties.

There was, however, great interest in the guard electrode log in Kansas, where the productive horizons of the Pennsylvanian are composed of thin beds. It allowed for a narrower A_0 current sheet than did Laterolog 7, actually a small advantage but enough for the competitors to advertise it *ad nauseam.* R. D. (Bob) Ford, then manager of the area that included Kansas, intervened strongly in Houston; his message, loud and clear, was persuasive to the extent that a Laterolog with guard electrodes, equivalent to the competitive model, was placed on the drawing board.

made it possible to have a log at a "compressed scale," that is, linear as to resistivity on one-half of the strip, and as to conductivity on the other.

This did not constitute plagiarism in any way since it was a mere resumption of the 1929 idea of the guarded monoelectrode.

Doll authorized the project without great enthusiasm. An electronic Laterlog with a narrow current beam ("Laterolog 3") was made, but with results far less satisfactory than those obtained with Laterolog 7. An improvement made in 1953 consisted of inserting small measure electrodes between current and guard electrodes, making it possible to nullify the potential gradient as in Laterolog 7. Laterolog 3 became, in fact, a Laterolog 7 in disguise.

However, Laterolog 3 offered a possibility that, strangely enough, passed unnoticed initially. The upper guard electrode was long enough to receive a gamma ray recording cartridge. This subject will be discussed again under nuclear logging; it suffices to say here that gamma rays, in any kind of mud, provide good detection of shale beds, and in this sense the gamma log prevails over S.P. whenever the latter is suppressed by the salinity of the mud. The combination of the Laterolog and gamma log quickly gained ascendancy in Kansas and in the Permian Basin of Texas. Still another factor made the study of Laterolog 3 fruitful: this technology was to prove indispensable for the development of the "Microlaterolog," the natural extension of the Laterolog. In fact—and this constituted a substantial practical advantage—an electronic cartridge was soon produced that could serve for either Laterolog 3 or the Microlaterolog.

The success of the Laterolog technique in the United States made it the standard resistivity log in salty mud. However, because it remained confined for such a long time to areas where these muds prevail, its commercial success did not match that of the induction log. Only much later, and initially in the eastern hemisphere, did the Laterolog regain its standing. No longer is the sole reason for its use the fact that it gives a significant resistivity log in saltwater mud, which neither the conventional nor the induction log can do: it is also suitable in situations where the resistivity of the invaded zone is lower than that of the undisturbed zone. In such cases an accurate evaluation of the latter requires measurement in series, which is precisely what the Laterolog provides.

The Microlaterolog Technique

It has been explained how the existence of the mud cake between the pad of the Microlog and a porous formation interferes with measurement of the resistivity of the invaded zone. To prevent the current from escaping through the conductive medium of the mud cake and to force it into the formation to be measured, a system similar to the Laterolog is needed. Seven electrode devices are placed on the pad. Since the problem is no longer to block only the vertical leaks, but rather to have a system electrically focused in any direction, the electrodes have a circular pattern (Fig. 76).[37] If, by modulating the output of the guard ring A_1, the potential difference created between M_1 and M_2 by the steady current emitted from electrode A_0 is balanced, this current is forced to flow through the mud cake into the formation. The spacing of the electrodes determines the outward shape of the beam emitted from A_0; for a correct measurement of the invaded zone, it must remain cylindrical for several inches, with a diameter somewhere between those of rings M_1 and M_2—about 2 inches—before opening up. It is assumed that the beam will flare at some 3 inches from the borehole wall, and this value, which represents the depth of investigation of the instrument, can be modified by changing the spacing of the rings.

With the focusing system of the Microlaterolog, the contact between pad and borehole wall is, basically, a less critical problem. A nearly perfect contact is positively required for the Microlog only; nevertheless the Microlaterolog has had the benefit of all the mechanical improvements added to the former, for example, the hydraulic pad and the powered sonde.

Whereas the Microlog and the Laterolog are systems in which the downhole electrical equipment consists merely of measure and current electrodes, the sonde of the Microlaterolog required electronics. Difficulties did not prevent its early realization. The first tests were conducted in 1950, and marketing was successful in Kansas in 1951. In October 1952 Doll disclosed its characteristics and potentialities to the American Institute of Mining Engineers in Houston.

Technically successful as it was, the Microlaterolog did not arouse the same enthusiasm as the Microlog had 2 years earlier, nor did it develop into the same profitable business. The reason was that, whereas its application was confined to quantitative interpretation, the Microlog had become the basic tool for the location of permeable beds in consolidated

[37] They are not actually continuous circles, but concentric circular strings of interconnected electrodes.

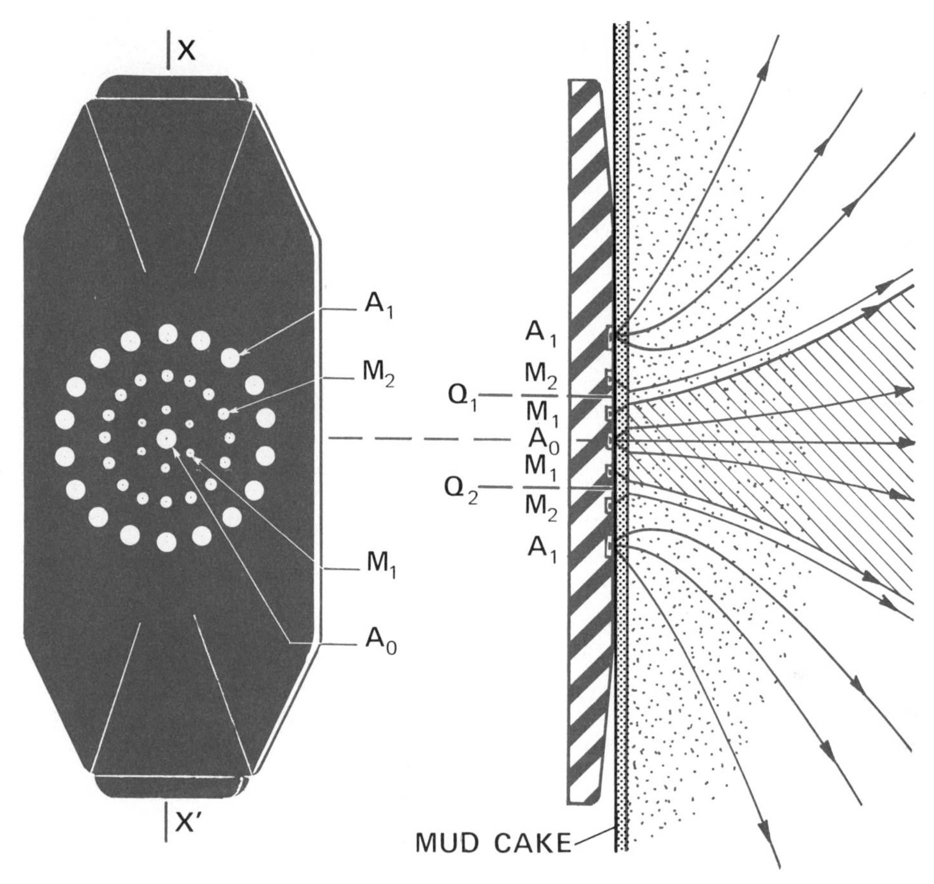

Figure 76. Pad of the microlaterolog and schematic distribution of the current lines. (From H. G. Doll, "The Microlaterolog" Journal of Petroleum Technology, *January 1953) (Courtesy SPE of AIME).*

formations. However, the Microlaterolog was to open new horizons to the specialists in interpretation, as it allowed a more thorough elaboration of the ideas put forward by Tixier in 1942 (see p. 203).

Instead of considering a rather vaguely defined invaded zone, the emphasis was now on the zone of maximum displacement of the original fluids by mud filtrate. In this 2 to 4 inch thick section—precisely that affecting the Microlaterolog—all the interstitial water has been replaced by the mud filtrate; as to hydrocarbons, if any, most have been flushed away, leaving in place only the amount which cannot be moved, usually termed irreducible. The resistivity of this zone, so effectively measured by the Microlaterolog, is called R_{xo}. If S_{xo} stands for the proportion in the pores of a mud filtrate of resistivity R_{mf}, the Archie equation becomes

$$R_{xo} = \frac{FR_{mf}}{S^2_{xo}}$$

whereas, applied to the undisturbed zone, it remains

$$R_t = \frac{FR_w}{S^2_w}$$

Therefore

$$\frac{S_w}{S_{xo}} = \sqrt{\frac{R_{xo}/R_{mf}}{R_t/R_w}} \qquad (1)$$

It is sometimes possible to obtain a fair idea of the residual oil proportion $(1 - S_{xo})$ by a laboratory analysis of various cores. Then S_w may be calculated through equation 1.

When S_{xo} is unknown, it is necessary to formulate a hypothesis based on an approximate empirical relation between S_w and S_{xo}. Such a relation will be similar to that observed by Tixier in the Rockies, except that, under the stricter conditions defined above, it will be closer to reality. Assuming that

$$S_{xo} = (S_w)^{1/5}$$

it follows that

$$S_w = \left(\frac{R_{xo}/R_{mf}}{R_t/R_w}\right)^{5/8} \qquad (2)$$

At this point the essential data for its evaluation, the oil and water percentages in the pores of the reservoir, are known.

Reverting to equation 1, if the term under the radical equals 1, that is, if

$$\frac{R_{xo}}{R_{mf}} = \frac{R_t}{R_w}$$

then

$$\frac{S_w}{S_{xo}} = 1 \qquad \text{and} \qquad S_w = S_{xo}$$

which generally means that

$$S_w = S_{xo} = 1$$

This will be the case for a water-saturated formation, where the invasion did nothing but substitute the mud filtrate for the original water. But even if S_w and S_{xo} differ from 1, their equality means that the water : oil ratio was not affected by the invasion—in other words, no oil has been flushed by the filtrate. The fact that there was no flushing of the oil indicates a virtually nonproductive reservoir. It will happen, for instance, with an oil so viscous that its recovery through conventional methods is not feasible.

The same method leads to an interpretation technique proposed as early as 1952. Its later application, however, was devised by Doll and published jointly with M. Martin (see note 24, p. 234). It is known that from the porosity (ϕ) a satisfactory value can be deduced for the formation factor (F):

$$F = \frac{1}{\phi^2} \tag{3}$$

The Archie equation then becomes, for the undisturbed and invaded zones, respectively,

$$R_t = \frac{R_w}{\phi^2 S_w^2}$$

$$R_{xo} = \frac{R_{mf}}{\phi^2 S_{xo}^2}$$

Here ϕS_w, the product of the porosity and the water percentage, represents the quantity of water by unit of formation volume. Likewise, ϕS_{xo} is the amount of filtrate in the invaded zone. The amount of oil displaced by the invasion is thus

$$Y = \phi S_{xo} - \phi S_w$$

by unit of volume, or, on the basis of the above equations,

$$Y = \sqrt{\frac{R_{mf}}{R_{xo}}} - \sqrt{\frac{R_w}{R_t}}$$

The interest in the formula will be readily understood as it informs not only on the amount of oil in the undisturbed reservoir, but also on its mobility or its "producibility."

It is still a puzzle as to why the Microlaterolog, with all these assets, was not an immediate commercial success. Without doubt its advanced technical features made it more difficult to operate than the Microlog; hence it lacked what the field people called "sex appeal." Yet, as early as 1950, quite a number of logging experts were using it profitably and were even able to elaborate novel methods of interpretation based on the Microlaterolog.

It must be acknowledged that the reason for its half-failure was the often dubious quality of the results. Serious practical difficulties existed. The first was the lack of dependability characteristic of the electronics of that period because they were being employed in a definitely hostile environment. A further obstacle was inherent in the Microlaterolog system: to read R_{xo} beyond the mud cake, focusing is required, but it must not be too deep lest the resistivity of the undisturbed zone be encompassed. Yet the thicknesses of both mud cake and invaded zone are variable. The former is contingent on mud quality: a good mud rarely forms a mud cake thicker than 0.2 inch, though at that time 0.8 inch and over was not uncommon; the latter, being a function of porosity, is reduced in a very porous reservoir. It is a tightrope exercise; a focusing is needed that suits every case even though it cannot pretend to apply to the whole range of thicknesses.

There was another and more subtle problem of design. The pad must be suitably curved to fit the borehole. Clearly, focusing, being determined by the tridimensional geometry of the whole system, is affected by this curvature; but since a borehole is never perfectly cylindrical, the pad is continuously changing its shape to keep in contact with the wall.

Aside from electronics, the Microlog has encountered the same problems but was content with being an excellent indicator of even very thin permeable layers and hardly pretended to quantitative results. The Microlaterolog, on the contrary, aimed at the highest accuracy. Its lack of favor with certain experts arose from its failure to fulfill all that it had promised.[38] Finally, there were almost simultaneous developments of porosity measurement processes other than resistivity, which made it

[38] At that time, R_t itself was not always above suspicion, and it is likely that a failing R_{xo}/R_t ratio has, sometimes incorrectly, been attributed to a faulty R_{xo}.

possible to calculate F through relations like equation 3 above. The problem of calculating F, or eliminating it between two equations with the R_{xo} of the Microlaterolog, became much less stringent.

Years passed before improved electronics, better focusing, and drilling muds of higher quality restored the Microlaterolog to its deserved place among the Schlumberger tools. The theory of interpretation set forth above and its application to the design of the Microlaterolog have been fully justified by experience.

The nuclear logs: gamma rays, neutrons

It may be safely stated that all the practical systems of resistivity logging originated with Schlumberger, and, furthermore, that the credit for these systems and also for S.P. goes to two men: Conrad Schlumberger and Henri Georges Doll.

However, geophysics is more than resistivity. When, after 1945, the accelerated development of petroleum exploration and production brought to the staffs of the major oil companies—Esso, Gulf, Mobil, Shell, Texaco—physicists assigned to the systematic study of such parameters as could lead to a better evaluation of the oil fields, the policy of the companies was not to keep a proprietary interest in the results of such studies, but to license their patents to service companies. When these patents were offered for licensing, Schlumberger, being best qualified to make a success of a new technique, was given priority but not exclusivity, since the latter would have been contrary to American law. This was the circumstance under which the array of Schlumberger techniques came to include some developed outside the Company.

Gamma Ray Logs

Natural radioactivity varies from one rock to another. It is very weak in pure sands or limestones and reaches a maximum in shales, where there is a concentration of radioactive uranium, thorium, and potassium-40 salts. The energy of the gamma rays the rocks emit is such that they pass through steel without a great loss: for instance, a typical well casing will absorb only 30 percent of the radiation.

The log of this radioactivity in an open-hole section will differentiate marls and sands as does S.P., but what is important is that radioactivity is not affected by the nature of the mud, whereas S.P. is generally useless in salt-saturated mud. Other substantial advantages are, in many cases, the capability to see through the casing and to record a log in a cased borehole.

The interest of geologists and geophysicists in radioactivity is nothing new. In 1909 J. Joly had published an article entitled *Radioactivité et Géologie (Radioactivity and Geology).*[39] In 1921 Richard Ambronn had measured the radioactivity of cores taken in an oil well at Celle, near Hanover, Germany. It is known that before World War II Russian geophysicists had built a logging device based on gamma rays, but no results were published. Finally, on October 29, 1938, natural radioactivity was recorded for the first time, under the name "Gamma Ray Log," in a well at Oklahoma City. The apparatus had been designed by a group of physicists[40] of Engineering Laboratories in Tulsa, Oklahoma. With the support of Socony Vacuum, the group established a service company, Well Surveys, Inc., which marketed the log from 1939 on.

The detector, based on a rather old technique, was an ionization chamber, a gas-filled enclosure made weakly conductive by the gamma radiation. Like all gamma ray systems, it required a number of electronic components in the sonde.

The results confirmed the theory: whatever the mud, even in a cased well the logs were comparable with the S.P. Practical applications were found immediately, especially for the resumption of production in old wells where no logging had ever taken place, and which could now be studied by correlating the gamma ray log with S.P. logs in surrounding wells. The same was true for the control of perforating depths.

Mention has already been made of how difficult it is to measure accurately the depth of a well and, in particular, to position the perforator at exactly the same depth where a bed has been identified by electrical logging under mechanical conditions (tension and elongation of the cable) quite different from those in the cased well. The necessary correction was now provided by the clear correlation between the open-hole S.P. and the gamma ray log in the cased well.[41]

[39] Reprinted in *Geophysics* (*Silver Anniversary Program*), Society of Exploration Geophysicists, 1955.

[40] W. G. Green, S. A. Scherbatskoy, R. E. Fearon, L. M. Swift, and J. Neufeld.

[41] The method is actually rather complex. Even when a horizon has been located with

Wartime personnel and equipment shortages delayed the full success of the gamma ray measurements and depth control technique. Well Surveys, Inc., merged with Lane Wells, specialists in perforation, and placed Schlumberger in an awkward position. True, in 1940 Doll had proposed marking the productive zones with radioactive bullets. Once the formation had been identified on the electrical log, and before the casing had been run, a bullet carrying a small amount of radioactive substance was fired into the formation, using a regular perforator. The bullet was very accurately positioned by means of an S.P. electrode lowered simultaneously, as for sidewall sampling. In a formation of average consistency the bullet penetrated deep enough not to be displaced during casing and cementing. It provided a marker that could be detected through the casing, and the detector used was simpler than that for measuring the natural radioactivity of rocks.

The method proved less valuable, however, than had been anticipated. A first problem was that the decision to use the gamma ray log could be postponed until the very time of perforating, whereas the firing of the radioactive bullet had to be scheduled before casing was run. The difference was, in effect, either doing two operations (firing and locating) or running a single gamma ray log. Also, some additional data could be expected from the gamma log, whereas the location of markers was limited to depth control. Finally, as the bullets contained a radium salt with a half-life of 1590 years, the zone surrounding the marker was made radioactive and no measurement (gamma ray) of its weak natural radioactivity could be made.[42] Although commercial prospects were never

---

gamma rays, a residual uncertainty over the depth arises from differences of weight, volume, and location between the sonde and the perforator. Such uncertainty would be eliminated if locating the horizon and perforating could take place in the same run. There are two reasons why this cannot be done: first, it is too hazardous to attach an electrically detonated pyrotechnical device and a sonde requiring a high voltage at the end of the same cable; and, second, the recoil of the perforator would destroy the electronic components of the gamma ray sonde. An intermediate reference is thus required, and it is provided by the collars of the casing. Since 1948 this reference has been a passive magnetic device, requiring no current and presenting no hazard. A first detector, fitted on the sonde, locates the collars on the gamma ray log, and their virtual location on the electrical log is deduced by correlation with the S.P. A second detector, mounted on the perforator, locates the collars again. To position the perforator, all that is needed is to interpolate depths between two consecutive collars, roughly 30 feet apart, with the cable measuring wheel. The accuracy reached is within an inch.

[42] With isotopes currently available today, this objection would no longer hold.

outstanding, there is one area where the system has prevailed. Whenever the problem is to detect with an accuracy within a fraction of an inch the ground subsidence that occurs or may occur in California, Japan, or Holland as a result of gas or oil recovery from thick unconsolidated beds, the radioactive bullet is the answer.

By the end of World War II it was obvious that gamma ray logging had a bright future, especially since a new radioactivity technique was emerging: "Neutron Logging" with its anticipated contribution to the evaluation of reservoir rocks. In 1945 an agreement had been concluded between Schlumberger and The Texas Company for the development of a gamma ray logging patent. The director of research for Texaco, G. R. Herzog, a Swiss physicist well known for his work on cosmic rays, was made available to Schlumberger. His specific plan was to use a Geiger–Müller detector for counting the gamma rays. The problem of creating an electromotive force of some 1000 volts in the sonde while maintaining good insulation at a bottom-hole temperature of 300 degrees F was solved, and the system, put on the market in August 1946, proved wholly satisfactory.

The absolute value of radioactivity has little quantitative use; it may sometimes be necessary to compare two readings at different levels, but this requires only proportionality between the deflection of the galvanometer and the natural radioactivity. Nevertheless, this was not good enough for Schlumberger, with its propensity for measurements of the highest quality.

Therefore the Company introduced a calibration technique that made the log independent of the particular features of the sonde or the control panel.[43] This calibration system was abandoned, however, when the American Petroleum Institute devised another, which became standard for all the service companies, even though the Schlumberger system seemed simpler and less artificial.

The Neutron Log

Together with the proton and the electron, the neutron is one of the elementary particles of the atom. It is usually associated with the proton in

[43] See A. Blanchard and J. T. Dewan, "The Calibration of Gamma Ray Logs," *The Petroleum Engineer,* August 1953.

the nucleus. Its mass equals that of the hydrogen atom, and it is electrically neutral.

How the neutron was discovered reads like a "whodonit." The race for the truth began at the end of 1930 with Bothe and Becker in Germany and ended on February 17, 1932, with Chadwick's triumph at the Cavendish Laboratory. To postulate the existence of the new particle, Chadwick had used, among others, the results published barely a month earlier (January 18, 1932) by Frédéric Joliot and Irene Joliot-Curie of the Institut du Radium in France.

No less extraordinary was the speed with which the adaptation of the neutron to lithological studies was proposed and achieved. Robert E. Fearon was granted a patent in 1938, another was granted in 1940 to Folkert Brons, and in 1941 the Italian physicist Bruno Pontecorvo published in *Oil and Gas Journal* an article entitled "Neutron Well Logging." Well Surveys, Inc., undertook the first tests that same year.

Meanwhile study of the new particle, whose role in nuclear weapons could already be foreseen by those with the requisite knowledge, had taken giant strides. It seems rather surprising that the neutron was considered for logging, but the highly penetrative character of a flux of these particles did augur well for logging through the casing. On the other hand, as a consequence of neutronic activity, high-energy[44] gamma rays that a gamma ray sonde can detect are emitted. As Bothe and Becker had demonstrated, neutrons are generated at the rate of several billions per second by bombarding a beryllium target with alpha particles provided by a naturally radioactive substance like polonium or radium.

However advanced nuclear physics may have been in 1945, its first application to logging was highly empirical. It was soon possible, with the log obtained, to distinguish shales from sandstones or limestones, but for a rather long time use of the neutron log was strictly qualitative, complementing the natural radioactivity log and enhancing lithological interpretation and well correlations. The gamma–neutron combination somehow recalled the S.P.–resistivity association, except that it was now possible to "see" through the casing.

The free neutron, a neutral particle, collides with atomic nuclei without interference from electrostatic forces. Ejected at a velocity of thousands of kilometers per second, it collides with various nuclei; these

[44] This property distinguishes them from the much less energetic natural gamma rays.

collisions slow it down to the velocity due to thermal motion at ambient temperature, or about 2 kilometers per second. It can then be absorbed by the nucleus of an atom, and this absorption results in the emission of a high-energy gamma ray, or "capture" ray. The fundamental interest of the neutron log lies in this slowdown, studied by Fermi as early as 1934. The neutrons sustain their highest deceleration when colliding with hydrogen atoms because the latter, having the same mass as neutrons, absorb rather than reflect energy at each collision. If, on the contrary, neutrons collide with much heavier atoms such as calcium or silicon, whose mass is about 25 times higher, they "bounce" without any significant loss of energy.[45]

In a hydrogen-rich medium the neutrons are thus slowed down and captured in the vicinity of the emitting source, and the gamma rays generated by this capture will be too far away to be sensed by a rather remote detector 1 to 2 feet away. On the contrary, if the medium is poor in hydrogen, the neutrons will proceed further before being captured, and a large number of gamma rays will be emitted in the vicinity of the Geiger–Müller counter. In other words, the number of gamma rays counted varies in inverse ratio to the hydrogen concentration.[46] In the formations drilled through by a borehole, the presence of hydrogen is almost solely caused by the fluids, and consequently the amount of hydrogen per unit volume of rock is a function of porosity. The latter can be evaluated by counting the gamma rays of capture if the pores are filled with oil or water, both of which have about the same molecular concentration in hydrogen. The same is not true, however, of gas, the hydrogen concentration of which is much lower; a gas-bearing horizon will therefore appear on the neutron log as less porous than if it were filled with oil or water. With the help of an additional porosity measurement (core, sonic or density log, etc.; see pp. 307 and 304), it thus becomes possible to distinguish not only between gas and water—which resistivity already does—but also between gas and oil—which resistivity cannot do.

In addition to being a significant parameter characterizing the reservoir, porosity makes possible the determination of the formation factor (Archie formula) and, provided that the resistivity (R_w) of the interstitial water is known, the deduction of the resistivity (R_o) of the 100 percent-water-saturated rock. If the electrical log gives a resistivity value of R_t

[45] This is only a rough mechanical image of a phenomenon belonging to nuclear physics.
[46] There are processes to evaluate the slowdown of the neutrons other than counting the gamma rays of capture (see p. 304).

higher than R_o, the rock must, in principle, contain hydrocarbons in a proportion that can be obtained from the ratio R_t/R_o.

The importance of the neutron log becomes clear from these remarks. The slowdown of free neutrons is a function of the hydrogen concentration, which in turn is a function of the porosity. Knowledge of this porosity at every point makes it possible to determine at any depth the excess of the measured resistivity over what it would be with 100 percent water saturation and thereby to establish the presence of oil and quantify it percentagewise within the pores. With this percentage, and with porosity once more intervening, the volume and value of the oil in place are known.

Those[47] with the necessary background were aware of such potentialities, but realization had to wait for several years. The practical difficulties were great. It soon became apparent that not only porosity, but also variations in the borehole diameter, the position of the sonde, and the nature of the mud substantially affected the results. The need to obtain identical logs irrespective of equipment mandated calibration, and since this had to allow for particular features of the sonde as well as the source, it was always a delicate and sometimes uncertain operation.

Schlumberger entered the field in 1949 with an ambitious $2,000,000 program. The problem was to catch up with the competition, design precise and dependable instruments, set up a first-class service, and integrate the results of the new log into the general technique of interpretation. A group of experts was hired (including John Dewan, Denis Tanguy, and Jay Tittman), who quickly made the basic decisions as to the source,[48] the type of detector, the spacing, and other relevant features.

[47] The relation between porosity and neutron slowdown had been considered in 1949 by Clark Goodman, a technical consultant for Schlumberger. In the same year R. Fearon referred to it in *Nucleonics* (June 1949). One of the first studies of Jay Tittman when he joined Schlumberger in 1951 was to tackle the theoretical basis of this relation.

[48] Schlumberger initially decided in favor of radium–beryllium, although this combination had serious shortcomings as compared with polonium–beryllium: the price was very high, and the gamma radiation required thorough protection of the operators. However, its half-life period is 1590 years against only 140 days for polonium. Whereas in the United States radioactive sources could be delivered within a few days, the situation was different when customs formalities interfered. The delays for polonium could then exceed one or two longevity periods, and a problematic recalibration was required, if the product had not become altogether useless. The selection of the source was therefore determined by the fact that Schlumberger's operations were international.

Special laboratories were established in Ridgefield and Houston for handling radioactive materials and designing the tools. A system of safeguards and tests had to be organized to comply with U. S. Atomic Energy Commission standards. Periodic inspections of the sources detected any possible leaks in the stainless steel block in which they were sealed. Instructions were given on fishing jobs, should the source be lost in a well.

Schlumberger's version of the neutron log was introduced in the United States in 1952. From the very outset it proved comparable in quality to competitive services, especially since it had the advantage of the field engineers' competence in interpretation. Observations made in 1953 in the Permian Basin provided practical data on porosity determination.[49] Concerned with giving the client the means to determine a physical parameter irrespective of the conditions under which the measurements were made, Schlumberger published correction charts that are to the neutron log what departure curves are to resistivity.[50] Moreover, by using two logs with different spacings between source and detector (which gave different depths of investigation), it became possible to identify gas-bearing horizons. The method was especially successful in eastern Venezuela, thanks to the outstanding teamwork of Pierre Majani, Frank Dennis, and Mike Grosmangin.[51]

Neutron logging developed steadily between 1952 and 1957, even though it did not fulfill the hope that it might become a universal porosity measurement method. In addition to the difficulties already listed, there is a rather large lithological effect, foreseeable but complicating interpretation, and in areas where the lithology was known empiricism often prevailed over rational procedure.

At Schlumberger, as elsewhere, nuclear physicists had much earlier proposed and tried to reduce the lithological effect by other methods than counting the gamma rays of capture to detect the slowing down of the neutrons. Such methods were developed later, and neutron logging realized its full value only when it became integrated into a set of mea-

[49] The linear relation between the logarithm of the porosity and the neutron log reading had been recognized by Shell's S. H. Rockwood, who had conducted many comparisons between porosity values measured on cores and neutron log readings.

[50] J. T. Dewan, "Neutron Log Correction Charts for Borehole Conditions and Bed Thickness," *Journal of Petroleum Technology,* February 1956.

[51] M. Grosmangin and E. B. Walker, "Gas Detection by Dual-Spacing Neutron Log in the Greater Oficina Area, Venezuela," *Journal of Petroleum Technology,* May 1957.

surements the interpretation of which was so complex as to require the help of the computer. In effect, from 1958 on the neutron log was no longer the only one of its kind: for a time the sonic log and then the density log rounded out the array of porosity tools.

Auxiliary operations

Dipmeter Surveys

The S.P. correlation dipmeter often gave excellent results. From the outset it almost always proved superior to the electromagnetic anisotropy dipmeter, as it indicated not only the azimuth but also the dip angle of the strata.

The operating technique was to select on the conventional log a number of sections, each about 30 feet long, such that the dips would be geologically significant and the S.P. curve would allow correlations. The selection was usually made in collaboration between the client's geologist and the Schlumberger engineer. Computation of the dip requires knowledge of the inclination of the borehole and the orientation of the instrument, and these data were given by the photoclinometer kept in fixed geometry with the dipmeter sonde. Once the tool had been lowered to the deepest level of a section selected for study, it was necessary to wait for about 1 minute until the needle of the compass and the ball of the inclinometer stabilized and could be photographed. While the sonde was then slowly pulled to the upper level of the section, the three dipmeter curves were recorded. There was then another stop, another wait, and another photograph. As a comparison of the photographs taken at the two levels almost always showed that the instrument had rotated during the recording, computation of the dip was possible only on the assumption that the rotation had been proportional to the vertical travel. This is the reason why the length of the sections was limited: a substantial rotation would have required an exceedingly hazardous interpolation.

Correlating the dipmeter curves was theoretically easy, but actually required great skill. Wherever the dipmeter operated, there were specialists analyzing the logs received from the field, correlating curves, making trigonometric calculations to determine the azimuth and dip angle, and preparing a graphic presentation of the results. These specialists were recruited from among draftsmen, assistants, and engineers removed from the field for reasons of health or especially hired for the job (there were a few women in the group), and some became outstanding experts. The quality of the service provided was due to their competence as well as to the proper selection of the sections and the accuracy of the field measurements. Such a strong organization, along with highly specialized teleclinometer equipment, gave Schlumberger a practical monopoly in dipmeter operations.

Southern Louisiana was the area of greatest activity and success for the S.P. dipmeter. In this region of salt dome tectonics, where knowledge of the dip is essential, the S.P. curves are excellent and highly diversified, and very sharp deflections yield unequivocal correlations.[52] Doll's study on the S.P. specified the conditions to be met for sharp deflections: what is needed is not only good resistivity contrast between drilling mud and interstitial water, but also permeable formations and shoulders of rather low resistivities. Such conditions also prevail in southern Texas, California, eastern Venezuela, Nigeria, and Indonesia. On the other hand—either because the drilling takes place with saltwater mud (Kansas, Permian Basin, offshore), or because the resistivities are high (northern Louisiana, northern Texas, Oklahoma)—there are many regions where the S.P. dipmeter works inadequately, or not at all.

There is also a geological factor that renders the S.P. inadequate as a universal dip-measuring parameter. The sharpest deflections (i.e., those giving the least equivocal correlations) are the ones marking the sand–shale contacts. Yet it often happens that these contacts, which are in a way stratigraphic accidents, are not in the plane of the structural dip. On the contrary, while nicely layered shales allow for excellent dip values, S.P. does not distinguish one layer from another in a typical shale stratum, so that the S.P. dipmeter is blind to the most significant sections. This is the reason why, despite its complexity and limitations, the electromagnetic dipmeter, extremely sensitive to the anisotropy of the shales, continued to operate for years in certain regions.

From 1945 on, the above considerations determined the path to follow: resistivity curves were necessary to obtain universally good corre-

[52] At the depth scale of 1:20 generally used in these recordings, the maximum theoretical curve displacement for a 5 degree dip in an 8 inch well is less than 0.7 millimeter. This already requires precision in evaluation, but the dips to be measured and the displacements are sometimes even smaller.

lations. Again, the solution came from Doll. He conceived an apparatus that combined three lateral sondes with a single common-current electrode. The measure electrodes *M* and *N* were placed in pairs on three arms at 120 degrees, with *MN* spacing of about 1 inch. The current electrode *A* was on the lower part of the mandrel, 3 feet from the plane defined by the centers of the three *MN* pairs. Three detailed resistivity logs, suitable for correlations, were thus anticipated.

When this sonde was designed, only the textile-braided quadricable was available, and it was necessary to work out simple[53] and dependable bottom-hole electronics so as to record the three resistivity signals independently. The results were immediately conclusive. Excellent correlations could be obtained from the resistivity curves in areas where nothing could be expected from S.P. In southern Oklahoma, where the initial tests took place in 1946, clients reacted positively to the new service. From January 1947 it was accepted throughout the region and its use was widespread. From there it promptly expanded first in the United States and then to Venezuela and Trinidad. Beginning in 1952, resistivity dip measurements became a standard, worldwide procedure.

More improvements were desirable because of the importance of selecting the proper sections and their boundaries. Should these selections be made *a priori,* or would it not be preferable to record the whole open hole and then select the sections after examination of the results? However, this was not possible at the time because of the stops required for the operation of the photoclinometer. This device therefore had to be replaced by a type of inclinometer that functioned continuously, if possible, or at least required no stops in the upward movement of the sonde.

From 1940 to 1942, during the German occupation, the Paris center had worked out a photoclinometer, fluid filled and consequently mechanically damped, whose elements could be photographed without interrupting the upward motion. A gas bubble moving inside a liquid under an upward-convex glass replaced the steel ball of the cup. Without interrupting the motion of the sonde, as many pictures could be taken as the film magazine permitted. In spite of such minor troubles as the formation of

[53] The inclination of the borehole and the orientation of the electrodes were still obtained by the photoclinometer, whose operation remained unchanged. A relay made it possible to switch from the dipmeter to the photoclinometer function. With electronics the two could have been performed simultaneously, but since the mechanical system required a stop to photograph the needle and the ball, simultaneity was not essential.

secondary bubbles, the instrument could have become operational except for the fact that it used film. Intermittent inclinometry data constituted no major drawback provided that the interval between photographs could be reduced. The major limitation was the sensitivity of the film to temperature. Whatever the precautions, it frequently happened that, after operating in a deep well, the film was found to be completely blackened; this was especially exasperating since it was discovered only after the operation. It was better to forego the film and transmit the inclinometry data to the surface, a practice that, in addition to actually yielding the required information permits monitoring the instrument behavior.

The first device meeting this requirement was realized in Ridgefield in 1951, and a small series of 10 was put into operation in the United States in 1952. Called the C.D.M., for "Continuous Dipmeter Microlog," it became known as the "de Chambrier dipmeter"[54] or "The Thing" by field engineers marveling at such complex and novel electronics. Indeed it contained a fluxgate compass borrowed from the aircraft industry, and an elementary electronic computer which, by combining the readings of the compass and the inclinations of two pendulums mounted at right angles, calculated the north–south and east–west components of the borehole deviation. As to the dipmeter curves, a considerable advance had been made by using three Microlog arrays, as their vertical resolution was much better than that of the old resistivity curves and permitted more correlations of higher accuracy. The system was further perfected in 1954 with the replacement of the Micrologs by Microlaterolog-like curves. Also, the mechanics of the sonde were improved: instead of fixed electrode supports selected according to the nominal diameter of the borehole, there were now three mobile pads applied to the borehole wall by springs, thus making it possible to record simultaneously the diameter variations of the hole.

However ingenious, the teleclinometer lacked dependability, was very difficult to maintain, and required bulky and complex surface equipment. These drawbacks did not prevent the de Chambrier dipmeter from being accepted by clients, proof that continuous dipmeter surveying was the solution of the future, and that discontinuous measurements were no

[54] Pierre de Chambrier was an engineer and electronics specialist who had expended great effort on dipmeter problems, especially during the several years he had spent in eastern Venezuela.

longer acceptable in regions where knowledge of the dip is particularly important.

At the same time, the Paris center had begun to develop a tool capable of similar performances, but mechanical in nature. The initiative came from Marcel. In a way it could hardly have been otherwise, for the device was so innovative that nothing less than his full technical and moral authority had been required to launch the project. The guiding idea was to measure the rotations of the deviation pendulum, the relative bearing pendulum, and the compass enclosed in the teleclinometer by having them actuate rotating potentiometers designed for minimal frictional torque. Imagine a short magnetized bar oriented in the terrestrial magnetic field and actuating a potentiometer: the torque is so weak that it does not seem able to overcome the friction of the cursor on the rheostat. Yet in 1950 the Swiss firm Ohmag was already making potentiometers that a compass needle could actuate. Conventional measurement of the resistances of the potentiometer easily gives the position of the cursor, and thus the angle between the magnetic north and the direction of reference of the instrument. What is true for the torque of the compass also applies to the torques of the deviation pendulum and the relative bearing pendulum when they are only 1 or 2 degrees off their positions of equilibrium, and thus their slightest angular displacements can be measured. These are the elements of the teleclinometer that was called the "Poteclinometer."

The instrument was tested in Pau in 1953 and then brought to the United States and Venezuela in 1954 for intensive trials. Edmond Boucherot and Bob Canup adapted to the poteclinometer the three-pad Microlaterolog sonde used for dipmeter curves. In October 1954 Rene Roussin in Paris resumed study of the mechanical design, which had made little progress since the death of Marcel in 1953. In cooperation with Ohmag, the potentiometers, under the name "Micropotentiometers," were improved and manufactured by Schlumberger under a licensing arrangement. The friction effect was further minimized by a small motor, giving a tacking motion to the whole mechanism. Its sensitivity was such that the compass had to be developed in the open countryside because the Paris subway trains deflected it by some 10 degrees.

From 1957 on the instrument was produced in its final form (Fig. 77). It not only replaced the de Chambrier instrument in the United States, but also, since it no longer required highly specialized operating per-

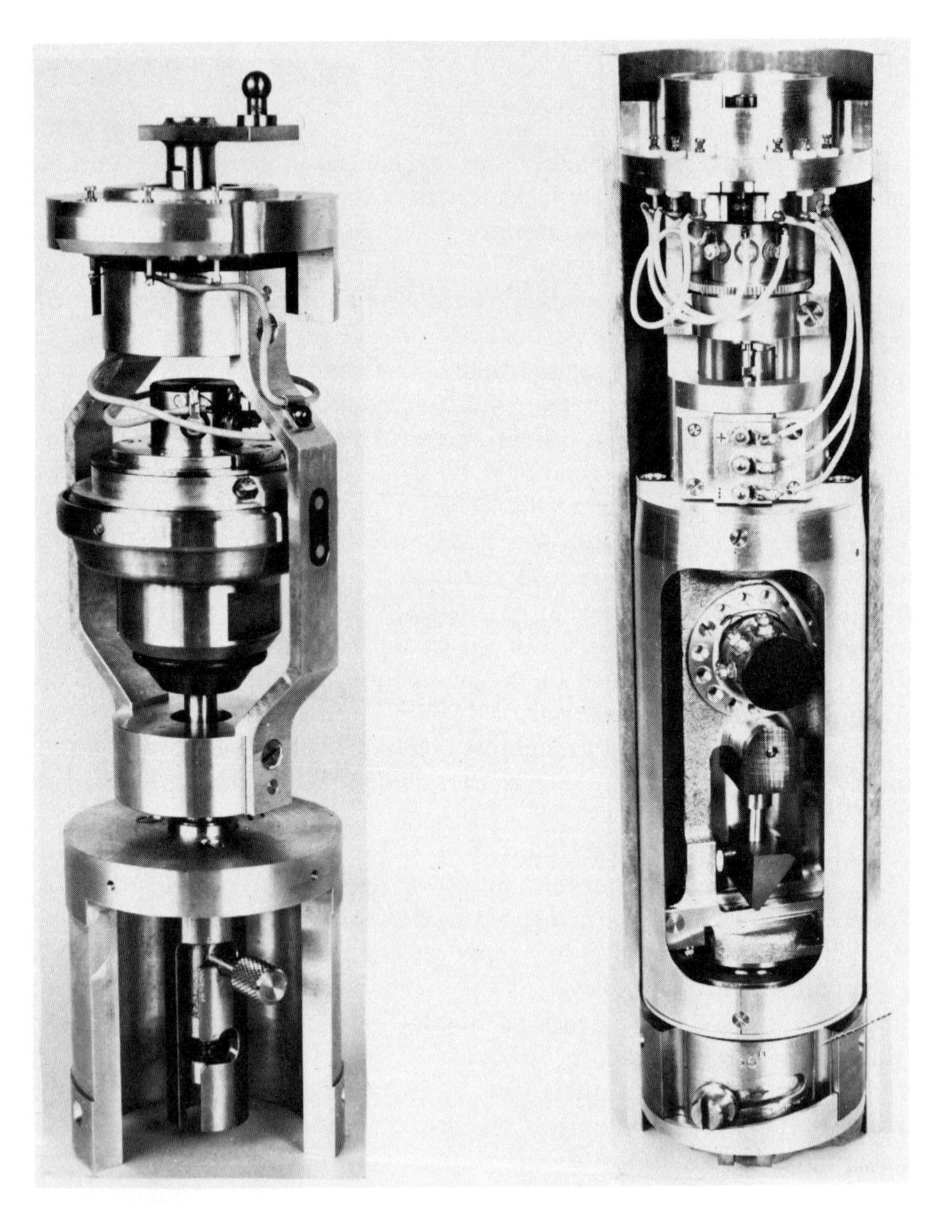

Figure 77. Poteclinometer.

sonnel, soon supplanted the photoclinometer and the resistivity dipmeter all over the world.

Thermometry

The battery cell thermometer had been employed in the field since 1938, mainly for thermal location of the cement tops in cased wells (see p. 179), and had given rather satisfactory results. Another application, less dependable because multiple intervening factors made any quantitative interpretation, even in broad figures, very difficult, was the detection of gas entries in the mud column. Certain cases demanded the study of minute temperature variations. However, the technique proved reasonably successful in the Permian Basin, where gas entries are important.[55] On the other hand, it was used to excess in Kansas to detect low-potential and low-gas : oil-ratio oil-bearing strata. Oddly enough for a thermometer, it had temperature limitations. The bottom-hole temperature induced a rather sharp drop in the voltage of the cells enclosed in the sonde, and in the potential variation measured at the surface it was difficult to distinguish between the change in thermometric resistance and the drop in battery voltage.

To achieve more dependable results, Boucherot undertook in 1948 the study of a "Deep Well Thermometer." He produced an instrument that withstood the highest bottom-hole temperatures and was more sensitive, quicker to operate, and less bulky than the previous model. The solution was a bold one. The thermometric wire was dipped bare into the drilling mud—hence a minimum calorific mass and a quicker response. The problem had been that, in order to render negligible the leaks through the mud, a low thermometric resistance was needed. As the trend of the time was to use resistivity logging circuits for all purposes, the resistance bridge constituting the thermometric sonde was fed with pulsated current. The signal was amplified by placing at the head of the sonde one of those unorthodox square current transformers already described in connection with the Laterolog. With no substantial change in operating method or interpretation of results, the deep well thermometer was put

[55] The problem was to produce a gas-free oil, as at that time gas had no commercial value and was flared on the spot. However, reservoir conservation rules strictly limited the gas : oil ratio. Only the reservoir energy of the gas was involved, but it remained essential to locate its entry.

into operation in 1950 and lived up to expectations. Even though the thermometric wire had to be insulated by a special coating for operations in highly conductive muds, its basic qualities were such that it led to consideration of a type of production logging to be discussed later.

The Borehole Caliper

It may seem strange that a tool to determine borehole diameter had not been produced by Schlumberger at the outset. To measure a geometric value by an electrical method would have been well within Marcel's talents. Perhaps he was too remote from the field, and the engineers did not call his attention to the significance of the problem. At any rate the first caliper was put into operation in 1938 by Halliburton, a leader in the cementing of casing. It had been proved by thermometric measurements that the level to which the cement rises in the annular space between borehole wall and casing is always less than it would be if the borehole diameter were equal to that of the drill bit. As such equality is nearly always prevented because the borehole is exposed to erosion by the circulation of the mud and the whipping of the drill pipe, the annular space must be accurately known to determine the amount of cement to reach the desired level. The war delayed the success of the Halliburton caliper until 1945.

The caliper put on the drawing board by Schlumber in 1946 reached the market in January 1947 under the name "Drill Hole Gauge." The sonde utilized three long, flexible spring bands that contacted the borehole wall, their lower ends fastened to a rod free to slide inside two coils on the sonde mandrel. The spring band diameter (hole diameter) was calibrated in terms of rod position, which, in turn, was a function of the coupling effect between the two external coils. Here again conventional resistivity logging circuitry was used. The primary coil was fed by a sawtooth alternating current produced at the surface from pulsated current by a combination of inductance coils and capacitors; accordingly, the output of the secondary coil was a pulsated voltage[56] easily accommodated by the conventional resistivity measurement system. Since the inductive system removed the need for any mechanical contact, the design

[56] The induced voltage is proportional to the derivative of the inductive current with respect to time, and the derivative of a sawtooth curve is a square-pulse curve.

was very simple. Furthermore, by varying the metallic assembly of the rod, one could make the signal proportional to either the diameter or its square. In the latter case, the log could be read in $\pi d^2/4$, in other words, in terms of the area of the supposedly circular cross section. A mere integration of the curve as a function of depth then gave the volume of the borehole. However ingenious this solution, clients preferred a linear scale. The presentation most in demand, the so-called "Mae West," consisted in tracing two symmetrical curves, which gave the impression of a cross section of the borehole by a plane passing through its axis.

Later experience demonstrated the need to record the borehole diameter simultaneously with certain measurements (Microlog, Microlaterolog, dipmeter, etc.) so as to better control their quality. Although this recording, which aimed solely at the sonde–borehole geometry, was less dependable than results obtained with the caliper proper, it often gave a first approximation of the volume of the hole and of the expected depth of the cement top. From that point on, special caliper measurements were reserved for problem cases: unusual borehole shapes, large diameter increases, the need for exceptional precision, and so on.

Sampling

In the southern United States and eastern Venezuela especially, the Sample Taker or Sidewall Sampler had been remarkably successful since its introduction in 1936.

The recovery rate had been considerably increased by the substantial improvements made over the following decade in the bullets, retrieving wires, and loading methods.[57] However, geologists and engineers were no longer content to look at the cores, smell or lick them, or even examine them with a steroscopic microscope or under ultraviolet light. They were now demanding a more systematic analysis of the samples, particularly for porosity and permeability. Certain laboratories provided such measurements, but these still had to be interpreted taking into account the variation sustained by both parameters under the impact of the bullet on the borehole wall. Furthermore, there was agreement among specialists that the samples were too small for a complete analysis that would include

[57] Up to 80 samples for every 100 shots fired was the recovery rate in favorable areas.

paleontology, clay percentage, oil saturation, oil composition, and other relevant factors.

The problem was tackled in Houston and Paris simultaneously. After many discussions that, incidentally, were indicative of the kind of competition existing between the two centers, the Paris model prevailed.[58] This was the last major contribution of Marcel, who had not conceived it, but had pursued its realization step by step. The C.S.T. ("Chronological Sample Taker") could take 30 samples in each run, against 18 for the old sidewall sampler, and their volume was 2.5 times greater. What was really new was the loading and igniting system, devised by Boris Schneersohn and Jean Planche (Fig. 78). Igniter needles, already tested on the bullet perforator and since extended to other models, allowed for the use of cartridges whose powder, carefully controlled as to weight and moisture, ensured a uniform fire power. Selectivity was achieved in an original manner: the firing of each shot armed the following one, thus eliminating any ambiguity as to the chronology of the firings. To this end, Schneersohn and Planche had first considered utilizing part of the deflagration gases somewhat like an automatic weapon; but Marcel raised the objection that the gas ports would be quickly eroded, and proposed a solution of his own: a steel wire, pulled by each bullet fired, armed the following one. Improvised as this solution appeared, it proved to be the right one.

A testimonial to the quality of the C.S.T. is the fact that it has undergone no major change since it was put into operation in 1951 in the United States and Venezuela. Only accessories—bullet bottoms, wires, and igniter needles—were gradually perfected to the point of ensuring optimum performance in areas best suited to sidewall sampling. However, despite displays of ingenuity and progress in steel metallurgy, sample taking remained as unsuccessful as in the past in hard formations such as sandstones or limestones. The conviction finally prevailed that percussion sampling was impossible in such formations, and that the problem had to be attacked from another angle.

As early as 1947 Schlumberger had offered the service of a mechanical sampler. It had been conceived, designed, and built in Houston by Maurice Mennecier, an enthusiastic and touchy perfectionist, whose basic idea was simple: a hollow-core barrel with a diamond drill crown was driven by an electric motor, and carved in a borehole wall was a core 1

[58] The Houston model had to be limited to small-diameter wells.

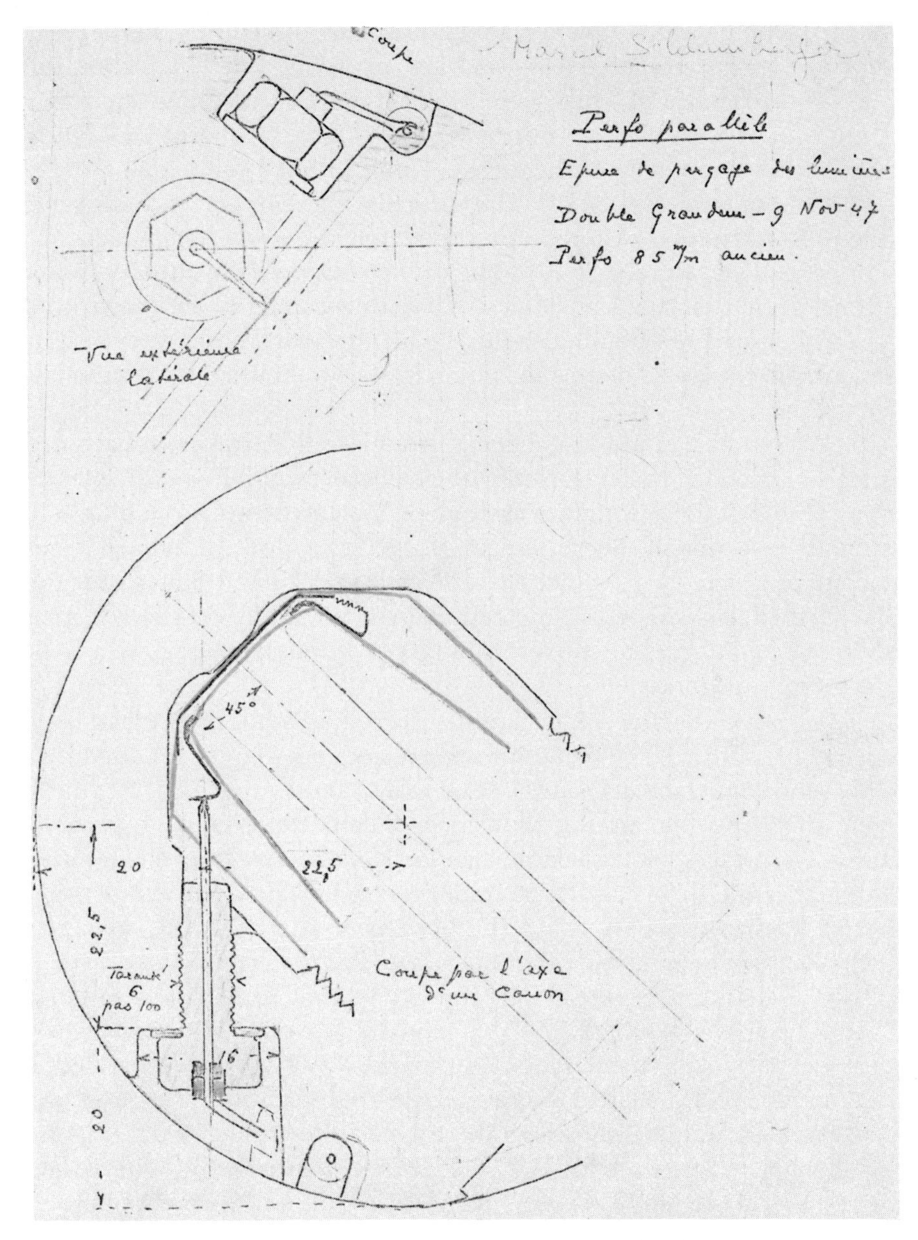

Figure 78. Sketch by Marcel Schlumberger of a proposed needle-igniting system.

inch in diameter and 2 inches long. Practical difficulties were great: anchoring, feed, core recovery, and so on. Only a few units of this ingenious mechanical prodigy were built; they met with some success in the sandstones common to northern Texas and southern Oklahoma. However, the complexity of the device, along with the skillful handling of the winch required, and chiefly the difficulty of sampling in limestones where porous samples disintegrate, prevented wider use. Manufacturing was discontinued, and about 1955 the service ceased. Not until 1970 was the concept of a mechanical sidewall sampler resumed in another form.

Another kind of sampling should be mentioned: the recovery of fluid samples from the pores of permeable rocks. Several methods are conceivable.

In 1951–1952, at Marcel's behest, study was undertaken in Paris of a "tester" or "suction" bullet. Fired by the sidewall sampler, it was to take a sample of the fluid through a system of instantaneously opening and closing valves actuated by the impact on the formation. In 1952 Jacques Delacour conducted a few successful tests in the United States, but the project was shelved when it appeared that the few cubic centimeters that could be tapped under the most favorable conditions were inadequate for any thorough diagnosis.

To increase the volume of fluid recovered, another device had been studied. This consisted of firing into the formation a bullet retained by a flexible tube connected at the other end to a container in which, through a system of valves, the liquid could accumulate. From similar projects undertaken in Paris and Houston, two instruments were born. The first, marketed between 1955 and 1957 under the name "Fluid Sampler," could take five samples of 330 cubic centimeters in a single run. This was deemed inadequate, and the other, designated as the "Formation Tester," prevailed. It took only one sample, but the volume exceeded 5 gallons.

The formation tester (Fig. 79) involved a flexible pad forced tightly against the permeable formation to be tested. Once it was well sealed, a bullet or the jet of a shaped charge pierced the pad and penetrated the formation, thus connecting it with the sampling container. All the stages of the operation were recorded at the surface, particularly the flow line pressure versus the sampling time. Thus, in addition to a good sample of the fluid, these data, which under certain assumptions permit calculation of the formation permeability with an acceptable approximation, are obtained.

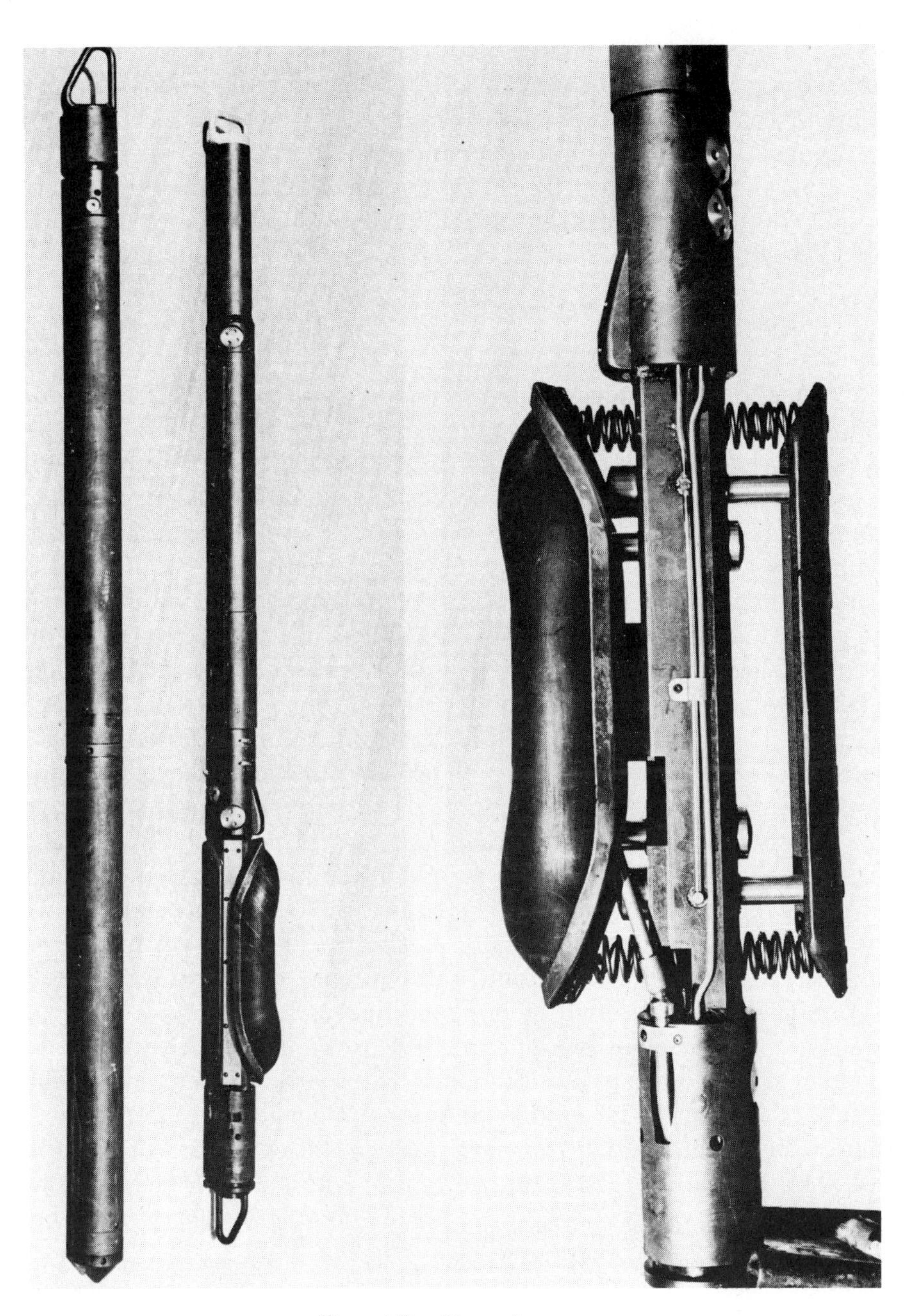

Figure 79. Formation tester.

Of all the instruments of its kind only the formation tester has survived to this day. However, in spite of many alterations and improvements, it remains of limited use because of its mechanical complexity, operational difficulties, and maintenance requirements. It has found certain application in southern Texas, in Argentina, and in Nigeria, but has in no way eliminated the drill stem test, which requires a more demanding and hazardous endeavor but yields much more comprehensive results.

Perforation

The perforation technique had barely evolved during the period from 1938 to 1945. The only noticeable change in Schlumberger field operations was the introduction of firing by bursts, which was more in the nature of a palliative of the shortcomings of the igniting system than actual progress. In 1948 Schlumberger, as well as its competitors, became equipped to lower and set bridge plugs[59] in the casing. Until then, the operating companies themselves had set the plugs, lowering them at the end of the tubing or the drill pipe. However, since the operation very often took place immediately before or after perforating, it seemed logical to take advantage of the cable to save rig time and improve setting depth accuracy. To enter the field, Schlumberger made an agreement with Baker Oil Tools, a manufacturer of bridge plugs and associated setting equipment.[60] In 1949 the encouraging reaction of clients led to augmenting this service with a more delicate operation: the setting of production packers.[61] Since in either case the problem is to lower an assembly almost as wide as the inside diameter of the casing, the operation can be jeopardized by the slightest (casing) burr. Unfortunately any gun perforation produced a burr; hence it was out of the question that a packer or a plug be run through a perforated section, and this made the service much less attractive. The problem was put to the Paris center, which was then in charge of all perforating developments. The solution was quickly found, but it took time to render it fully operational. It was to place in the gun, in front of the bullet, a small "no-burr" plate, thanks to which the perforations were as neat as if they had been made with a drill. Penetration was

[59] A plug to seal a borehole at a given depth.
[60] Sealing results from submitting the packing of the plug (hardened elastomer) to very high stresses, generated in the borehole by the combustion of a special slow powder.
[61] A packing gland between casing and tubing.

not affected, because the energy consumed to perforate the plate was compensated for by an increase in pressure build up and muzzle velocity.

More serious problems, however, soon arose. One was that the duration of operations involving hundreds or even thousands of shots had to be shortened; more importantly, the bullet penetration in consolidated sandstones and limestones was inadequate, and there was need for increased firing power. The first problem was one of accessories, of proper organization of the loading shop, and of better personnel training, rather than of the perforator itself. This aspect concerned Kuwait, Iraq, and Venezuela more than the United States, where large shooting jobs were few and, in any case, not handled by Schlumberger. The operational speed was first increased in Kuwait. A truck-mounted shop equipped to reload the guns on the spot as shooting went on was driven to the well site. The layout was excellent, and every convenience was available: guns were assembled and disassembled in record time by pneumatic tools. The tightest safety measures were enforced, for with perforators firing in burst any accidental ignition could have meant catastrophe. This equipment had been designed and built in Paris by Marcel, but its operation in the field was conducted by, among others, Louis Bordat, Mike Grosmangin, Joe Moffet, Jacques Priotton, and Ahmud, the Kuwaiti foreman who was unanimously credited for the liaison achieved with the local manpower and for the high efficiency obtained.

The second problem, penetration, became especially acute in 1948. The inadequate penetration in consolidated rocks was revealed not only by tests made on cores, but also by the fact that, in bailing the bottom of the hole, drillers would retrieve parts of bullets, and even whole bullets, that had not penetrated the casing. On the contrary, in sand–shale series, perforation at the correct depth resulted in oil production if the cementing had been properly done, and in water production otherwise. The efficiency of the perforator was obvious. Yet in 1948 Gulf developed a new cementing process through which the seal between casing and formations was substantially improved, with the result that even in sand–shale series penetration did not always suffice to bring the well in.

By that time the service companies had already undertaken the study of more powerful perforators. Schlumberger had been working since 1945 on a new gun, and in 1947 entered the field of shaped charges—an area where others had a substantial lead. The oil companies themselves were increasingly unhappy with conventional tools and submitted the various perforators to surface tests. These were not very meaningful in absolute value, since surface and bottom-hole conditions were quite unrelated. Nevertheless, they did allow a rough comparison of the various tools. During one of these tests Schlumberger had one of the poorest scores, and something spectacular had to be done. Even though its new gun perforator was not in final shape, it was advanced enough for a surface demonstration, and in December 1948 Planche was dispatched to Houston for this purpose.

The new perforator, to be known later as the "Supergun," was technically highly elaborate. Theory had suggested, and experience confirmed, that during the fraction of a second that followed firing the stresses developed during propagation of the shock wave were above the elastic limit of the alloy steels from which the body of the perforator and the gun barrels were built, but without any resulting damage. The threads had been given a special profile; the powder load density was very high; the seal plugs were integrated with the bullets; and the firing was actuated by the needle system of the sidewall sampler gun, a design that had made it possible to reinforce the bottom of the chambers. Whereas the bullet velocity was formerly 750 meters per second, with the supergun it reached 1000 and ultimately 1500 meters per second.

The supergun was an outstanding success, first in Venezuela, where it was introduced in 1949, and then in the United States in 1950. It was initially used as a special tool in conjunction with the conventional perforator but gradually supplanted the latter throughout the world.

Indeed, it was finally surpassed by shaped charges, but not before the 1960's. As to penetration, for a long time it could successfully compete, especially in sand reservoirs, where it had the advantage of fracturing the cement behind the casing and thus enhancing production.[62] Operationally, however, shaped charges were much more convenient, and the supergun was actually the ultimate in what could be achieved with a bullet perforator. Only marginal improvements resulted from further intensive engineering work, whereas with shaped charges the prospects seemed unlimited.

The technique of shaped charges derives from the armament industry. Its origins lay in an 1856 discovery by Munroe that a cavity in an

[62] If perforation takes place close to the water–oil contact, such fracturing may also cause water intrusion. But if the firing is positioned much above that contact, the advantage is such that it justifies the development of fracturing shaped charges.

explosive charge concentrates the energy of deflagration (the Munroe effect). How to multiply this effect was discovered a few years later: when a conical cavity is dug at the end of a cylindrical charge and lined with a thin metal, the hole in the target will have a narrow opening, but a much greater depth. Analysis of this phenomenon shows that the shock wave disintegrates the cone, and that the particles of the metallic lining act like a jet with a velocity of 6000 to 10,000 meters per second, a velocity at which the developing pressures are far above the mechanical resistance characteristics of the best steels. Under the impact the target behaves like a plastic body.

Great enthusiasm was aroused by the discovery: the armor of iron-clad vessels seemed obsolete. Disappointment followed; under water, shaped charge torpedoes did not work because the jet formed only when the cavity was filled by a compressible fluid such as air. Not until World War II did the proliferation of armored vehicles bring about, on both the Allied and the German sides, the shaped charge projectiles, known as *Panzerfaust* or bazookas, that could pierce the heaviest armor. Thinking then turned naturally to a similar solution for the perforation of casings. The only difficulty was the need to maintain air in the cavity and hence to use sealed containers. This feature was tightly protected by patents.[63] Preliminary tests were conducted in Texas in 1946, but not until 1947 did a major development take place with the founding by Robert H. McLemore and R. C. Armstrong of the Welex Jet Perforating Company and the manufacture of the charges by du Pont de Nemours. The first commercial perforation took place in Mississippi to the advantage of Gulf Oil Corporation.

Great improvements in the charges themselves have since been made, but their operation remains essentially unchanged. The charge carrier is a steel cylinder (about 6 feet long and 4 inches in outer diameter), in which threaded openings have been machined at intervals. The charges are housed in the cylinder, the base of the cone facing the opening, into which is screwed an aluminum or steel plug that will be perforated and destroyed by the firing. The charges are generally distributed on a helix so as to have four perforations per foot in four azimuths at

[63] The first patent of Clyde O. Davis (1942), licensed to du Pont de Nemours, did not include this feature; but that of Muskat (Gulf, 1945) specified that the cavity of the cone, lined with a ductile metal, is filled with a compressible fluid.

90 degrees from each other. They are simultaneously ignited by a detonating cord, which is itself primed by a standard blasting cap (Fig. 80). Each firing is a burst of 24 to 30 shots. As nearly all the energy of the explosive is concentrated in the perforating jet, the stresses sustained by the charge carrier are much lower than those in the bullet perforator. The result is much less weight and easier handling, loading, and access to the well. Provided that the operation requires a substantial number of holes and that the spacing of the openings in the charge carrier is suited to the type

Figure 80. Four inch shaped-charges carrier.

of production contemplated, the shaped charge perforator is an almost ideal tool.

Remarkable performance was attained as soon as the shaped charges became operational. Surface demonstrations, much easier than with the conventional perforator, showed that mild steel targets 4 inches thick were pierced through and through, and that penetration in hard sandstones or limestones was at least 2.5 times that of a bullet. Under the circumstances the bullet perforator, in spite of such advantages as flexibility of bullet-by-bullet firing, fracturing of the cement, and lesser cost, was doomed.

The development of shaped charges had caught Schlumberger flat-footed. At a time when the Paris center, still in the process of postwar reorganization, was undertaking responsibility for the whole perforation technique, new choices were already emerging in the United States. However, while still concentrating its efforts on the supergun as a tool suitable for difficult operating conditions, Schlumberger had the sense to keep abreast of what was going on elsewhere in the field. In 1946 a study was commissioned from the French Direction des Poudres, which includes leading experts in the field of explosives, and somewhat later Schlumberger entered into an agreement with the Brandt Company. These steps enabled Boris Schneersohn, Alexis Venghattis, and Claude Baks to realize, in 1947, their first shaped charges. In 1948 Schlumberger, in collaboration with the French–German Laboratory in Saint-Louis (near Basel, Switzerland), undertook basic research from which it acquired the capacity to develop and make its own explosives.

The first shaped charge perforations by Schlumberger took place in Kuwait in 1949. The results were so conclusive that bullet perforation soon became obsolete. This experience was repeated in 1949 in Venezuela, where competitors with their newly imported shaped charges were giving the supergun a tough time. Finally, in 1950, once the question of patents had been settled, Schlumberger was the only company in the United States to bypass du Pont de Nemours, the exclusive supplier of all other perforating companies.

Schlumberger's original contribution was all the more appreciated since a special engineering section had just been established in Houston with one of its first objectives that of adapting perforation to a new technique which the major oil companies, Esso in particular, were developing under the name "Permanent Well Completion." This production technique consisted of arranging, from the outset, for all completion and work-over operations involved in the lifetime of an oil well (water shutoff, recementing, reacidization, reperforation) to be feasible without interrupting production, that is, without "killing" the well by pumping heavy mud, and then removing the tubing. To make perforation an integral part of this new technique required not only a small-diameter perforator but also a cable small enough to pass through the packing gland of the blowout preventer. The quick solution was to scale down the $3\frac{3}{8}$ inch shaped charge so that it could be used in a $1\frac{11}{16}$ inch expendable aluminum tubing carrier. The effort resulted in a charge which shot upward at a 45 degree angle with a shot density of five charges per foot. Later on, a string of sealed frangible aluminum capsules, each with its own shaped charge, replaced the tubular gun. These were of particular interest in operations outside the United States because of their convenience and portability.

The $1\frac{11}{16}$ inch overall guns were able to pass through most production tubings; as to the cable, it was a single-conductor, armored type about $\frac{7}{32}$ inch in diameter, wound on a small truck-mounted winch. Gillingham deserves the credit for having recognized the appeal of such equipment and taken the initiative in its manufacture. Even though widespread commercial success was not immediate, Schlumberger had led the way in a new and revolutionary technique.

Phase Two: 1957–1974

A general view

To set 1957 as the beginning of the contemporary history of the Schlumberger processes is no arbitrary decision. Even though no major technical developments took place during that year, it was marked by a fundamental change in the legal setup of the Company: 1957 saw the consolidation into a holding company, Schlumberger Limited, of Société de Prospection Électrique (Europe and Africa), Schlumberger Overseas (the Middle East and Far East), Schlumberger Surenco (Latin America), and Schlumberger Well Surveying Corporation (North America). This restructuring of hitherto more or less autonomous companies into an international holding company required lengthy and laborious negotiations with French and American fiscal authorities. Once achieved, it put the group on a rational organizational basis, no longer tied to family interests. A few years later the Schlumberger stock was listed on the New York Stock Exchange. Although members of the family still hold a substantial number of the shares, they no longer make kinship requisite for top management. Jean Riboud, the president of the holding company since 1965, is not related to the Schlumbergers.

Technical and commercial development was deeply affected by the reorganization. Intensive diversification ensued, especially in the field of electronics.[1] On the other hand, Paris, after some slowdown following the death of Marcel in August 1953, resumed a leading role, as shown by the

[1] This activity is mentioned here for the record only, as it does not fall within the scope of this work.

establishment of a new center at Clamart in the outskirts of Paris. Although less extensive than the Houston center, it handles in a modern plant an increasing share of the joint research and development program and manufactures a substantial part of the equipment. A number of problems of coordination and standardization arise from the distribution of activities among Ridgefield, Houston, and Paris. It often happens that a field crew receives equipment manufactured partly in Houston and partly in Paris, and its assembly in the field is expected to proceed without a hitch. On the other hand, as regards both design and workmanship, there is no doubt that the friendly competition between Houston and Paris contributes to the success of many projects.

The research and development budget ($11,000,000 in 1970)[2] allows for the most ambitious studies. Electronics and informatics give Schlumberger direct access to highly advanced technology and specialized know-how. Although basic contributions were perhaps less frequent than in the postwar years, a definite increase in efforts directed toward improving the quality of instruments and services followed in the wake of the 1957 restructuring.

Time, accuracy, and environment are of the essence. The cost of standby time in drilling, especially offshore, is such that time becomes the overriding factor. This, in turn, creates a demand for equipment that can be set up quickly, is foolproof, and combines an array of tools whereby several measurements can be made in a single run. The increasing demand for accuracy in interpretation calls for redesigning both the electronics and the downhole mechanics of the equipment. Finally, each component requires special study when the equipment must function under extreme conditions, such as pressure and temperature at depths of 30,000 feet, polar cold, and storms at sea.

Schlumberger benefited greatly from advances in electronics and, in the first place, from the development of transistors. Especially for sondes, their value is obvious. With transistors, volume reductions become possible that even miniaturized tubes would not allow. This is an essential asset for a tool housed in a cartridge whose overall diameter never exceeds 4 inches. Transistors are much less fragile than tubes, sensitive as the latter are to shocks and vibrations, not only because of the glass, but mainly

[2] This figure covers only traditional Schlumberger activities. An almost equal amount is earmarked for other undertakings of the holding company.

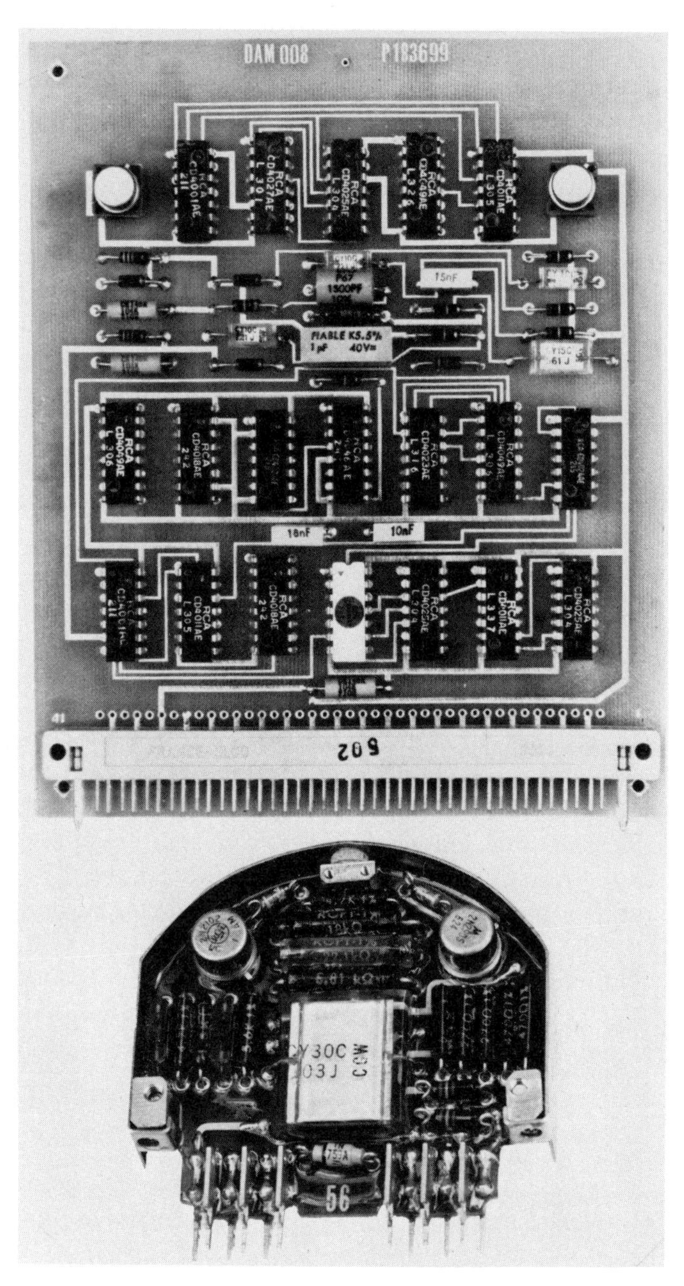

Figure 81. Printed circuit and modular element of cartridge.

because their components (cathode, grid, and anode) have a rather high inertia and easily become deformed. Furthermore, the transistor needs very little power, a valuable asset for circuits operating at the far end of a cable several miles long and ill suited for power transmission. In the latter case it was necessary to await the advent in 1954 of silicon transistors, which can withstand temperatures of 350 degrees F.

The printed circuit technique (Fig. 81) was, in its turn, put to ample use. In this case, unlike the audiovisual industry, dependability, together with easier troubleshooting and, mainly, volume reduction, was more important than cost cutting. On the other hand, the first integrated circuits, in which all the elements are assembled on a crystal of a few millimeters, were developed in 1960, and only their early inadequate resistance to temperature delayed for a time their incorporation into Schlumberger instruments.

The modern tools

Resistivity Tools

By 1957 the development of resistivity measurements had almost attained its high point. Schlumberger's unique experience, plus the basic improvements that had been added, left little room for further changes except to increase the dependability of the equipment through better selection of the components, to refine the features of the instruments, and to combine them in order to obtain maximum logging capacity in a single run.

The last pulsated current systems gradually passed out of use in the 1960's. The 20 year old chronological system was superseded by theoretically superior and at the same time equally convenient operational techniques.

A new 6 FF 40[3] induction sonde was produced with a greatly increased depth of investigation. The short normal was replaced by Laterolog 8, which had the same shallow depth of investigation and, through appropriate focusing, was less affected by variations in borehole

[3] The code name for a six-coil, vertically and laterally focused sonde with 40 inch spacing between main transmitting and receiving coils.

diameter. Later, a single tool combined two induction sondes, one the 6 FF 40 and the other much less deep, as well as the Laterolog 8. In conjunction with the Microlaterolog, this tool proved useful in the qualitative interpretation as well as the quantitative study of resistivities.

Francis Perrin had based his calculations on the assumption that concentric elementary currents had a negligible mutual effect (see p. 247). This is not always correct because, at the frequency used, a reaction resulting from the skin effect can be observed in low-resistivity zones. Although appropriate charts made it possible to correct the reading, an automatic correction at the time of recording was more rational, and the surface apparatus was modified accordingly.

The year 1971 saw the integration into a single tool of an induction sonde, a sonic device (see p. 307), and a very shallow resistivity instrument, the "Spherical Focused Log" (S.F.L.). The primary use of the S.F.L. is to replace the Laterolog 8. Its focusing system forces the current emitted by the central electrode to spread as in a homogeneous isotropic medium. The equipotential surfaces are then spheres. The borehole effect is eliminated without any increase in the depth of investigation, thus providing an accurate resistivity measurement of the invaded zone. The induction–sonic–S.F.L. combination was a distinguished technical achievement; the tool was particularly suited to the characteristics of the Texas and Louisiana coasts.

In regard to the measurement of R_t, it is worth putting on record that the advocates of Laterologs 3 and 7, up to recent years, could not agree as to which was the better system. Although Laterolog 3 is easy to combine with radioactivity instruments, Laterolog 7 gives a better determination of the true resistivity, especially in cases where R_t is unusually high. One might say that, roughly, Laterolog 3 is better suited to the conditions of American fields and Laterolog 7 to those of the eastern hemisphere. This is the reason why Houston refined Laterolog 3, and Paris Laterolog 7.

Laterolog 3 became a conductivity measuring instrument. The problem was not merely to convert the ohm scale into mhos, which could have been done by a reciprocator, but to obtain a measurement signal linearly proportional to conductivity: the current emitted by the central electrode being constant, voltage varies as resistivity; conversely, if the voltage is kept constant, conductivity is measured by the current variations. The latter system is used in Laterolog 3.

From 1964 Laterolog 7 included a long-contemplated downhole

amplifier. Next came the analysis of the so-called "Delaware effect," first observed in a western Texas basin on a sand overlain by a thick anhydrite bed of practically infinite resistivity. The effect is eliminated by a special arrangement of the electrodes, unless mutual induction generates perturbations in the opposite sense:[4] the "anti-Delaware effect."

During this period detailed theoretical studies were available, directed at defining a Laterolog suitable for conditions, frequent in the eastern hemisphere, of a true resistivity much above that of the invaded zone. André Poupon was the first to become fully aware of the problem and was able to enlist the interest of Doll, who lent his support and original views. Jean Dumanoir was put in charge of the mathematical treatment, and after many transatlantic discussions specifications for the tool were set. Two systems were combined: the deepest possible Laterolog, and another one with guard current return electrodes close to the transmitting electrodes, an arrangement that focuses the current throughout the significant part of its flow through the invaded zone. This shallow investigation system is called a "Pseudolaterolog" (Fig. 82). However attractive the combination, it appeared that two resistivity measurements are not always enough to give a good definition of the invasion profile and to determine R_t accurately; an additional Microlaterolog-type, low-penetration measurement was needed on the borehole wall. As the design of a tool able to make these three measurements simultaneously seemed difficult, a phased approach was adopted. With the measurement of R_{xo} left to the conventional Microlaterolog, an instrument was created that, in two successive runs and without leaving the borehole, recorded both the Laterolog and the pseudolaterolog curves. As a wide range (0.2 to 40,000 ohms) was desired, a technique had to be devised that would associate the features of both measurement systems: conductivity and resistivity. Under the name "constant power technique," it consists of maintaining the constancy, not of the voltage (V_o) or the current (I_o) of the central electrode, but of their product, that is, the power.

The results, bolstered by those of the Microlaterolog, proved quite satisfactory. The presentation of the three logs in logarithmic scales allowed in particular for the determination of their ratios, which are

[4] *Schlumberger Documentation, Log Interpretation,* Vol. I: *Principles,* Houston, 1972; J. Suau, "L'Outil Dual Laterolog R_{x0}," *Revue de l'Association Française des Techniciens du Pétrole*, No. 223, January–February 1974.

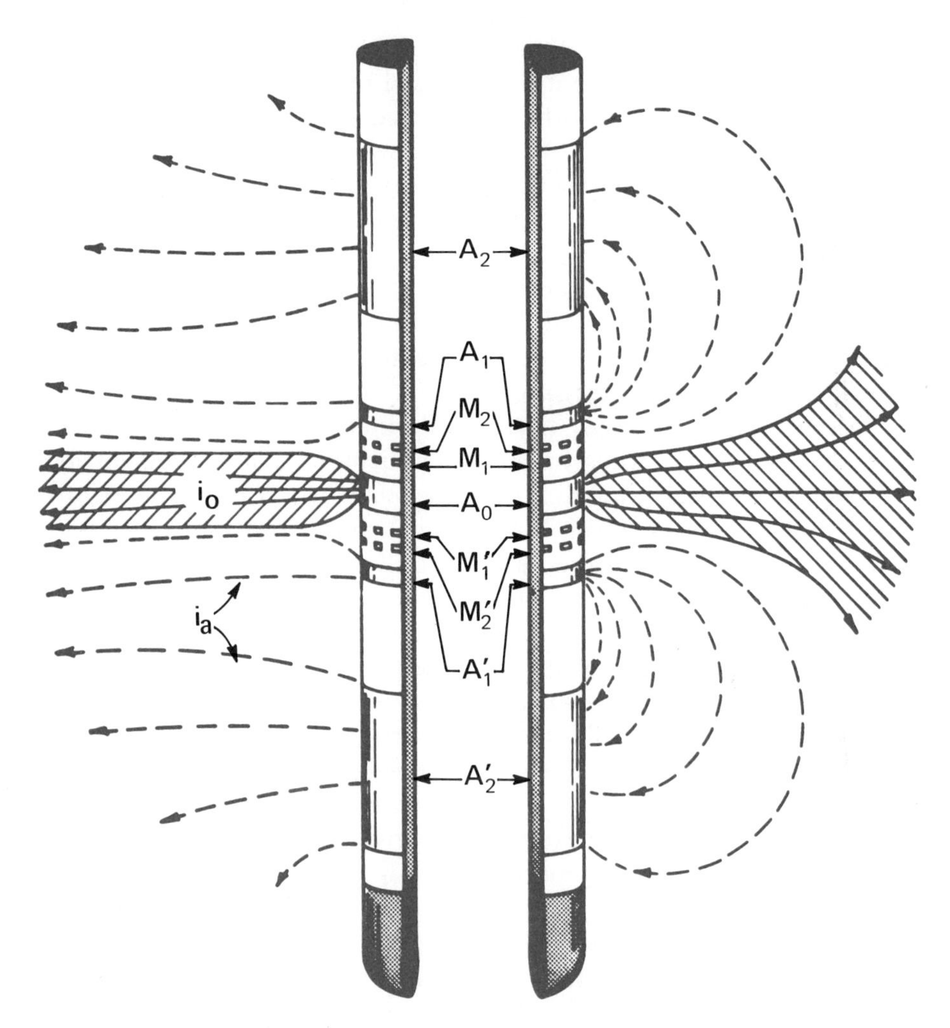

Figure 82. Principle of the Double Laterolog. Left: *Deep Laterolog.* Right: *Pseudolaterolog.*

essential data for the "quick look" (rapid qualitative interpretation), as well as for a complete quantitative analysis.

The final version, the "Dual Laterolog-R_{xo}," was completed in 1972. Three resistivities are recorded simultaneously: deep, medium, and very shallow. The last of these is no longer dealt with by the Microlaterolog, which is excessively affected by thick mud cakes, or even by the Proximity log,[5] which operates satisfactorily only if the invasion is rather deep; rather, it is measured by a miniaturized, pad-mounted "Spherical Focused Log" (Micro S.F.L.), which is hardly affected by the mud cake in spite of its very shallow depth of investigation. Moreover, by combining the values of the focusing and measuring currents, the Micro S.F.L. provides an indication of the thickness of the mud cake and a quality factor for the R_{xo} measurement.

The fact that in a Dual Laterolog the two systems use in common the electrodes A_0, M_1, M'_1, M_2, and M'_2 entails equal thicknesses of the current sheets, and therefore of the vertical definition of both measurements; even in thin layers the responses are in complete concordance, a great advantage in getting clearer results and exploiting them quantitatively. However, as had been foreseen, since the Micro S.F.L. measurement gives a much finer vertical resolution than the other two, it results in an often unwelcome degree of detail. This led to "smoothing" the Micro S.F.L. curve by a weighted average, automatically computed over an interval with a thickness equal to that of the Laterolog–Pseudolaterolog current sheets.[6] The result is equal vertical resolution for the three curves, which, in addition, are recorded at the same depth by means of a memorizing system. Finally, the tool also measures the S.P. through an electrode located far above the metallic mass by which it could be affected. This S.P. is also memorized and recorded at the same depth as the three resistivity logs. When the S.P. reading is unsatisfactory, as often happens under conditions in which the Laterolog is the most effective tool, a gamma ray log[7] is substituted, the appropriate cartridge then being housed inside electrode A_2.

The simultaneous recording of multiple parameters, the logarithmic scale presentation, and the smoothing and memorizing are all operations made possible by modern electronics. Perhaps less spectacular, though equally essential, is the mechanical sophistication of the equipment. Like

[5] Through a focused pad-mounted system, the Proximity log measures the resistivity of the zone fully invaded by the mud filtrate, but is less affected by the mud cake than the Microlaterolog, from which it is derived. It was put into operation in 1958 and has gradually replaced the Microlaterolog wherever there are thick mud deposits.
[6] What is actually computed is an average of conductivities.
[7] Another solution is the recording of a neutron log (C.N.L.; see p. 304) or a compensated density log (see p. 304–305), but the quality of the pseudolaterolog is affected by the excentralization these tools require.

all tools using pads, the Micro S.F.L. requires tight contact with the borehole wall. Correct alignment is achieved through a system of four hydraulically collapsible legs, which also ensures the centering of the bottom of the sonde and the measurement of the borehole diameter.

The Dual Laterolog-R_{xo} has confirmed in practice the theoretical considerations on which it was based, thus justifying the effort invested in its design.

Nuclear Tools and Techniques

Only after 1957 did nuclear logging techniques begin to show promise of successful development. Every service company concentrated its efforts in this new field, either on instrumentation or on interpretation. Although not a pioneer, Schumberger built on its experience and remained in the leading ranks.

One of the first objectives, as was true for the whole nuclear industry, was to design better detectors. As radioactivity phenomena are of a random nature, the number of events occurring within 1 second—emission of natural or induced gamma rays, neutrons, or other particles—is variable. A valid measurement requires an average over a time span that, according to the theory of probability, will be longer as the events are more widely spaced. All the counters used heretofore, whether ionization chambers or Geiger–Müller counters, were of poor efficiency as they counted only a minor part of the rays passing through them. To obtain significant measurements, logging required either overlong counters, which worked against the degree of resolution, or slow logging speeds that extended the duration of the operation. The scintillation counter (a photoluminescent crystal coupled with an electron multiplier) was welcomed as the ideal detector for logging, but the question of downhole temperature had to be dealt with first. Without waiting for the manufacturers' improvements, Schlumberger tackled the problem by undertaking an exhaustive study of the components that could serve in such an instrument. In 1960, with the assistance and advice of André Lallemand,[8] Jean-Pierre Causse designed and built a scintillation counter combining

[8] A French astronomer, the inventor of cells with high-sensitivity electron multipliers, which he developed at the Paris Observatory for the electronic camera he had conceived in 1937.

good performance under high temperatures with excellent mechanical strength, which proved highly suitable for logging. Its standards are so high that it has found wide application outside logging, especially in space technology; it is part of the equipment on a number of American satellites.

The neutron logging technique evolved along several lines. A basic change was the use of a criterion other than the number of gamma rays of capture to measure the neutron population in the vicinity of the detector, with the result that the present systems measure the epithermal neutrons, that is, those with an energy level slightly above 0.4 electron volt (eV), or the thermal neutrons, 0.025 eV. A further step was in the direction of lessening the effect of the mud. To this end the whole tool was initially excentered so as to rub against the borehole wall, an arrangement soon improved by introducing a neutron source and a detector into a kind of elongated skid hydraulically pressed against the wall. In an even more recent system the "Compensated Neutron Log" (C.N.L.) eliminates the mud effect by making simultaneous measurements on two counters located at different distances from the source; the relation between the readings, automatically established by the recorder, determines a porosity index that is practically independent of the borehole. The C.N.L. utilizes very powerful sources—up to 16 curies—such as plutonium–beryllium or americium–beryllium, which allow for increased spacing between source and detector. Thus, while maintaining a counting rate adequate for a satisfactory logging speed, the investigation is deeper.

Together with improved techniques, new methods were elaborated. As measurement of density by gamma ray absorption was used in various industries, it seemed logical to apply it to logging. The idea was that the porosity of a rock could easily be deduced from its global density, provided that the densities of the matrix and of the pore-filling fluid were known.[9] The matrix density is well known for various formations: for

[9] If ρ_b is the measured global density, ϕ the porosity, ρ_{ma} the density of the matrix, and ρ_f the density of the pore-filling fluid:

$$\rho_b = \rho_{ma}(1 - \phi) + \rho_f\phi$$

Hence

$$\phi = \frac{\rho_{ma} - \rho_b}{\rho_{ma} - \rho_f}$$

typical rocks it varies between 2.65 and 2.87. As to the fluid—since theory contemplated a shallow depth of investigation involving the invaded zone—it usually is a mud filtrate with a density in the vicinity of 1. A small amount of residual oil, with a density between 0.8 and 1, hardly affects the results; only the possible presence of gas will lead to an exaggerated porosity estimate. From thereon the road seemed clear, and Lane Wells was the first to tread it. Its tool was a sonde containing a radioactive source and a detector, pressed against the borehole wall by a spring blade. Shielded against the direct flux of the source, the detector counted the gamma rays after their diffusion into the surrounding formations. The theoretical basis was found to be correct, but mechanical trouble was serious inasmuch as the mud cake effect was important, and loss of contact with the wall led to aberrant results.

Schlumberger tackled this type of logging in 1957. On the instigation of Majani, then manager of the Maracaibo Division, the first tests were conducted by Tittman, a nuclear physicist of the Ridgefield center. One year later, the "density" or "gamma gamma" log appeared. In the beginning it had to compete with the sonic log on the Texas and Louisiana coasts, but its intrinsic qualities finally won out. Much of its success was due to Schlumberger's special lead in hydraulically controlled power sondes. Another 6 years passed, however, before a final solution to the mud cake problem was found.

As for the previously mentioned C.N.L., the locating of a second detector close to the gamma source provides a double reading from which, in the course of recording, the correct density values can be automatically computed. For reservoirs filled with liquid, the subsequent porosity determinations are entirely satisfactory. Although this is not true for residual gas, the difference between the porosity given by the neutron log and that provided by the density log, which is overestimated, is in itself a valuable indicator of the presence of gas, a criterion widely used in interpretation.

Another innovation in the nuclear field required a much more ambitious program. For a long time there had been interest in the rate of decay of thermal neutrons as a function of time: this rate increases with the amount of chlorine in their path. The reason for this is that, of all common elements, chlorine has, by far, the greatest neutron absorption power. Thus the rate of decay of thermal neutrons depends on the amount of sodium chloride or, ultimately, salt water.[10] The striking analogy with other measurements also dependent on the amount of salt water in a nonconductive medium gave reason to hope that a log could be obtained that would be similar to the resistivity log but could be recorded through the casing. It was not for Western scientists to lead the research in this field: in 1958 the Russians published their first experiments.[11]

Conventional sources cannot be used to measure the rate of neutron decay as a function of time; a discontinuous source that generates a large number of neutrons within a few microseconds is needed. The decay is measured within a few milliseconds following the interruption of the emission. The process is repeated a great number of times within a second. The measurement is simplified because of the fact that, the decay being exponential, two values suffice in principle to give the time constant. A log is obtained with a pulsated source and a detector for gamma rays of capture or thermal neutrons. From 1964 Lane Wells used this system in the United States under the name "Neutron Lifetime Log." Schlumberger's version, put into operation in 1966, was called the "Thermal Decay Time Log" (T.D.T.).

The great difficulty here lies in "manufacturing" neutrons in large quantities within a very short time span. The most suitable method is to use a particle miniaccelerator to bombard a tritium target with deuterium atoms. Yet to generate neutrons the nuclear reaction calls for energy involving a potential difference of some 100,000 volts, an apparently insuperable obstacle considering the reduced space and high temperatures at the bottom of a borehole. The problem was solved, not only for standard sondes about 4 inches in diameter, but also for the $1\frac{11}{16}$ inch sondes required to go through the tubing. This was a remarkable feat, particularly for the latter, which are used to run through the tubing under pressure and log the formations behind the casing while the well is flowing. One interesting application is the repeated observation at scheduled intervals of the saltwater level in a producing horizon.

Schlumberger's contribution to the pulsated neutron technique was

[10] The conventional neutron log studies the *distance* covered by the particles before their collisions with hydrogen atoms render them thermal or epithermal. In the present case what is at stake is the life *duration* of the thermal neutrons before absorption.

[11] B. G. Erozolimski, R. L. Voitsik, N. V. Popov, and A. S. Chkolnikov, "New Methods of Logging Bore Holes Using Pulsed Neutron Sources,"*Neftyanoie Khozyaistvo,* Vol. 36, 11 (1958).

important not only for the instrumental progress achieved, but also for the practical utilization of the results. The Ridgefield and Houston centers manufactured small series of instruments dependable enough to be put into service in the Middle East, but their correct operation was made possible only by the international technical organization. An original method had been devised for precision measurement of the neutron rate of decay;[12] interpretation was supported by Schlumberger's long experience.[13]

The function of the T.D.T. is not to evaluate a reservoir before casing is set. In the current state of the technique, its depth of investigation is not sufficient to make it competitive with the various resistivity logs. The quantitative interpretation of the results, however, makes it an outstanding instrument for production control insofar as it determines drainage efficiency and assists in optimizing production methods. Such is its role today in the most productive fields of the U.S.S.R., the United States, and the Middle East.

The Sonic Log

The origin of the sonic log is somewhat curious. It was developed as an adjunct to seismics; initially it contributed nothing to log interpretation, although it was to become one of its essential elements. However, its history begins much earlier.

In this connection a return to surface prospecting is appropriate. Schlumberger's field activities emphasized electrical methods, but since 1935 seismic prospecting had enjoyed the lion's share in the study of deep structures. It was known that seismic (or sound) waves propagate at velocities varying with geological formations. The thought then came naturally of an acoustical log, similar to the resistivity log, and on June 1, 1934, a patent was filed by Schlumberger under the title *Procédé et Appareillage pour la Reconnaissance des Terrains Traversés par un Sondage* ("Process and Apparatus for the Reconnaissance of Formations Drilled Through by a Borehole") (Fig. 83).

[12] J. S. Wahl, W. B. Nelligan, A. H. Frenttrop, C. W. Johnstone, and R. J. Schwartz, "The Thermal Neutron Decay Time Log," *Society of Petroleum Engineering Journal,* December 1970.

[13] C. Clavier, W. Hoyle, and D. Meunier, "Quantitative Interpretation of Thermal Neutron Decay Time Log," *Journal of Petroleum Technology.* June 1971.

RÉPUBLIQUE FRANÇAISE.

MINISTÈRE DU COMMERCE ET DE L'INDUSTRIE.

DIRECTION DE LA PROPRIÉTÉ INDUSTRIELLE.

BREVET D'INVENTION.

Gr. 8. — Cl. 1. **N° 786.863**

Procédé et appareillage pour la reconnaissance des terrains traversés par un sondage.

Société dite : SOCIÉTÉ DE PROSPECTION ÉLECTRIQUE (Procédés SCHLUMBERGER) résidant en France (Seine).

Demandé le 1er juin 1934, à 16 heures, à Paris.
Délivré le 17 juin 1935. — Publié le 11 septembre 1935.

[Brevet d'invention dont la délivrance a été ajournée en exécution de l'art. 11 § 7 de la loi du 5 juillet 1844 modifiée par la loi du 7 avril 1902.]

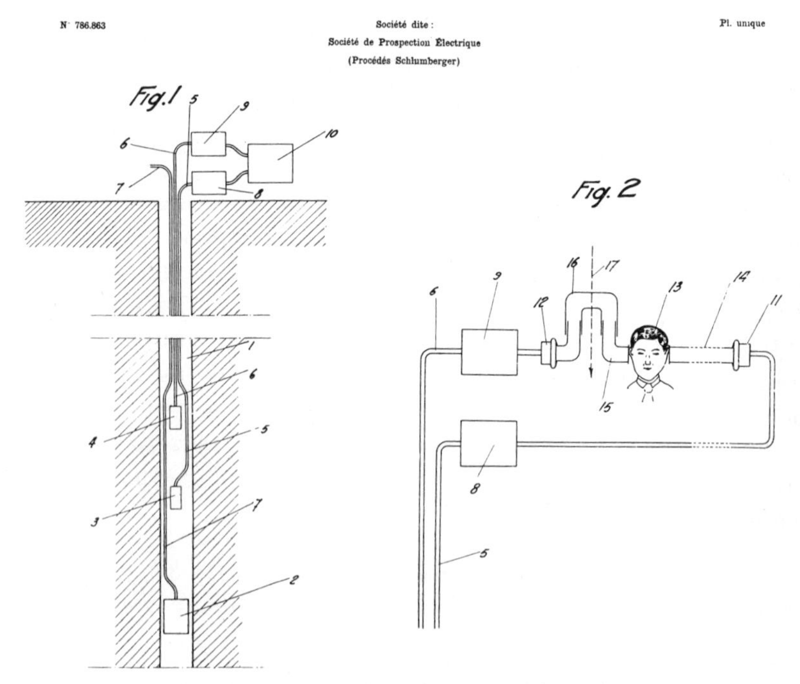

Figure 83. The Schlumberger sonic log patent.

On the basis of acoustic velocities of 1000 meters per second observed in sands, and 5000 in hard limestones, the patent covered "the reconnaissance process of the formations drilled through by a borehole . . . which . . . consists in determining the velocity of sound propagation in the various formations penetrated, such velocity being measured at various depths in the open-hole section of the borehole" (translation from the French). Also covered by the patent was the possibility of measuring in the same operation the intensity of the sound received, which would give "an indication on the absorption caused by the various formations drilled through" (translation). In fact the patent covered all the sonic techniques used today. The sonde included a sound generator and, close to the source, two microphones appropriately spaced vertically. If the acoustic signal is a sinusoid vibration, the propagation velocity in the formations is determined by the phase difference between the microphone signals; if it is not, a highly picturesque method comes to the rescue, based on the biauricular technique. It consists in listening, both ears working separately, to two earphones, each one fed by its own downhole microphone. The operator, all ears, moves a kind of trombone slide, thereby lengthening or shortening one of the channels, until he perceives the sound coming from straight ahead of his honest face (see Fig. 82, lower right). The position of the slide determines the phase difference, and the rest of the operation is pursued as in the case of sinusoid vibration.

With an automobile horn for sound, the device was tested in Pechelbronn in 1935. The biaural system proved too inaccurate, and the only remaining record of that episode is of a practical joke;[14] thereafter, despite a July 1947 research note from Doll, the matter rested until 1952.

However, the surface geophysicists felt an increasing need to know accurately the velocity of seismic waves in various formations, and it often happened that, as soon as a borehole had been drilled in an unexplored area, this parameter was measured point by point.[15] A geophone was lowered to the desired depth at the end of a logging cable, a dynamite charge was detonated at the surface, and the time was measured between

[14] At a certain moment the operator heard, instead of the usual cacophony, these distinctly spoken words: "Trilobites speaking. Go to hell!" (Trilobites are fossil crustaceans of the Paleozoic.)

[15] There is here a similarity with one of the first goals of downhole resistivity measurements: to obtain parameters for the interpretation of surface results.

the explosion and the arrival of the seismic wave at the geophone. The operation was repeated at some 20 different levels in the hole; the process was slow and costly, however, and there was too great a discontinuity between results for the desired accuracy.

At the instigation of Frank Kokesh, a field engineer transferred to Houston, better systems were designed. In the first the arrangement was reversed: with the geophone at the surface, a standard perforator fired 24 blank cartridges at as many depths in the borehole. Much time and many explosives were saved, but the energy of the shots did not reach beyond a few hundred meters. The few units built were used in 1952 in shallow stratigraphic reconnaissance coreholes. In 1955 new tests were made with the geophone being lowered downhole about 100 feet below the perforator, making it possible to measure, regardless of depth, the propagation time over 24 intervals in a single run. This "Long-Interval Velocity Logging" system was marketed for several years without any great success. One deterrent was the fact that two instruments 100 feet apart on the same cable meant a difficult fishing job, should either get stuck or lost in the hole.

The advantage of all these measurement methods was their relative facility. Propagation times were short, though not much more so than those of the conventional seismic method. But the processes were clumsy, and the results lacked continuity. The geophysical services of the major companies, Magnolia (Mobil), Humble (Esso), and Shell, resumed the search for a continuous propagation velocity log and undertook the development of more advanced techniques. Although reverting to spacings of a few feet between transmitter and receiver allowed for a much weaker source, it entailed measurements in microseconds, a unit not current in geophysics. Electronics easily solved this problem in 1952. Two systems prevailed. In the first, propagation time between transmitter and receiver was measured; in the other, as in the 1935 Schlumberger patent, the measurement took place between two vertically spaced receivers, which in principle eliminated the travel time in the mud and hence the effect of the borehole diameter.

The first operational equipment (March 1954) was produced by Magnolia, which licensed its patent rights to the Seismographic Service Corporation. In 1955 Schlumberger acquired the patent rights of Humble and thereupon developed a two-receiver system, which was marketed in 1958.

In addition to the essential contribution of the sonic log to seismic interpretation, more applications were soon found. Tom Hingle, a petrophysicist of Magnolia in the Mid-Continent, had observed correlations between formation factors and the first sonic log readings. Experience was soon to confirm that the more porous a formation, the lower is its sound velocity. M. R. J. Wyllie, already known from his study on S.P., deserves the credit for having, in collaboration with A. R. Gregory and L. W. Gardner,[16] expressed this correlation in mathematical terms. Their "time average formula" indicates that the time for a sound wave to cover a certain distance in a porous medium equals the sum of the time spent in the liquid and that spent in the mineral grains.[17]

These efforts to measure porosity lent considerable impetus to the development of the sonic log. It gave excellent results in sands and, as shown by Tixier, was remarkably compatible with the induction log in shaly sands.[18] In conjunction with the neutron and density logs, and integrated into the system of parameters fed into the computer for the complete processing of data, the sonic log has become a basic factor of reservoir evaluation.

In October 1965 two experts of Shell Oil Company, C. E. Hottman and R. K. Johnson, presented to the Society of Petroleum Engineers a paper entitled "The estimation of Formation Pressures from Log-Derived Shale Properties," in which they pointed out an interesting possibility regarding sonic log interpretation. While the travel time in shales evolves steadily with depth, distinct anomalies are observed when the bottom of the borehole is approaching a high-pressure porous bed. This is a most useful warning for the driller since it signals the danger of a blowout.

As anticipated in the 1935 Schlumberger patent, by then fallen into both oblivion and the public domain, velocity is not the only significant factor: another source of information is the intensity of the sound. It could not be measured with the initial sonic instruments, but their ab-

[16] M. R. J. Wyllie, A. R. Gregory, and L. W. Gardner, "Elastic Wave Velocities in Heterogeneous and Porous Media," *Geophysics*, Vol. 21, pp. 41-70 (1956).

[17] Their formula is

$$\frac{1}{V_m} = \frac{\phi}{V_f} + \frac{1-\phi}{V_r}$$

where ϕ represents porosity; V_m, measured velocity, V_f, velocity in the fluid; and V_r, velocity in the rock grains.

[18] Although both logs are affected by shale, a first approximation of the oil content is gained by proceeding as for nonshaly sands, without either log requiring correction.

normal behavior signaled zones of very low intensity. It soon became manifest that such intensity drops occurred precisely where high acoustical absorption, caused chiefly by horizontal fractures in the rock, could be expected. Despite the temptation to look here for a fracture indicator, experience showed that the phenomenon was complex and needed much closer scrutiny before an unequivocal diagnosis could be formulated.

Far from laboratories and engineering offices, there was an unexpected development. The sonic log had just been introduced in Venezuela, where the Maracaibo crews were pursuing a suggested experiment. Until then all measurements had been in the open hole; would a sonic log recorded inside the casing have any significance? Pierre Majani and Alan Rushton were struck by certain recurring anomalies. Intuition, based on a thorough knowledge of drilling, casing, and cementing processes, led them to the conclusion that the cause of these anomalies was the condition of the cement rather than the cased formations: a good bonding between steel and rock dissipates the signal, of which only a minor part reaches the detector. Conversely, in the case of poor bonding, the wave travels mainly in the casing, and the detector registers a strong signal. Should this assumption be confirmed, it would have momentous consequences: whereas cementing control then rested on a temperature log that had to be recorded within days after the setting of the casing, it would henceforth be possible to do the recording months or years after the well had been completed.

Day-to-day experience seemed to provide confirmation of this view, but for lack of any other substantiation it was accepted only with reluctance. Proof required that the equipment be modified so as to record the amplitude (intensity) of the signal. Again the Maracaibo team did what was needed: its electronics specialist, Jean Bouffard, traveled to Houston to work out the change he had proposed. A new service was born: "Cement Bond Logging" (C.B.L.), which after 1961 superseded thermometric cementing control.

Even though earlier tests had shown the difficulty of detecting fractures by sonic logging, the importance of the objective justified further effort. The analytical approach was to differentiate in the acoustical wave train between the compression and the shear wave, from which differentiation might come the ability to locate horizontal as well as vertical fractures. Geographically limited successes resulted. Although a number of specialists hold that the key to the problem is at hand, it seems doubtful that a solution is imminent.

An altogether different application of the acoustical phenomena was proposed in 1956 by Dowell, specialists in oil well cementing, acidizing, and fracturing. It was a sonar-type device, designed to measure the volume of the huge caves obtained by dissolving rock salt in order to provide space for hydrocarbon storage. On the basis of the same idea, Mobil Oil, in 1967, took out a patent covering an instrument having the rather misleading name "Borehole Televiewer." The target was no longer caves but conventional boreholes, and what was measured was not the propagation time but the intensity of the reflected pulses. It was like observing the borehole wall in monochromatic light, through the mud and even the mud cake if any: if it were hard and smooth, it would appear as shiny, whereas fractures would show in dark.[19]

Schlumberger, having acquired the license for the Mobil patent, made it operational, notably in western Texas, Libya, and Iran. A continuous log is obtained when pulling the sonde up the hole, while a rotating acoustic beam scans a low-pitch helix all along the borehole wall. In compact limestones, nonhorizontal fractures emerge sharply, as do the perforations in a cased well. This is the only method known to locate them unequivocally.

However important these side applications, the main use of the sonic log remains the measurement of sound velocity and, through the Wyllie formula, the computation of porosity. The tools were greatly improved, and in 1963 the Schlumberger sonic equipment was thoroughly redesigned. The present sonde includes two ultrasound generators and four receivers. An average is taken between two measurements, one on an ascending, the other on a descending, wave. This eliminates errors arising from diameter variations of the borehole or from the inclination of the sonde in relation to its axis.

[19] The question has often been raised as to how useful it would be to televise, or take color films, in boreholes. However interesting, a picture of a borehole wall would be less informative than its resistivity, density, sound velocity, and so on. Moreover, a borehole is usually filled with mud or crude oil, and since the real subjects of interest (i.e., the porous and permeable borehole walls) are plastered with mud cake, nothing much would be revealed. The television cameras that have been designed for highly specific purposes have found little application.

Miscellaneous Techniques and Tools

The continuous dipmeter with its three Microlaterolog pads had quickly achieved general use since its creation in 1955 and had opened the way to a new interpretation technique. Thorough studies in southern Louisiana showed that valuable data could be obtained not merely from the dip itself but, above all, from its variations relative to depth. Further developments included the identification of the structural dip, the detection of fault crossings, the location of old channels in deltaic complexes, and even the determination of the direction of the current at the time of sedimentation. Such interpretations were due to the collaboration of oil company geologists and Schlumberger engineers, of whom Al Gilreath was the first to formulate the basic principles.[20] Ray Campbell, then at the Ridgefield Research Center, participated in working out the theory, according to which a more thorough interpretation of the dips required a much higher number of determinations per meter than heretofore.

Rather than perfecting the existing dipmeter, the need was for a system with the highest possible resolution. Michel Gouilloud was put in charge of the project and achieved an outstanding new correlation dipmeter. Its telemetry includes the poteclinometer; a fourth pad confirms the results of the other three while ensuring a better positioning of the tool in the borehole. The microresistivity curves are as detailed as possible, and since five of them are recorded simultaneously (the fifth being used for the evaluation and correction of the speed variations of the sonde as it is being pulled out), the number of data transmitted by the cable created a problem that only sophisticated electronics could solve. The sonde being powered, the opening, the closing, and even the pressure of the pads on the borehole wall are controlled from the surface. After the instrument was put into operation in 1966 under the name "High Resolution Dipmeter" (H.R.D.), its mechanical and electrical dependability was enhanced by several changes. An important step forward was the manufacture by Schlumberger in 1970 of the first sonde made entirely of titanium, a metal preferable to the conventional stainless steel because of its better mechanical behavior, its light weight, and its corrosion-resistant and non-magnetic qualities (Fig. 84).

[20] J. A. Gilreath and J. J. Maricelli, "Detailed Stratigraphic Control Through Dip Computations," *Bulletin of the American Association of Petroleum Geologists,* Vol. 48, No. 12, pp. 1902–1910 (December 1964).

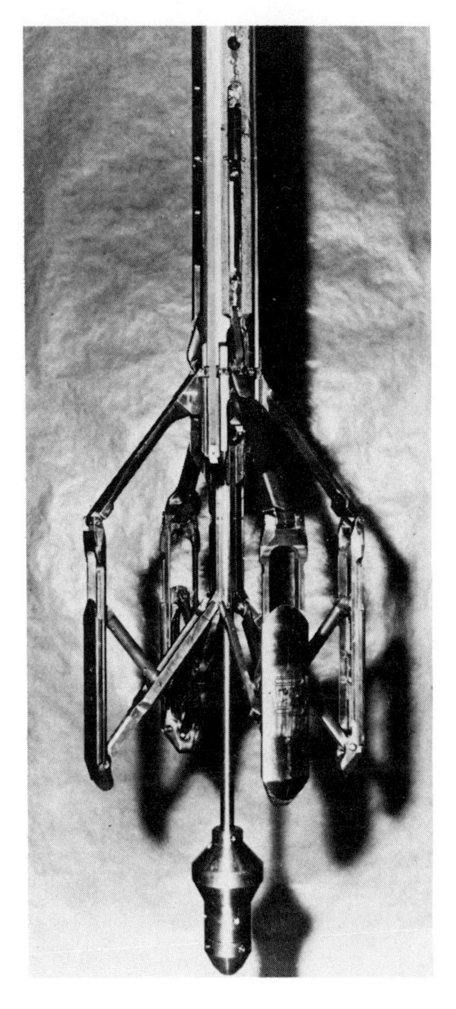

Figure 84. H.D.T. dipmeter sonde.

So many correlations can be derived from the extremely detailed microresistivity curves that they are beyond the scope of conventional analysis. From the outset the high-resolution dipmeter was conceived for computer processing, which required recording on magnetic tape and adequate programming. Magnetic tape recording was not limited to the dipmeter; in 1960 it was already clear that data processing was to play a leading role in the interpretation of logs of all kinds. Programming was and continues to be the subject of intensive research from log correlations proper to the analysis and presentation of results. A few magnetic recorders were operating as early as 1961, but they came into general use only with the advent of the H.R.D., and from that time forward the optical log of the dipmeter merely monitored the operation. The dipmeter has become today an indispensable tool for the geologist in the exhaustive study of sedimentation and structural problems.

An interesting outgrowth of the dipmeter is the use, since 1973, of the four-arm sonde as a borehole caliper under the name "Borehole Geometry Tool." As the cross section of a borehole is more frequently elliptic than circular, the four arms allow for the measurement of the two axes of the ellipse, and consequently a more accurate computation of the volume. The (optional) use of the poteclinometer gives the azimuths of these axes and reveals the distortion and rotation of the ellipse relative to depth; heretofore the geometric representation of the borehole had never reached such a degree of accuracy.

Whereas the C.S.T.-type sidewall sampler has undergone few changes in recent years, the formation tester has been substantially improved. Fitted with a second pad and shaped charge, its sampling of small intervals (1 foot) is more representative than a single-hole test. In addition, the two-pad tool can be used for taking a fluid sample through the casing, as is desirable in some cases. A drawback is that the perforations left in the casing provide openings through which—at inopportune times—water, oil, or gas can enter the borehole. The conventional remedy is to plug them by a squeeze job involving a special truck equipped with powerful pumps, although often only a few quarts of cement is required. Schlumberger therefore designed a system whereby the tester, once anchored in place, discharges into the perforations the amount of quick-setting cement necessary to seal off the holes. With this design the tool becomes more complicated, longer, and heavier. The operation is delicate and the choice of cement critical; ultimately, success is contingent mainly on how good the initial cementation between casing and formation was. In any case, the operation is worth trying; if it fails, it is not too late to summon the cementing truck.[21]

[21] This same Schlumberger tool is used for an altogether different purpose: the injection into incompetent sand of resinous compounds that bind the grains without excessively reducing the permeability of the reservoir, and prevent them from being carried by oil or gas through the production string and eroding it.

The coring of hard formations reverted in 1970 to diamond tools, but instead of the Mennecier system (see p. 284) two small circular saws, slightly offset at an angle from one another, travel vertically along the borehole wall and carve a 4 or 5 foot triangular prism with a concave base. The mechanism is delicate and complicated; much know-how is needed for its anchoring in the well, the automatic feed of the saws, the recovery of the sample, and so on. The elegant mechanical concept, however, would have made Marcel happy.

From 1957 to 1970 perforation made substantial progress. These advances derived from the technology of explosives and shaped charges and also from better adaptation of perforators to completion methods.

Today bullets have been completely superseded by shaped charges. With powerful, high-temperature-resistant explosives and conventional carriers, penetrations of 20 inches through casing, cement, and formation can be obtained that, in almost any case, exceed the depth of the invaded zone. Different liners are used in the shaped charges in order to ensure penetration, fracture the cement, or even crack the rock, according to what is needed. Perforation below the tubing, as worked out for the permanent well completion technique, is now frequently used, not so much to avoid killing a well under production as to optimize its output from the beginning. Indeed, a conventional perforation entails the risk of a blowout unless the well is filled with a mud of such high density that its hydrostatic pressure overbalances that of the reservoir. The trouble, however, is that this heavy mud may invade the porous formation and reduce production. On the contrary, when an oil well is fitted with wellhead tubing and packers, it can be perforated safely even if the reservoir pressure is very high. This significant advantage explains the existence of the many "through tubing" perforator models, using small-diameter shaped charges rendered increasingly powerful by constant engineering efforts. They represent an important contribution to the improvement of well completion methods.

Multiple completion, that is, the simultaneous production of several beds in a single well with one tubing for each reservoir, has given rise to specialized techniques. Such wells may include the exploitation of up to five pay zones, each with its individual tubing. The problem is then to perforate the casing at the desired level without damaging the tubing or tubings that carry the production from the lower beds. To determine the azimuth of the firing, a radioactive pellet is lowered inside each tubing to the desired level; the pellets are located from inside the casing by a directional detector coupled with a swiveling perforator firing in a single direction. Another system comprises a gamma ray densimeter, which is also directional and locates the tubings by the increased density read in their direction. These are all exceptional operations, difficult to execute and requiring an extremely complex apparatus, along with a flawless interpretation of measurements.

Traditionally, Schlumberger's activity in the field began with the drilling of an oil well and ended with its completion. As soon as the well was connected with the pipeline, the crews had to withdraw without the opportunity to put their experience and versatile equipment to further use. There seemed to be a loss of potential business in this routine. One possible direction for such expanded activity was analysis of the flow at the various producing levels of an oil well. Optimum production of one or several reservoirs mandates that, when necessary, the behavior of a well be corrected. This, in turn, requires information on the amount and nature of the fluids flowing into the well at every level. The greater the production, the more important the problem, as is typically the case in the Middle East. However, it was in 1957, in Brunei, that at Shell's instigation, and thanks to the perserverance of Simon Noïk, Schlumberger became interested in the question. On his own initiative, Noïk had built, with locally available facilities, rudimentary tools promising enough to convince Paris that the lead was worth following. It soon became apparent that, in addition to instruments, the primary need was for a theoretical and experimental study of the flow of fluids in vertical tubes. A laboratory equipped for the simulation of the whole range of diphasic flows as they occur in petroleum production was built—probably the only one of its kind in Europe.

A flowmeter suffices in the case of monophasic production: the total flow is measured point by point, the inflow at each level being determined by the derivative of the flow curve as a function of depth. The situation is no longer simple when the flow becomes diphasic or triphasic (water–oil-gas) because the flowmeters are affected by the nature of the fluids driving them. A system of measurements then becomes necessary in which the percentage of each phase is determined level by level and from which, in turn, the amount of each fluid entering the well can be calculated at each level. One difficulty is that in order to lower the tools below the tubing small diameters (about $1\frac{11}{16}$ inches) are required; another is that the measurements must not interfere with the phenomena being evaluated.

Schlumberger flowmeters are of the spinner type. The percentage of each phase is obtained by also measuring the density and the dielectric constant of the mixture: the density reveals the ratio between gases and liquids, and the dielectric constant that between water and hydrocarbons. Once more, the research program proved more complex than anticipated, especially as the solution to the problem calls for two series of instruments, one for outputs exceeding 500 barrels per day, another for lesser productions. In the latter case, fluid velocity is boosted by narrowing the flow section with an inflatable bag. The measurements are translated into simple physical parameters: electrical capacitance, revolutions per second, frequencies, and so on. However, there is another original method, also credited to Noïk: determination of the density on the basis of the hydrostatic pressure differential between two levels (Gradiomanometer). Other tools complete the array, in particular a special caliper, and a high-resolution thermometer which locates gas leaks in the tubing and fluid flows behind the casing.

In spite of the importance of production logging a long time was required for it to prevail. The interpretation technique, which had to be created from scratch, showed the possibilities and limitations of the system. Although its advantages are no longer questioned in the case of large producers, the fact is that most oil wells are small producers for which such a service, however beneficial, remains too complicated and costly. Yet it seems logical to think that Schlumberger's initiative, immediately followed by the activity of its competitors, has marked the beginning of an irreversible trend and that with the benefit of new measurement techniques, among which the Thermal Decay time log is already operational, this type of logging will become a key element of hydrocarbon production.

A Note on Offshore Drilling

Even to a public not especially aware of petroleum problems, there is something exciting in offshore drilling with its great economic investment, its quality of adventure, and its use of bold techniques. Consequently, the question may be asked why a special chapter has not been devoted to it. Indeed, for Schlumberger the problem was not new: by

1930 drilling in southern Louisiana had moved from the "bayous" to the coast, and from there to the shallow waters of the Gulf of Mexico. In Venezuela, too, in the 1950's drilling platforms could be found in the middle of Lake Maracaibo. Solution of the problem had simply required a switch from the truck of the O.S.U.-C (see p. 224). This unit was normally installed permanently on the drilling platform and included the equipment common to all services: cable, winch, cabin, optical recorder, and generator. As a radio network and air or sea transport were available, no time was lost in moving the crews and their equipment to the site.

Far greater difficulties arose with deep-sea drilling, and to master them a new generation of platforms was needed. On the other hand, logging was not affected by any major new problem. Technical features were gradually adjusted. There was, first, the question of the optimum location for the O.S.U.-C. The prohibitive cost of standby time on offshore platforms required as many measurements as possible in a minimum of logging runs; combination tools had to be designed to meet this objective. Along the same lines the first radio transmissions of logs were studied to devise the fastest possible delivery of logging results Finally, specialized crews had to be set up to meet the specific requirements of the platforms and their environment—a minor task for a company whose motto was "Wherever the Drill Goes, Schlumberger Goes." Though there are colorful elements in the ways to reach the platforms and in the work and life on them, logging services—whether in the North Sea or off the shores of Alaska, Nigeria, or the Sunda Islands—continue to offer the same routine efficiency as they do in Iran or Oklahoma (Fig. 85).

Presentation and interpretation of logs

Despite a wealth of new and improved services, the outstanding achievement of the 1960's was the progress in interpretation.

Any log reading involves two stages: an immediate and a deferred interpretation. The former determines the decisions to be taken forthwith. This is the "quick look"; it usually gives only approximate results, but takes few computations and can be applied at the well site. The second interpretation seeks a thorough reservoir engineering study and requires long and difficult calculations. As a matter of fact, it has been a problem for all log users to distinguish what must be solved on the spot from what

Figure 85. Schlumberger unit on offshore platform.

deserves elaborate processing. Only when such priorities have been established, can substantial progress be achieved.

The quick look had been the basis of Schlumberger's reputation and success on the Texas–Louisiana Gulf Coast. In this region of sands with low shale content and uniform porosity, where there is little variation in the salinity of the interstitial water, the S.P. and the two conventional resistivity logs gave quick and safe diagnoses. In other regions, however, variable porosity or salinity and a higher shale content blurred the effect of the hydrocarbons on the resistivity—a fact that no interpretation, however summary, may overlook. Although no good quick-look method has yet been devised that accounts for the shale content, it has been possible to eliminate the effects of porosity and interstitial water to the point that, by the 1960's, some logs presented in the field lent themselves to immediate interpretation.

The first of these logs, the so-called R_{wa} log, followed shortly upon the introduction of the sonic log as a porosity tool. The method consists in considering all the formations as being entirely filled with water and calculating the apparent resistivity (R_{wa}) of this water on the basis of the true resistivity (R_t) and the formation factor (F) provided by the sonic log.[22]

The apparent resistivity (R_{wa}) will be the true resistivity (R_w) of the interstitial water wherever it fills the formation at 100 percent. If, however, R_{wa} is higher that R_w, and if, furthermore, the S.P. maximum deflections are approximately constant, indicating that R_w is the same over

[22] The following relations are used:

$$F = \frac{0.62}{\phi^{2.15}}$$

where

$$\phi = \frac{\Delta t - \Delta t_{ma}}{\Delta t_f - \Delta t_{ma}}$$

Δt_f and Δt_{ma} are the interval travel times in microseconds per foot (known in principle) in the fluid and the rock matrix, respectively, and Δt is the time measured by the sonic sonde; and

$$R_{wa} = \frac{R_t}{F}$$

where R_t is given by a deep investigation resistivity device, preferably an induction sonde.

the whole section logged, then these abnormally high values of R_{wa} are indications of oil or gas.

Although a sonde has recently been built that records sonic and induction logs in a single run, to obtain an R_{wa} in the field in the 1960's required the combined results of two runs. On the first run the sonic log was recorded optically and on punched tape.[23] On the second run, while the induction was being recorded, the punched tape was played back through a simplified computer that calculated R_{wa} from R_t and Δ_t, the constant values Δ_{tma} and Δ_{tf} being dialed in; and as it was being calculated, R_{wa} was put into analog form and also recorded optically. This was a complicated exercise, requiring equipment so bulky as to call for an additional pickup truck. The operation put a strain on the equipment, as well as on the nerves of the crew because of the devilish noise of the paper tape puncher.

This service was operative for a few years but was active only in southern Texas and Louisiana.[24] It had serious limitations; for example, the variations of R_w had to be small, the invasion could not be too deep, and the sonic log was expected to give a good porosity value. But the major practical difficulty was to obtain a perfect depth matching of both R_t and Δ_t logs. In brief, there was hardly any justification for such complexity when the result was merely to indicate hydrocarbon potentialities without even giving their approximate percentages. More could be achieved at less cost.

To avoid these inconclusive calculations, Doll thought of comparing the various logs by overlaying their transparencies. Sliding one log over another would readily give the optical depth correlation; and by using logarithmic coordinates for the various resistivities, an immediate evaluation of the ratio between two values (especially R_{xo} and R_t) should result from measuring the separation between curves. It then takes the mere translation of a log with respect to a fixed grid to multiply or divide all its ordinates by a constant. This makes it possible to multiply all the values of

[23] A conventional recording in binary code with 8 bits, that is, 128 values, a resolution that is adequate for a sonic log.

[24] More specifically, it involved also a so-called F_R/F_S method consisting of calculating the formation factor on the basis first of the resistivities (F_R) and second of the porosity as provided by the values of the sonic log (F_S). In principle, $F_R/F_S = 1$ in aquifers, and any value above 1 indicates hydrocarbons.

an R_{xo} log (Microlaterolog, Proximity log, micro S.F.L.) by the ratio R_w/R_{mf}. According to equation 2 on p. 264:

$$S_w = \left(\frac{R_{xo}/R_{mf}}{R_t/R_w} \right)^{5/8}$$

the new R_{xo} log will coincide with the R_t log (induction or laterolog) in water-bearing horizons where $S_w = 1$, whereas its noncoincidence will be a good hydrocarbon indication: where the curves are separated, it will suffice to measure the separation with a special rule (allowing for the ⅝ exponent), and a first approximation of S_w will be obtained without further ado.

As only comparables may be compared, the method demands that the vertical resolution be the same for both logs, which is not generally the case. To be valid, therefore, the method requires obtaining the average log R_{xo}—actually a "sliding average" of conductivities—a routine operation easily carried out during recording.

The logarithmic scale has other merits, timidly acknowledged in the early days of the hand recorder. It amplifies differences in the low-resistivity zones where they are of interest, but compresses them where they are less so. It thus allows for the clear recording on a single curve of a considerable range of values (from 0.2 to 2000 ohms), whereas the log would be difficult to read on an linear scale. What had doomed prior efforts was the fact that to geologists the shapes of the curves were more important than their numerical values.

However satisfactory quick-look methods have become, and however skillful geologists, geophysicists, and petroleum engineers are in their use, they are only a beginning for an industry capable of assimilating huge amounts of information. Quantitative interpretation had long been reserved for specialists: it consists in the application of the proper formulae to translate physical parameters into petroleum production language. As computations made by slide rule and charts were slow and arduous, their practice was limited to zones that logs and local stratigraphy indicated as being promising. Yet a real danger existed that a productive zone might be overlooked for lack of a systematic study of the log over its whole length. Computers were now available, however, and most of the laborious routine work seemed to lend itself to programming. Automated interpretation was the preocccupation of the day, but far from being easy, as some believed, it involved certain thorny problems that required several years of research to master. The first thing necessary to make the log usable by the computer was to digitize it, that is, to substitute for each curve a series of numbers recorded on punched cards, punched tape, or magnetic tape; programs that varied with the logs, the geological features, and the information sought were then drawn up.

Digitization had two aspects, depending on whether existing or future logs were involved. The first group, available in optical form only, was usually processed by specialized organizations operating more or less automated equipment; Schlumberger had recourse to these services, as did all the oil people. As to the future, the Company had been gaining experience by recording data in digital form together with some conventional logs since the mid-1950's. As a result of this experience, the dipmeter log, recorded on magnetic tape and processed by computers, became operational in 1961.

From that time on, abetted by the fad for computers, most oil companies tried to set up their own automated interpretation programs. Many interesting publications resulted from their research. But as long as digital recording in the field remained confined to the dipmeter, progress continued to be of limited scope. The oil companies were most eager to see digitization spread to all kinds of logs, but the environment of the oil well is such that it took Schlumberger years to succeed.

Various decisions, both as to recording technique and format of magnetic tape, were preceded by prolonged studies, with the participation not only of the Houston, Ridgefield, and Paris staffs, but also of recognized interpretation experts. The service started in the United States in 1965 and after 1968 became widespread with the introduction of a new digital recorder. Although the optical log remains in use for quick estimations on the well site, the magnetic tape is becoming the principal document. In addition to being suited for computer processing, it has substantial advantages. When it is played back, not in the field but in the silence of the office, it is possible with the help of special devices to reconstitute an optical log, whose scales and presentation can differ from the original. On the other hand, the optical recorder with its nine galvanometers can record nine parameters at best, whereas the magnetic recorder has up to 22 channels, thus considerably enlarging the scope of combinations that can be achieved with the downhole equipment.

The magnetic tapes recorded on the well cannot be directly processed by the computer; they first need "editing," that is, the depth and unit (ohms/m, grams/cm^3, microseconds/ foot, etc.) scales must be reconstituted, and then "normalizing," in other words they must be corrected for environmental factors[25] (borehole diameter, mud cake, invasion, etc.). Only after the tapes have been normalized by Schlumberger is actual interpretation possible; it is for the client to decide whether to undertake the programming himself or entrust it to Schlumberger. This programming[26] has been the subject of a vast amount of research and is today the bailiwick of a highly specialized staff. Yet experience reveals that, in spite of more and more automation, the operator's intervention remains inescapable when it comes to basic choices. However, since the computer willingly accommodates a mass of calculations, one can invoke complex relations and repetitive processes heretofore unthinkable. By using the whole gamut of logs, it usually becomes possible not only to evaluate the porosity, the oil saturation, and the amount of oil flushed from the invaded zone, but also to estimate the shale content and the density of the hydrocarbons, and even to draw up a "permeability index" from relations between capillary pressure, permeability, irreducible water saturation,[27] and porosity.

The end product may be of several types. It can be a magnetic tape called CERT (Computer Evaluation Result Tape), which contains not only the global results, but also edited and normalized basic data enabling the client to proceed with his own computations. Most frequently it is an optical log materializing the results of computer processing, and obtained through a highly sophisticated electronic camera developed in Ridgefield (Fig. 86). The conventional aspect of the log is maintained, except that the parameters provided are no longer couched in the language of ohms and millivolts, but are expressed in terms of barrels of oil. Today only Houston and Paris (Clamart) have the requisite equipment and personnel for this kind of processing, but local centers with a capacity for increasingly complex operations are being developed all over the world. With the

[25] See p. 227–228. This is the "first phase" of interpretation.
[26] SARABAND: A. Poupon, C. Clavier, J. Dumanoir, R. Gaymard, and A. Misk, "Log Analysis of Sand–Shale Sequences, a Systematic Approach," *Journal of Petroleum Technology*, July 1970.
CORIBAND: A. Poupon, W. R. Hoyle, A. W. Schmidt, "Log Analysis in Formations with Complex Lithology," *Journal of Petroleum Technology*, August 1971.
[27] This is the minimum percentage of water in a reservoir.

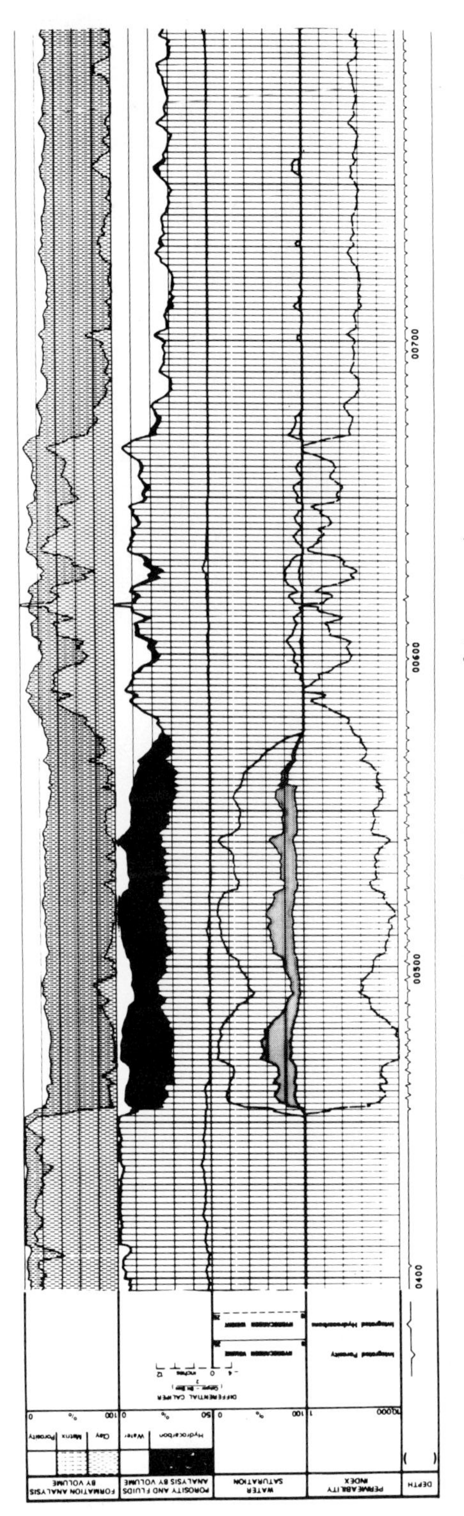

Figure 86. Computer-processed interpretation.

proliferation of computers and constantly improved programming, they are likely, within a few years, to offer a complete service of automated interpretation.

Ample advantage has already been taken of the adaptability of magnetic tape to radio communications for the nearly instantaneous transmission of offshore logs to the decision centers with an appreciable saving of time. With the advent of worldwide satellite communications, it is probable that this method, too, will serve for the transmission of logs and of their interpretations.

Conclusion

Though playing only a limited role in overall geophysical activities (except in the U.S.S.R.), surface electrical prospecting is still commonly used in the search for minerals other than petroleum. Many of its methods have remained basically—but with frequent improvements—those that Conrad invented during the first quarter of the century. However, his contribution was made too long ago to find an honorable niche in recent literature. Had it not been for logging, the name Schlumberger would hardly emerge from the shadow of bibliographical dictionaries.

Logging was a major success. To be sure, Schlumberger did not invent all the logs, but what is essential is that the Company brought logging to the petroleum industry. There was nothing new for physicists in marking the depths as abscissae and the values of a certain parameter as ordinates; but the curve that resulted and that gave the measurements concrete form was a true reflection of stratigraphy, and this was something genuinely new. To the Texas driller, every log is still a "schlumberger."

If logging had remained confined to the measurement of resistivities, it would have become nothing but a remarkable correlation tool. It achieved its full impact with the contribution of S.P. in 1930, and a highly productive method was created when the latter was plotted on the same diagram as resistivity. This was, unknowingly, the first "Synergetic" system: the information provided by the whole exceeded the sum of the data contributed by each single curve. Similarly, combinations of sondes with various depths of investigation were soon to provide, in the very frequent case of sand–shale series, precise answers to the problems of the operators.

At this point the future of logging was no longer in question. It had been good fortune that a technique developed for surface measurements met the needs of deep exploration. But how would a small company located in Paris, despite all its pioneering laurels, become aware of, let alone solve, the many problems that the petroleum industry would encounter in those far-off poles of its activity that were then the U.S.S.R. and the United States? The challenge was met because its leaders were not only researchers but also men of action. They could very rapidly avail themselves of a vast experience gathered from all over the world; but it had to be digested, evaluated, disseminated with discretion, and used wisely for the technical and commercial guidance of the young Company. During this phase of intense learning and important decisions great help came from clients prompt to visualize the potentialities of logging, and appreciative of the strenuous efforts of the Schlumberger international staff, whether in the field or at the engineering office. Schlumberger's place in the petroleum industry was earned by the conception and development of a series of basic inventions fostered by such a climate. It should be mentioned incidentally that, perhaps for lack of any other choice, Schlumberger, in delegating almost full autonomy and responsibility to its field engineers, was a few decades ahead in the application of certain tenets of modern management.

Throughout its development Schlumberger earmarked an ever-increasing share (sometimes deemed excessive) of funds for the design of tools of the highest standard. The conviction prevailed that, however reluctantly, petroleum operators would give priority to quality over economy, and experience confirmed the wisdom of this policy. In keeping with it and allowing for specific conditions predominant in boreholes, original concepts shaped the design of every basic tool. Perturbations arising from the environment were gradually studied so as to eliminate their effect on the measurements. This took place in the first stage through departure curves and in the second through the development of tools that made the corrections automatically and read the real parameters directly, thereby extending the range of logging to conditions heretofore unfavorable.

Thus, starting from three cable strands wound on a wooden reel, a potentiometer, a notebook, and a pencil, logging took less than half a century to reach the stage of a complex association of mechanics, electricity, electronics, and automation. Indeed, there have been other such cases, but few techniques have encountered more obstacles strewn across their paths. Born in a country and at a time when specialists in oil production were very few, tested at Pechelbronn under adverse geological and archaic drilling conditions, promoted in the world by a small group of engineers who, initially, had only the faintest knowledge of the problems and the language of their clients, logging had as its assets the guiding light of brilliant concepts, the keen perceptiveness of its creators, and a high sense of teamwork.

Logging—whether by Schlumberger, its competitors, or its clients—is today a vast industry employing thousands of specialists. One cannot conceive of any oil company, whatever its size, operating without the support of its daily harvest of various logs. They are the foundation of a document that contains every possible bit of information on a well, the "terminal log," a wide paper strip showing, side by side, the overall depth of the well, the parameters recorded as well as the results of interpretation, and the data on casing, cementing, perforating, conventional cores and sidewall samples, cuttings, production tests, and so on. The modest hand-drawn curves of Pechelbronn have evolved into thick folders filling drawers and cabinets, from the prefabricated hut in the African jungle or the Alaskan tundra to the top floor of the skyscrapers where the oil companies have their executive offices.

Each company disseminates copies of the logs to its geological field offices, regional offices, central services, and sometimes to its subsidiaries and partners. Other copies go to the technical secretariat of top management, the drilling and production departments, the research centers, the economic, statistical, and financial divisions, and finally, in compliance with the law of most countries, to the authority in charge of mineral resources conservation. These copies are the basic tool of geologists and engineers for the computation of reserves and the location of aquifers and gas caps. The logs are likewise consulted in planning the location and output of development wells. Drillers refer to them in order to draw up the technical drilling program for new wells (depth, bit sizes, casing lengths). They support decision-making in high-level policy meetings of a company.

When a concern is interested in drilling in a new area, the first thing it does is to scrutinize the logs of that area or, if there are none, those of the nearest wells. In the United States most of them are published within a few weeks and can be obtained in the trade. When an organization is in the process of resuming operations in a territory already prospected by another, the technical arguments put forward in the financial discussions rest on seismic surveys for surface work and on logs for deep exploration. When an independent American producer seeks financing, he submits the logs to bankers to stress the potential return on the investment. In addition, the banks have their own log libraries (often on microfilm), where their petroleum advisers seek the documentation they need for investment decisions.

The innumerable uses of logging should not obliterate what has perhaps been its greatest contribution to the development of the petroleum industry. Closely tied to the rotary drilling system, in fact designed and developed for its needs, logging and its auxiliary activities have become intrinsic elements in its progress. Needless to say, logging is not the sole source of information for deep exploration, and, especially in wildcatting, coring remains crucial. Nevertheless the driller, relieved of the constant burden of evaluating the formations through which he is drilling, is able to concentrate on the technique and efficiency of the drilling proper, in the certainty that methods exist through which all the formations can be detected and correlated with those of nearby wells, the potential reservoirs evaluated, and the formations and fluids they hold sampled. It is fair to say that, by relieving drilling of the disruptive operations for which it was ill fitted, logging has contributed to the remarkably high rates of penetration of modern wells.

At the service of an industry that still has vast areas open for exploration, notably the barely touched resources of the deep seas and the Arctic regions, Schlumberger must actively continue its research and development efforts along the same general lines as were pursued for the past two decades. The development of tools for the recording in a single run of the greatest possible number of parameters will be carried on, the ultimate objective being to record them all (i.e., about 15) under the increasingly severe constraints of ever-greater drilling depths. Magnetic tape recorders will be provided to all crews; in addition, they will be equipped with computer systems for the simplification and automation of otherwise complex adjustment processes, a heavy burden on field engineers.

With the ever-expanding use of computers, interpretation has reached a high level of efficiency over the past 15 years. However, the intervening factors are so many and so diverse that it will take a long time to complete current research and realize mathematical programs the output of which will clearly indicate the relative impacts of the various factors involved. It will be recalled also that in many cases the parameters measured are incomplete sources of information (shale content of reservoirs) or are exceedingly vague (permeability, location of fractures before completion). However uncertain the prospects of success, the possibility should not be excluded that these gaps will be filled some time in the future by inventions as significant as S.P. or induction logging.

Further progress in telemetry will make possible the transmission of logs between wells, especially offshore, and a growing number of centers equipped for computerized interpretation. Nonetheless, whatever the speed and flexibility of such transmissions (in any case it may be some time before they are introduced into a number of developing countries), field interpretation will remain indispensable for such urgent decisions as where to core or take fluid samples, when to lower a casing, how to select intervals for perforation, and, more importantly, when to continue or discontinue drilling. This is the reason why the effort is made to provide the crews with processing systems that give more thorough interpretations than the quick look.

Whereas the extrapolation of current trends gives some hint of short-term developments to come, any longer range forecast would be risky. Twenty-five years ago, before the discovery of transistors withstanding high temperatures, it seemed unlikely that Schlumberger would some day switch from pulsated current to electronic solutions. Likewise, nobody could foresee from the first-generation computers the role that informatics would play in measuring and interpretation techniques. There are many similar examples. What can be safely stated is that all the technological developments of the future will be drawn upon by Schlumberger to maintain its lead in the field.

In conclusion, it may be fitting to add that Schlumberger's technical and industrial success is also, in a way, a human achievement. From the outset, a certain spirit motivated a staff of young graduates fresh out of school: fellowship, the attraction of distant horizons, the feeling of participation in an effort where a common task did not stifle individualism. This rather exceptional atmosphere, although undoubtedly somewhat attenuated by years of growth, remains alive. Over the half century since Schlumberger became a leader in geophysics, it has grown without labor disputes, while avoiding paternalism in the relationship between management and staff. It is to be hoped that the future will preserve this human factor, without which no technique can make history.